Microorganisms for Sustainability

Volume 50

Series Editor
Naveen Kumar Arora, Environmental Microbiology
School for Environmental Science, Babasaheb Bhimrao Ambedkar University
Lucknow, Uttar Pradesh, India

Microorganisms have been in existence since the origin of life on earth and can survive the most extreme habitats or conditions on earth. Microorganisms are involved in regulating biogeochemical cycles, maintaining plant and animal health, and sustaining the global food chain. Moreover, they play crucial roles in addressing the challenges of climate change and achieving the targets of Sustainable Development Goals (SDGs).

This multidisciplinary book series captures the role of microbes towards building a sustainable world, while encompassing cutting-edge technologies and current needs across various fields such as agriculture sustainability, bioremediation, restoration of degraded habitats and wastelands, and food security. Additionally, this series explores microbial applications in industries, and building their utilization in clean and green energy solutions. Furthermore, themes like microbial secondary metabolites, extremophilic microbes and modern omics, including next generation sequencing and metagenomics, are also covered in this series.

With contributions from researchers across the globe, this series addresses the important call of 'One Planet-One Health-One Future'. It comprises a collection of diverse volumes that provides insights for scientists, young researchers, educators and decision-makers in the government, private sector, and non-governmental organizations, empowering their efforts to achieve the global goals.

The series invites, evaluates, and accepts book proposals to ensure a diverse, inclusive, and evolving program. The final decision regarding acceptance rests with the series editor.

Peer-review

This book series follows a stratified review process. Proposals for individual volumes are reviewed by the series editor and then the editorial board members. On a case-to-case basis, external reviewers are also invited for further evaluation of the book proposal. Review of the chapters is the responsibility of the volume editor(s). A manuscript submission platform has been recently made available to the authors, volume editors and the series editor.

Ravi Kant Bhatia • Abhishek Walia
Editors

Advancements in Microbial Biotechnology for Soil Health

Springer

Editors
Ravi Kant Bhatia
Department of Biotechnology
Himachal Pradesh University
Summer Hill Shimla, India

Abhishek Walia
Department of Microbiology
College of Basic Sciences, CSK Himachal Pradesh Agricultural University
Palampur, India

ISSN 2512-1901 ISSN 2512-1898 (electronic)
Microorganisms for Sustainability
ISBN 978-981-99-9481-6 ISBN 978-981-99-9482-3 (eBook)
https://doi.org/10.1007/978-981-99-9482-3

This Springer imprint is published by the registered company Springer Nature Singapore Pte Ltd.
The registered company address is: 152 Beach Road, #21-01/04 Gateway East, Singapore 189721, Singapore

Paper in this product is recyclable.

Foreword

The advancements in the field of microbial biotechnology have revolutionized our understanding of this hidden treasure. Today, we stand at the forefront of a new era, where microorganisms can be used for improving soil health and agricultural sustainability beside other applications. Untapped potential of microbial biotechnology for soil health offers a promising avenue not only to enhance soil fertility, mitigate the detrimental effects of chemical inputs but also to promote sustainable agricultural practices.

I am delighted to write the foreword for the book *Advancements in Microbial Biotechnology for Soil Health* which brings together a wealth of knowledge from esteemed experts who have dedicated their professional career to studying the soil microbiome and sustainable agriculture. Soil is the foundation of agriculture providing vital nutrients and support necessary for plant growth. However, intensive agricultural practices, such as the overuse of chemical fertilizers and pesticides, have led to the degradation of soil quality, loss of biodiversity, and ultimately negative impacts on ecosystem health which calls for urgent action to restore and rejuvenate our soils. Microbial biotechnology offers a powerful tool to achieve this goal through leveraging the power of microorganisms resulting in enhanced soil fertility, ecological balance, and promote sustainable agricultural practices. This book gives a detailed account of the remarkable progress made in understanding the functions and interactions of microorganisms in soil ecosystems, the diverse strategies employed to harness their potential, for developing biofertilizers and bio-control agents including genetically modified microbes for targeted soil remediation.

The book gives a comprehensive overview of the key aspects of microbial biotechnology important for soil health including the intricate mechanisms by which microorganisms interact with plants and how these interactions can be manipulated to enhance crop productivity. As we face the challenges of feeding a growing global population while preserving our natural resources, microbial biotechnology offers a pathway towards a sustainable future. This book serves as a valuable resource for scientists, educators, policymakers, and practitioners who seek to deepen their understanding of soil health and explore innovative approaches to enhance agricultural productivity, especially at a time when governments all across the globe have started planning alternatives for improving soil health.

I would like to commend the editors and authors for their very useful contributions to this book. Their expertise and knowledgeable inputs will serve as an indispensable resource for researchers, students, policymakers, and farmers besides anyone interested in the future of sustainable agriculture. I hope that this book will inspire further research and innovation in microbial biotechnology, and that the idea/suggestions of the experts will be both utilized to foster a more sustainable and resilient agricultural system and economy besides healthy world. This book will certainly help in our journey towards healthier soils, a greener planet, and a brighter future for all the generations to come.

Department of Biotechnology
Himachal Pradesh University, Shimla, HP, India
Planning and Teachers Matters
HP University, Shimla, HP, India
Departments of Biotechnology, Microbiology &
Forensic Science, HP University, Shimla, HP, India
Institute of Integrated Himalayan Studies (IIHS)
HP University, Shimla, HP, India
UIIT, HP University, Shimla, HP, India

Arvind Kumar Bhatt

Preface

In recent years, there has been a growing recognition of the critical role that soil health plays in sustainable agriculture and environmental conservation. Microbial biotechnology has emerged as a transformative field in the quest for sustainable agricultural practices and soil health improvement. Soil is not just a medium for plant growth; it is a complex ecosystem teeming with microorganisms that form intricate networks and interactions between microorganisms and their environment has opened new doors for harnessing the power of microbes to enhance soil fertility, nutrient cycling, and overall ecosystem resilience.

Microbial biotechnology, the science and application of harnessing the power of microorganisms, has emerged as a promising approach to enhance soil health and agricultural productivity. These microorganisms drive key processes, including nutrient cycling, organic matter decomposition, disease suppression, and the formation of beneficial symbiotic relationships with plants. Researchers and practitioners have been exploring innovative ways to understand, manipulate, and utilize the diverse microbial communities present in soil ecosystems. The goal is to leverage the natural functions of these microorganisms to promote sustainable agricultural practices and mitigate the negative impacts of conventional farming methods.

This book, *Advancements in Microbial Biotechnology for Soil Health*, aims to provide a comprehensive overview of the latest developments and breakthroughs in this exciting field. It brings together a diverse group of experts, including scientists, agronomists, and biotechnologists, who have made significant contributions to the understanding and application of microbial biotechnology for soil health. The chapters in this book cover a wide range of topics, starting with an introduction to the importance of soil health and the role of microorganisms in soil ecosystems. The book also explores the application of microbial biotechnology in various aspects of agriculture, including biofertilizers, biocontrol agents, plant growth-promoting rhizobacteria, managing soil-borne diseases, promoting plant growth and resilience remediating polluted soils, including soil remediation strategies.

Furthermore, this book aims to bridge the gap between scientific research and practical application. It provides case studies and examples of successful microbial biotechnology interventions in various agricultural systems, demonstrating their efficacy and economic viability. It also addresses the challenges and limitations that need to be overcome for the widespread adoption of these technologies. Additionally, we shed light on the challenges and opportunities associated with translating scientific knowledge into practical applications for farmers and land managers. Understanding the functioning of soil microbial communities and their dynamics has allowed us to develop novel approaches to enhance soil fertility and mitigate the negative impacts of agricultural practices. We believe that the insights and knowledge shared in this book will contribute to the advancement of microbial biotechnology and its integration into sustainable soil management practices. By harnessing the power of microorganisms, we can unlock the full potential of our soils, ensuring food security, environmental sustainability, and the well-being of future generations.

We extend our gratitude to all the contributors who have shared their expertise and insights in this book. Their dedication and passion for advancing microbial biotechnology for soil health have made this endeavour possible. We would also like to express our appreciation to the editorial team and the publisher for their support throughout the publication process. It is our hope that this book will serve as a valuable resource for researchers, students, and professionals working in the fields of soil science, agriculture, microbiology, and biotechnology. By promoting a deeper understanding of the interactions between microorganisms and soil ecosystems, we aim to foster sustainable practices that enhance soil health, protect the environment, and ensure food security for future generations.

Summer Hill, Shimla, India — Ravi Kant Bhatia
Palampur, India — Abhishek Walia

Contents

About the Editors

Ravi Kant Bhatia presently working as Assistant Professor in the Department of Biotechnology, Himachal Pradesh University, Shimla-05, India, is actively engaged in research activities in the area of environmental biotechnology. Dr. Bhatia received his Master's in Biotechnology in 2007 from HNBGU, Garhwal, Uttarakhand, and Ph.D. from Himanchal Pradesh University, Shimla. He has vast experience in industry and academia and is continuously engaged in teaching and research activities. He has qualified UGC-NET, ICAR-NET, GATE and obtained UGC-PDF fellowship to pursue his Post-Doctorate. He has also received the SERB-ITS fellowship award to visit Scotland to present his research findings in an international conference. He has published more than 75 research publication, 18 chapters and 2 books. He serves as a reviewer for various prestigious national and international journals and received projects from various funding agencies related to environmental microbiology and biotechnology. Besides his teaching and research assignments, he is also looking after Biotechnology Incubation Centre (BIC-HPU) to promote bio-entrepreneurship in the field of environment, agriculture and industrial biotechnology.

Abhishek Walia is an Assistant Professor in the Department of Microbiology, College of Basic Sciences at CSK Himachal Pradesh Agricultural University, Palampur, Himachal Pradesh, India. He has a total of 8 years of teaching and research experience after obtaining his doctorate degree. Dr. Walia has published 38 international and national research papers, 5 review articles, 14 book chapters and 2 conceptual articles. He has received the DST-Inspire Fellowship from the Government of India. He is an Editorial Board Member and Review Editor in international and national journals of repute. He has also headed the Department of Microbiology at DAV University, Jalandhar, Punjab. Dr. Walia is working in the area of Agricultural and Environmental Microbiology, currently focussing on agro-residue waste management and transformation into various valuable products for the hilly regions of Himachal Pradesh.

Chapter 1
Soil Microbiome as a Key Factor in Soil Health

Chayanika Putatunda, Nyaton Kitnya, and Preeti Solanki

1.1 Introduction

The term "soil microbiome" encompasses all the microbial flora and fauna inhabiting the soil including prokaryotes—archaeans and bacteria and eukaryotes—protistans and fungi, as well as acellular entities like viruses (Fierer 2017). The soil microbiome is absolutely essential for carrying out agriculturally significant processes like nitrogen fixation, nitrification, solubilization of phosphates, secretion of phytohormones, generation of siderophores and antibiotics, etc.; hence, the significance of soil microbiome in soil health and plant growth cannot be overemphasized (Narveer et al. 2014; Jacoby et al. 2017). Additionally, the soil microbiome is involved in the catalytic breakdown of organic and inorganic waste matter, as well as carbon sequestration. These processes allow the cleaning up of the environment from plant-based wastes, animal excreta, and carcasses as well as anthropogenic wastes and are responsible for nutrient recycling (Nannipieri et al. 2017). In fact, all the biogeochemical cycles heavily rely on the soil microbiome; thus, the soil microbiome tends to impact all living organisms in a direct or indirect manner (Bender et al. 2016).

Carbon content in the soil microbial biomass may reach even greater than a thousand kilos per hectare of soil (Serna-Chavez et al. 2013). Although a lot of information is available about the soil microbiome and its role, as per many estimates,

C. Putatunda
Department of Microbiology, Om Sterling Global University, Hisar, Haryana, India

N. Kitnya
Department of Zoology, Himalayan University, Itanagar, Arunachal Pradesh, India

P. Solanki (✉)
UCRD-Biotechnology, Chandigarh University, Mohali, Punjab, India
e-mail: preeti.e15024@cumail.in

R. K. Bhatia, A. Walia (eds.), *Advancements in Microbial Biotechnology for Soil Health*, Microorganisms for Sustainability 50,
https://doi.org/10.1007/978-981-99-9482-3_1

approximately 90% of soil microbiome is yet unexplored (Dubey et al. 2019). However, with the help of newer techniques especially the molecular biology and genomics-based approach, gaining an improved comprehension of these uncharted microorganisms, along with their phylogeny, interactions, and roles within the soil, is becoming increasingly feasible (Nesme et al. 2016; Mitter et al. 2017).

The constitution and organizational structure of soil microbiome cannot be uniform across different types of soil and is dictated by intricate interactions between biotic and abiotic factors. The multifaceted interactions among the different types of microorganisms themselves and the interactions between the microbes and the plant roots are some of the major biotic factors (Bahram et al. 2020). On the other hand, the type and composition of the soil, soil temperature, pH, nutrients, moisture, etc. are some important abiotic factors impacting the soil microbiome (Islam et al. 2020). Also, these factors are in turn impacted by anthropogenic activities like waste disposal, the addition of xenobiotic compounds, and agricultural practices such as crop rotation, application of fertilizers, biofertilizers, pesticides, biopesticides, manure, etc. (Kraut-Cohen et al. 2020; Li et al. 2020; Santos and Olivares 2021). A better comprehension of the factors impacting the soil microbiome can help us modify it and use it for bettering soil health, crop productivity, and bioremediation of contaminated soils (Wang et al. 2021; Santos and Olivares 2021). Enhancement of crop productivity is absolutely essential for feeding the ever-increasing population. However, overexploitation of our soil would tend to affect the chances of survival of future generations. So, the manipulation of soil microbiomes offers an avenue for sustainable growth and development (Omotayo and Babalola 2021).

1.2 Structure and Characteristics of Soil Microbiome

The soil offers a great diversity of microenvironments. The physicochemical factors like temperature, pH, moisture, soil texture, soil composition, etc. and the biotic factors like the interaction and activities of soil macrofauna, mesofauna, the microbial communities along with human activities go a long way in shaping soil microenvironments, which in turn significantly influence the structure, characteristics, and functions of the soil microbiota. The environment within the soil may vary even within a range of millimeters, and micrometers and the microorganisms are not evenly distributed (Fierer 2017). The microbial biomass present in soil's organic matter generally constitutes from 1% to 5%, which is equivalent to microbial biomass carbon of 280–1940 kg/ha (Nannipieri 2020). However, soil bacteria only occupy <1% of the total volume of the soil. Certain microsites like rhizosphere or detritusphere which are enriched in nutrients host a much higher abundance of microflora (Nannipieri et al. 2003; Orgiazzi et al. 2016). Lesser occupancy in other portions of the soil may be attributed to unfavorable conditions like pH extremes, scarcity of water, lack of appropriate level of oxygen, predation, presence of toxic compounds or antibiotics, etc. (Dini-Andreote and Van Elsas 2019). Thus, not only does one encounter variation in soil environments across the soil horizons or water flow paths but also intrinsically or extrinsically to the rhizospheric soil aggregates, or the bulk soil (Islam et al. 2020). Consequently, the microbial community structure varies as a result of the variation in the soil microenvironment. Blagodatskaya and Kuzyakov (2013) have reported that ~95% of the soil's total microbial biomass exists in an inactive form in addition to variation in the level of microbial activity with respect to space and time. This observation further supports the fact that the conditions in soil greatly influence the soil microbiota.

Even though there are unfathomably uncountable species of viruses, bacteria, archaea, protists, and fungi in the soil microbiome, only a minuscule of them have been thoroughly characterized (Sokol et al. 2022). Although there is no "typical soil microbiome" (Fierer 2017) and there is a lot of spatial-temporal dissimilarity in the microbial groups as per different reports, as little as a gram of surface soil can sustain $>10^9$ bacterial and archaeal cells, >100 m fungal hyphae, >10,000 protistan cells, and $>10^{12}$ viral particles (Tecon and Or 2017; Leake et al. 2004; Stefan et al. 2014; Williamson et al. 2017; Sokol et al. 2022). But, on comparing biomass from different microorganisms, the bacterial and fungal biomass are reported as the most dominant constituents (Fierer 2017; Islam et al. 2020). Also, as per several investigations based on 16S sequences, undescribed lineages of bacteria and archaea comprise more than 80% of the prokaryotic soil microbiota (Ramirez et al. 2014). Dassen et al. (2017) also made similar observations in the case of soil fungi and protists.

Another observation that has been made by several researchers is that just a small number of bacterial and archaeal phyla exhibit higher abundance than others which tend to display rarer occurrences (Crowther et al. 2014; Lynch and Neufeld 2015). Yurgel et al. (2018) analyzed *Vaccinium angustifolium* (wild blueberry) for its soil microbiome (bulk soil, root, and rhizosphere) and reported niche-specific functional distinction based on metagenomic data of identified 16S rRNA reads of the most relatively abundant top 15 soil/root bacterial classes, with ~94%, ~83%, and 89% from root, bulk soil, and rhizosphere samples, respectively. They also reported that the most profuse bacterial classes constituting the root samples were *Acidobacteria*, *Actinobacteria*, *Bacteroidetes*, *Chloroflexi*, *Ktedonobacteria*, *Proteobacteria*, *Saprospirae*, *Spartobacteria*, and *Verrucomicrobia*. A recent investigation involving shotgun metagenomic assessment of *Lactuca sativa* (lettuce) rhizosphere soils from South Africa was carried out by Babalola et al. (2023). They found the community analysis to comprise bacteria (97.65%), eukaryotes (1.36%), archaea (0.95%), and viruses (0.04%). *Acinetobacter*, *Pseudomonas*, *Streptomyces*, *Candidatus solibacter*, *Burkholderia*, *Bradyrhizobium*, and *Mycobacterium* were the predominant genera.

Delgado-Baquerizo et al. (2018) assessed surface soil samples collected from 237 sites from 18 different countries and 6 continents, for soil microbes and reported through gene sequence analysis of 16S rRNA the presence of undescribed taxa. The analysis showed that 2% of bacterial phylotypes were the dominant ones. They reported the dominant phylotypes as *Acidobacteria* (*Candidatus* sp. and *Salibacter*),

Actinobacteria (*Mycobacterium* sp., *Salinbacterium* sp., *Streptomyces* sp.), *Alphaproteobacteria* (*Bradyrhizobium* spp., *Devosia* sp., *Kaistobacter* sp., *Rhodoplanes* sp., *Sphingomonas* sp.), *Betaproteobacteria* (*Methylibium* sp. and *Ramlibacte*), and Plancttomycetes. In another investigation, Bahram et al. (2018) carried out a study encompassing 189 topsoil samples across the terrestrial regions of the globe. They analyzed the soil microbiome of these samples with the help of metagenomics and DNA metabarcoding, soil chemistry, and biomass assessments. In the case of bacteria, both the taxonomic and gene functional diversity was highest in the mid-latitudes and became lesser toward the polar and equatorial regions. This finding contrasted with the latitudinal diversity gradient seen in higher plants, where diversity tends to decrease at the poles and increase as one moves closer to the equator. However, in the case of overall fungal taxonomic diversity (but not in fungal functional diversity), the latitudinal gradient diversity was observed by the researchers.

Delgado-Baquerizo et al. (2021) also examined the soil microbiomes of surrounding natural ecosystems and urban greenspaces from 17 different nations on 6 different continents. They have reported that at the global scale, the microbiome from the soils of urban greenspaces tended to be more homogeneous as compared to the soil microbiomes from their natural ecosystem counterparts. The pattern was seen in the case of the community composition of not only archaea and bacteria but also fungi and protists. The urban greenspaces were found to harbor greater proportions of *Amoebozoa*, *Ascomycota*, *Bacteroidetes*, *Chlorophyta*, *Deltaproteobacteria*, *Gammaproteobacteria*, and *Gemmatimonadetes*. They also concluded that the chief factors that influenced the soil microbiota of these urban green spaces included city affluence, management practices, and climatic conditions. Although the analysis of any given soil sample provides a glimpse into its soil microbiome, a more holistic approach involving temporal and special analysis along with the integration of taxonomic and functional analysis will show a clearer composition of the soil's microbiome (Nannipieri et al. 2019; Geisen 2021).

1.3 Functions of Soil Microbiome

The soil microbiota is a diverse and complex community of microorganisms that inhabit the soil environment. These microorganisms (including bacterial, fungal, protozoan, and viral components) interact with each other and with the soil to perform critical functions that regulate nutrient cycles, organic matter decay, and formation of soil structure (Bastida et al. 2015; Delgado-Baquerizo et al. 2016; Peiffer et al. 2013). The following paragraphs describe the various functions of the soil microbiome in more detail, drawing from relevant studies to support the statement.

1.3.1 Nutrient Cycling

One of the pivotal roles of the soil microbiota is nutrient cycling. Microbes help in the transformation of organic and inorganic nutrients found in the soil, into elemental nutrients such as nitrogen (N), phosphorus (P), and potassium (K), critical for plant growth, and are easily taken up by plants (Bergersen 1997; Bardgett and van der Putten 2014).

1.3.1.1 Nitrogen Fixation

Nitrogen cycling is a critical function of the soil microbiome. Because it is often present in the soil in an inert form that plants cannot utilize, the element nitrogen, which is essential for plant growth, is frequently a limiting factor in agricultural systems. Nitrogen-fixing microorganisms, like *Rhizobium*, *Azotobacter*, and *Bradyrhizobium*, transform atmospheric nitrogen gas (N_2) into ammonia (NH_3) or other plant-useful forms (Sánchez-López et al. 2021). Further, numerous studies have also shown that soil bacteria, like *Azospirillum*, *Azotobacter*, and *Rhizobium*, have a vital function in the nitrogen cycle, i.e., fixing atmospheric N_2 and changing it into a plant-usable form (Bergersen 1997; Herridge et al. 2008).

1.3.1.2 Phosphorus Solubilization

Phosphorus, another critical plant nutrient, is often bound to soil particles and unavailable to plants. Phosphate-solubilizing bacteria and fungi break down these compounds, making phosphorus available to plants (Adesemoye and Kloepper 2009). The mycorrhizal fungi can increase the uptake of phosphorus by plants, which is often limited in agricultural soils (Smith and Read 2008; Vallino and Stephan 2018).

1.3.1.3 Carbon Cycle

Soil microbiomes play a significant role in carbon cycling. Soil microbes help to decompose organic matter, releasing carbon dioxide, which contributes to atmospheric carbon. However, microbes also convert organic matter into stable forms of soil organic carbon, which can remain in the soil for hundreds of years (Lal 2004). The formation of stable organic matter in the soil is influenced by the balance between inputs of organic matter and the microbial decomposition of that organic matter (Cotrufo et al. 2013). Thus, a diverse group of soil microorganisms plays a vital role in carbon sequestration: reducing the amount of carbon dioxide in the atmosphere and mitigating climate change in the soil.

1.3.2 Organic Matter Decomposition

The soil microbiome is also responsible for the decomposition of organic matter, which releases nutrients into the soil and helps to maintain soil fertility. Microorganisms such as bacteria, fungi, protozoa, etc. break down organic matter into simpler compounds that can be taken up by plants (Anderson et al. 1983).

A study by Anderson et al. (1983) found that soil bacteria were responsible for the breakdown of 60–90% of the carbon in plant residues. Nevertheless, fungi are also essential decomposers of organic matter in the soil. Lehmann and Kleber (2015) and Sinsabaugh et al. (2016) have shown that fungi can decompose complex organic compounds such as lignin, which is resistant to degradation by bacteria. Microbial diversity has been found to play a critical role in the rate of organic matter decomposition, with higher microbial diversity resulting in increased rates of decomposition (Delgado-Baquerizo et al. 2016). These microorganisms secrete enzymes that break down complex molecules like cellulose, lignin, and chitin, into simpler forms like carbon dioxide, water, and nutrients (Bardgett and van der Putten 2014).

1.3.3 Soil Structure Formation

The soil microbiome is responsible for the formation of soil structure, which refers to the way in which soil particles are arranged and can influence water infiltration, aeration, and nutrient availability (Tisdall and Oades 1982). Microorganisms such as bacteria and fungi can produce extracellular polymers such as polysaccharides, glycoproteins, and other substances that bind soil particles together, creating aggregates that improve soil structure (Tisdall and Oades 1982; Herrmann et al. 2008; Kaiser et al. 2015). These aggregates help to create pore spaces in the soil, improving soil water and air movement, increasing nutrient availability, and supporting plant growth.

1.3.4 Plant Growth Promotion and Pathogen Suppression

The soil microbiome plays a crucial role in promoting plant growth. Microorganisms in the soil can interact with plants in various ways, including nutrient cycling, hormone production, and disease suppression (Glick 2012). These relationships are crucial for maintaining soil health and promoting plant growth, with recent studies suggesting that plant-microbe interactions may be more critical for plant growth than previously thought (Berendsen et al. 2012).

Soil bacteria such as *Bacillus* and *Pseudomonas* have been found to promote plant growth through the production of plant growth-promoting hormones such as auxins and cytokinins (Zhang et al. 2016); in particular, *Bacillus subtilis* and *Pseudomonas fluorescens* produce antimicrobial compounds that can inhibit the growth of plant pathogens (Compant et al. 2005; Glick 2012). Likewise, Mycorrhizal fungi form symbiotic relationships with plants and can increase nutrient uptake, improve water use efficiency, and protect plants from pathogens by inducing systemic resistance and producing antibiotics (Smith and Read 2008). Also, microorganisms can suppress pathogens through various other mechanisms, including competition for resources and the production of secondary metabolites (Berg et al. 2005).

Though plant roots release organic compounds that attract and support microbial communities, providing a source of carbon and energy for soil microorganisms (Kuzyakov and Razavi 2019), they also exert selective pressure on soil microbiomes, favoring certain microbial groups over others. For example, some plants may selectively recruit microbial communities that can help them acquire nutrients, such as phosphorus or iron, from the soil (Kuzyakov and Razavi 2019). This is achieved by releasing specific root exudates that stimulate the growth and activity of these microbes, which in turn can solubilize or chelate these nutrients and make them more available to the plant. Other plants may prioritize the recruitment of microbes that can protect them from soil-borne pathogens, such as by producing antibiotics or eliciting systemic resistance (Kuzyakov and Razavi 2019). In these cases, the root exudates may contain compounds that attract these beneficial microbes and help establish a protective microbial barrier around the roots.

1.3.5 Bioremediation

Microorganisms can degrade or transform various pollutants like pesticides, heavy metals, and hydrocarbons, making them less harmful to the environment. For instance, certain bacteria like *Pseudomonas* and *Bacillus* can degrade pesticides like atrazine and chlorpyrifos, reducing their impact on the environment (Arora and Bae 2014). Fungi like *Trametes versicolor* can degrade polychlorinated biphenyls (PCBs), which are toxic to the environment (Osono et al. 2009).

1.4 Factors Affecting Soil Microbiome

The factors that affect the soil microbiome can be broadly categorized into biotic and abiotic factors. Abiotic factors are nonliving components, while biotic factors are living components of the soil ecosystem that influence the soil microbiome.

1.4.1 Abiotic Factors

1.4.1.1 Soil Type

It is a well-known fact that the soil microbial community is influenced by soil type. Such as Ushio et al. (2015) reported that the composition and diversity of the soil microbiome varied significantly across different soil types. The physical and chemical properties of the soil can affect the availability of nutrients and water, which can in turn affect the growth and activity of microorganisms in the soil. For example, sandy soils tend to have low nutrient content and low water-holding capacity, which can limit the growth and activity of microorganisms. In contrast, clay soils tend to have high nutrient content and high water-holding capacity, which can promote the growth and activity of microorganisms (Lehmann and Kleber 2015). Furthermore, soil type can also influence the distribution of different microbial groups in the soil. A study by Fierer et al. (2012) found that the soil microbiome of arid soils was dominated by bacteria, while the soil microbiome of wet soils was dominated by fungi. Hartman et al. (2017) found that bacterial communities in soil were more strongly influenced by soil type than by plant species. This suggests that soil type plays a key role in shaping the composition and diversity of the soil microbiome.

1.4.1.2 Soil pH

No doubt, microorganisms in soil play a vital role in nutrient cycling, organic matter decomposition, and soil formation. However, different microorganisms have varying pH requirements for growth and activity. For instance, acidophilic microorganisms thrive in acidic soils, while alkaliphilic microorganisms thrive in alkaline soils (Boddy 2011). Changes in soil pH can significantly affect the microbial community composition and activity (Fierer and Jackson 2006). Moreover, soil pH can also impact the availability of nutrients to microorganisms. Some nutrients are more readily available at certain pH levels, and changes in soil pH can alter the availability of these nutrients. For instance, Brady and Weil (2016) reported that in acidic soils, the solubility and availability of aluminum and manganese increase, whereas the availability of calcium and magnesium decreases. Conversely, in alkaline soils, the availability of phosphorus, iron, and zinc decreases. Therefore, soil pH is a crucial factor that influences the availability and activity of the microbial community in the soil.

1.4.1.3 Moisture

The importance of moisture for soil microbial activity and survival has been well-documented. Mau et al. (2015) reported that moisture plays a crucial role in facilitating microbial metabolic processes, influencing nutrient availability and soil structure. Adequate moisture is essential for the growth and activity of microorganisms, while waterlogged or saturated soil conditions can limit microbial activity and survival (Sui et al. 2019). Dry soil conditions can also restrict microbial activity, as the lack of moisture can limit the diffusion of nutrients and substrates needed for microbial metabolism (Schimel et al. 2007). They also reported that dry soils can lead to the formation of physical barriers that prevent microbial movement, further limiting microbial growth and function.

1.4.1.4 Temperature

Soil temperature is a critical factor that affects the growth and activity of microorganisms in soil. Different microorganisms have varying temperature requirements for growth and activity, and changes in soil temperature can significantly impact the composition and diversity of the soil microbiome (Singh et al. 2019).

Cold temperatures can limit microbial activity, leading to reduced nutrient cycling rates and organic matter decomposition. Conversely, in warm soils, microbial activity increases, which can lead to increased rates of nutrient cycling and organic matter decomposition (Wardle et al. 2012). However, prolonged exposure to high temperatures can lead to the death of microorganisms, thereby reducing soil biodiversity and ecosystem functioning (Singh et al. 2019).

1.4.2 Biotic Factors

1.4.2.1 Plants

According to several studies, plants play a significant role in shaping the soil microbiome by providing nutrients and organic matter that support microbial growth and activity (Bais et al. 2006; Chaparro et al. 2013; Philippot et al. 2013). The roots of plants release exudates, which are organic compounds that provide a food source for microorganisms (Badri et al. 2009). The composition and diversity of the soil microbiome can be influenced by the different types and amounts of exudates released by different plant species (Mendes et al. 2013; Bakker et al. 2018). Furthermore, plants can alter the physical structure of the soil, which can impact the distribution and activity of microorganisms (Jin et al. 2010; Lal 2004).

1.4.2.2 Soil Fauna

Soil fauna, including earthworms, nematodes, and arthropods, can significantly influence the soil microbiome by their activities. They can consume and excrete organic matter, physically mix the soil, and contribute to nutrient cycling in the soil. Earthworms, for example, consume soil and organic matter, which passes through their gut and is excreted as castings. Castings are rich in nutrients and microorganisms, which can promote the growth and activity of other microorganisms in the soil (González et al. 2019). Nematodes and arthropods can also consume organic matter and contribute to nutrient cycling in the soil (Bradford et al. 2002).

Furthermore, soil fauna can also physically mix the soil, which affects the distribution of microorganisms and nutrients in the soil. For instance, earthworm burrows can increase soil porosity, water infiltration, and oxygen availability, which can promote microbial activity and nutrient cycling in the soil. Nematodes and arthropods can also create pores and channels in the soil, facilitating water movement and nutrient diffusion (Bradford et al. 2002).

1.4.2.3 Microbial Interactions

The interactions between microorganisms in soil are diverse and can influence the structure and function of the soil microbiome. According to Philippot et al. (2013), these interactions can involve competition for resources, predation, parasitism, commensalism, and mutualism. For instance, some microorganisms produce antibiotics that can inhibit the growth of other microorganisms, while others can feed on or infect other microorganisms (Philippot et al. 2013). Conversely, mutualistic relationships, such as mycorrhizal associations, where plants exchange nutrients with fungi, can enhance plant growth and nutrient acquisition (Smith and Read 2008). These interactions can significantly influence the composition and diversity of the soil microbiome, which, in turn, affects soil fertility, nutrient cycling, and plant growth (Bonkowski et al. 2019).

The competition for resources among microorganisms can result in the development of diverse metabolic pathways and functional traits, which can enhance the efficiency of nutrient cycling and ecosystem functioning (Philippot et al. 2013). On the other hand, predation and parasitism can regulate the abundance and distribution of microorganisms in the soil, affecting the composition of the microbial community (Bonkowski et al. 2019). Furthermore, the mutualistic relationships between microorganisms can facilitate the exchange of nutrients and metabolites, promoting the growth and survival of the participating microorganisms and the plants they associate with (Smith and Read 2008).

1.4.2.4 Land Use

The impact of land use practices on the soil microbiome has gained significant attention in recent years. Land use practices, such as agriculture, forestry, and urbanization, can significantly alter soil microbial communities and associated functions. Agriculture, for instance, can lead to changes in the physical and chemical properties of soil, as well as alterations in soil microbial communities (Griffiths et al. 2016). The use of fertilizers, pesticides, and other chemicals in agriculture can also affect the composition and diversity of the soil microbiome (Philippot et al.

2013). These chemicals can alter the microbial community structure, reduce microbial diversity, and negatively impact soil functions (Philippot et al. 2013).

Similarly, forestry practices, such as clear-cutting, can lead to changes in soil microbial communities and nutrient cycling (Hartmann et al. 2018). Clear-cutting can alter soil moisture, temperature, and pH, leading to changes in microbial communities and functions (Hartmann et al. 2018).

Urbanization can also have a significant impact on the soil microbiome. The removal of vegetation and soil compaction can lead to changes in the soil's physical properties and nutrient cycling (Ramirez et al. 2014). Urbanization can also increase soil metal and organic contaminants, which can negatively impact microbial communities (Ramirez et al. 2014).

1.5 Techniques for Analyzing Soil Microbiome

Conventional analysis of soil samples by microbial culture and staining techniques is advisable for primary screening, but the complex diversity of microbial communities can be elucidated only through molecular techniques analyzing the nucleic acids—DNA and RNA. The frequently employed method of molecular analysis involves the isolation of DNA from the collected sample of soil or rhizosphere, followed by PCR-based amplification of genes (16S rRNA from bacteria and ITS regions from fungi), and subsequent sequencing of the PCR products obtained for microbial species identification and characterization.

The population dynamics of the soil microbiome can be analyzed through high-throughput omics procedures and modern biotechnology tools, like next-generation sequencing and genome shotgun sequencing, allowing the evaluation of biological and functional diversity in soil microbial communities. Approaches like metabolomics and genome sequencing have shown promising results in increasing plant and fruit yields, environmental stress tolerance, and production of phytohormones, through enhancing biodiversity in rhizospheric and native microbiomes and development of biofertilizers (Bertola et al. 2021). Metaproteomics and rDNA gene sequencing of bacterial genomes have identified the microbial species possessing biostimulant potential for benefitting plant growth (Mattarozzi et al. 2020).

Precise identification of microbial diversity in the soil sample allows simultaneous identification and comparison of the various microbial species present in the soil sampled from various sites. Additionally, functional biodiversity can be evaluated by newer molecular methods like metagenomics and shotgun sequencing, improving the comprehension of methodologies that are present in soil microbes and empowering them to survive under various environmental stress conditions (Bertola et al. 2021). The genomic and taxonomic evaluation through a referenced database also allows the designing of genes encoding beneficial gene products that can enhance the potential of microbes for bioremediation of soil and also have positive effects on plant and soil health and productivity.

Protein profiling of the microbial communities, also known as the "soil metaproteomics" approach, is another functional biodiversity analysis method that is a useful indicator for comprehending the intricate relationship of different microbial species present in the soil ecosystem (Starke et al. 2019). Enzymomics is an emerging omics strategy that has yet to pave its path in the field of agricultural microbiology. Hydrolytic microbial enzymes present in the soil like phosphatases, glucosidases, cellulases, chitinases, proteases, etc. have a critical role in the decomposition of soil organic matter as well as its transformation and mineralization; hence, soil organic matter is progressively processed by the soil microbiota and helps in regulating and supporting the soil ecosystem leading to a regenerative and healthy agroecosystem (Nannipieri et al. 2017). Soil enzymomics hence offers the potential to analyze through multivariate data sets of enzymes and provide befitting advice for the selection of samples that may be sequenced with other omics techniques.

Metabolomics-based approaches include the primary and secondary metabolites produced by the soil microorganism in response to an environmental stimulus. These approaches are useful to define the relationship between microbial communities and rhizospheres providing a map of food webs that regulate the soil's organic and inorganic chemistry and further influence inter- and intraspecies interactions and gene expressions among the microbial species of the soil. The metabolomic profiles are created with the help of liquid chromatography-mass spectroscopy-based data and comparing them with existing databases using bioinformatics tools. These strategies can help identify metabolic pathways in soil microorganisms and the corresponding genes that may encode beneficial enzymes which may be employed in bioremediation or as bio-control agents and biostimulants (Bertola et al. 2021; Hu et al. 2018).

Although omics and genome sequencing have improved the specificity in the identification and characterization of microbial taxa, nonetheless offering targets of gene manipulation and identifying novel microbial strains with unique environmental advantages, the major drawback is the high-cost and complicated computational analysis associated with these technologies. Additionally, the spatiotemporal heterogeneity in soil microcosm requires exhaustive statistical techniques to comprehend the dynamics of the factors involved in regulating microbial communities.

1.6 Manipulation of Soil Microbiome for Improving Soil Health and Productivity

Modulating various parameters of the soil may be aimed at enhancing the availability of nutrients and productivity in plants. Specifically, the constitution of the rhizosphere's microbial community may be targeted, being the most effective factor. Every biotic constituent of the soil works intricately with the complete microbiome affecting the soil's health and productivity. Efficiently managing soil is a critically significant task that enables us to meet the ever-growing agricultural demand in a sustainable manner. The organization of soil microbial communities in the agricultural ecosystem primarily depends on the interaction of various biotic and abiotic factors. These factors play a pivotal role in shaping the composition and functioning of soil microbiota.

Additionally, agricultural practices and anthropogenic activities on fertile soils also alter the composition of the soil microbial communities, directly or indirectly. Applications of contemporary agricultural practices excessively and injudiciously have led to deleterious effects on the environment and human health, besides being toxic to soil microbiota (Table 1.1; Fig. 1.1).

Manipulating the soil microbiome in and around the plant roots offers a favorable approach to improving food crop produce, its nutritive quality, and safety benefitting both human and environmental health. Additionally, the complicated intricacy of the soil biodiversity and ecosystems needs to be well understood and established before attempting manipulations that may lead to resilience from microbial ecosystems. Most appropriately, functional diversity, and not just biodiversity, offers a more befitting approach for considering soil's biome status. In recent times,

Table 1.1 Types of agricultural practices and their effects on soil microbiome (Bertola et al. 2021; Bhat et al. 2018; Gupta et al. 2022; Hermans et al. 2023)

Agricultural practices	Impact on soil microbiome
(a) Fertilizers and plant protectors	
Use of agrochemicals: inorganic fertilizers, pesticides, insecticides, weedicides, fungicides	Deteriorated biological and physicochemical characteristics of soil and hence the microbial communities
Green manures and biofertilizers, biostimulants, vermicomposting, biological control agents—biopesticides, bioherbicide	Enhance biodiversity in the soil, offer bioremediation, prevent resistant pathogens, and is safer for microbial ecosystems
(b) Cultivated plant varieties	
High-yielding cultivars	Deplete nutrients from the soil leading to loss of microbiota
Wild varieties	Improved biodiversity and environmental sustainability
(c) Tillage	
Traditional practices: increased tillage leads to better harvest	Beneficial short-term effects but long-term effects include soil erosion and loss of microbial species
Modern practices: no/zero tillage with enhanced residue management	Benefits the soil microbiota on a long-term basis
Conservational practices: organic inputs with minimal tillage	Benefits the soil microbiota both on a short-term and long-term basis
(d) Cropping practices	
Mono-cropping	Depletes soil nutrients from and leads to a decrease in microbial species and their number
Crop rotation and mixed cropping or inter-cropping, agro-forestry, landscape diversity	Nutrient recycling leads to increased microbial species and population

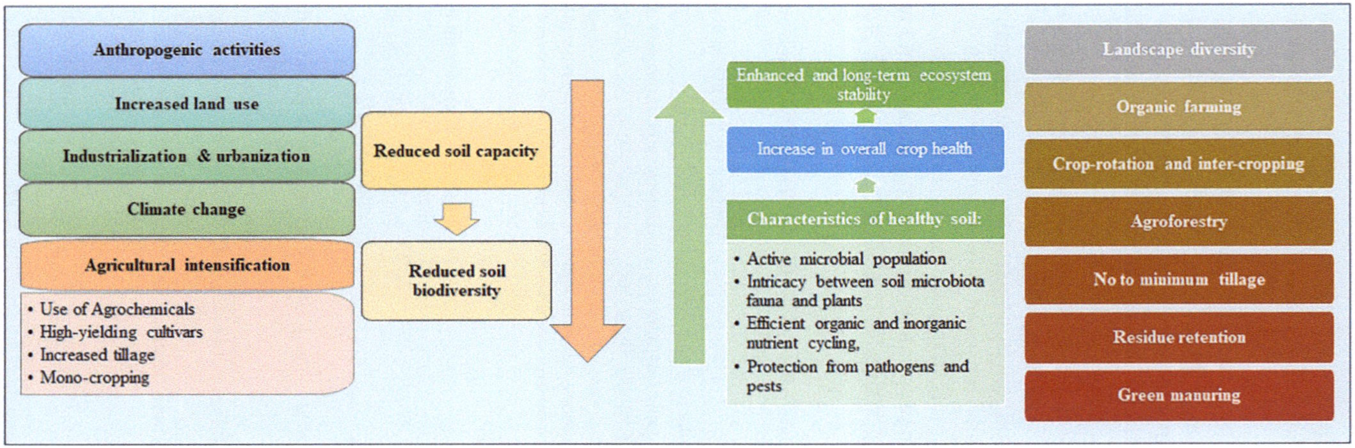

Fig. 1.1 Agricultural and anthropogenic factors affecting soil health and microbiota

"omics" methodologies have shed light on the intricate mechanisms underlying the mutual manipulation and influence of individual components, employing omics for the high throughput study and characterization of the complete microbial biological and functional diversity with genomics (DNA), transcriptomics (RNA), proteomics (proteins), or metabolomics (metabolites) (Bertola et al. 2021).

The biological and functional diversity of soil microbiome can be manipulated by indirect methods and management practices as explained in Table 1.1 or direct modification of the soil microbial communities. The latter includes the introduction of beneficial microbial inoculants in the soil ecosystem or rhizosphere. Although these are challenging in terms of implementation, they do offer commercially viable options with the potential to enhance soil fertility and act as biocontrol agents (Hermans et al. 2023). Both direct and indirect methods require deeper insight into the relationship between plant productivity and soil microbiome.

1.7 Conclusions and Future Perspectives

For developing self-regenerating agroecosystems, further studies deciphering the specific role and interactions of microbial communities with the soil ecosystem are required. A holistic understanding of the interconnected system of biotic and abiotic components of soil in coordination with agricultural practices, geographical or habitat-specific factors, and the history of soil usage is required to comprehend and offer a solution that is better for both the environment and the agriculture industry. A combination of both traditional and modern-day methods is also being studied aiming at environmentally sustainable and economically viable agricultural practices. Given the extensive research that has been carried out in the field of soil and agricultural microbiology and soil bioremediation in the last two decades and the advent of omics and AI technologies, an integrative approach needs to be developed based on bio-physicochemical characteristics of the soil, crop variety(s) to be grown, supply-chain logistics, and geographical/political factors. Practices of "regenerative agriculture" may be adopted encompassing methods of preventing soil degradation and improving soil fertility and ecosystem through biological means. Large-scale spatiotemporal studies on manipulating soil microbial communities, directly or indirectly, need to be conducted to observe the patterns and changes in soil fertility and dynamics. They may further be empowered with computational predictive models to offer solutions to the complexities of natural and anthropological factors affecting agricultural land.

References

Adesemoye AO, Kloepper JW (2009) Plant–microbes interactions in enhanced fertilizer-use efficiency. Appl Microbiol Biotechnol 85(1):1–12

Anderson TH, Domsch KH, Gajdos E (1983) Physiological methods for studying microbial biomass in soil. Soil Biol Biochem 15(6):657–663. https://doi.org/10.1016/0038-0717(83)90134-4
Arora PK, Bae H (2014) Bacterial degradation of chlorpyrifos: present and future perspectives. J Hazard Mater 267:50–59
Babalola OO, Akanmu AO, Fadiji AE (2023) Dataset of shotgun metagenomic evaluation of lettuce (*Lactuca sativa* L.) rhizosphere microbiome. Data Brief 48:109214. https://doi.org/10.1016/j.dib.2023.109214
Badri DV, Quintana N, El Kassis EG, Kim HK, Choi YH, Sugiyama A, Verpoorte R, Martinoia E, Manter DK (2009) An ABC transporter mutation alters root exudation of phytochemicals that provoke an overhaul of natural soil microbiota. Plant Physiol 151:2006–2017
Bahram M, Hildebrand F, Forslund SK, Anderson JL, Soudzilovskaia NA, Bodegom PM et al (2018) Structure and function of the global topsoil microbiome. Nature 560:233–237. https://doi.org/10.1038/s41586-018-0386-6
Bahram M, Netherway T, Hildebrand F, Pritsch K, Drenkhan R, Loit K, Anslan S, Bork P, Tedersoo L (2020) Plant nutrient-acquisition strategies drive topsoil microbiome structure and function. New Phytol 227:1189–1199
Bais HP, Weir TL, Perry LG, Gilroy S, Vivanco JM (2006) The role of root exudates in rhizosphere interactions with plants and other organisms. Annu Rev Plant Biol 57:233–266
Bakker MG, Chaparro JM, Manter DK, Vivanco JM (2018) Impacts of bulk soil microbial diversity on plant microbiomes and organic matter decomposition in a wheat agroecosystem. Sci Rep 8:16911
Bardgett RD, van der Putten WH (2014) Belowground biodiversity and ecosystem functioning. Nature 515(7528):505–511
Bastida F, Moreno JL, Nicolás C, Hernández T, García C, Montesinos-Navarro A (2015) Soil microbial diversity—a review of the main ecological factors involved in soil microbial diversity. Soil Biol Biochem 80:29–37
Bender SF, Wagg C, van der Heijden MGA (2016) An underground revolution: biodiversity and soil ecological engineering for agricultural sustainability. Trends Ecol Evol 31:440–452. https://doi.org/10.1016/j.tree.2016.02.016
Berendsen RL, Pieterse CM, Bakker PA (2012) The rhizosphere microbiome and plant health. Trends Plant Sci 17(8):478–486
Berg G, Smalla K, Kinkel L (2005) Plant-microbe interactions and disease suppression. In: Karpinski DD (ed) Ecological communities: plant mediation in indirect interaction webs. Cambridge University Press, Oxford, pp 187–208
Bergersen FJ (1997) Nitrogen fixation. Academic Press, Cambridge, MA
Bertola M, Ferrarini A, Visioli G (2021) Improvement of soil microbial diversity through sustainable agricultural practices and its evaluation by -Omics approaches: a perspective for the environment, food quality and human safety. Microorganisms 9(7):1400. https://doi.org/10.3390/microorganisms9071400
Bhat SA, Singh S, Singh J, Kumar S, Bhawana, Vig AP (2018) Bioremediation and detoxification of industrial wastes by earthworms: vermicompost as powerful crop nutrient in sustainable agriculture. Bioresour Technol 252:172–179. https://doi.org/10.1016/j.biortech.2018.01.003
Blagodatskaya E, Kuzyakov Y (2013) Active microorganisms in soil: critical review of estimation criteria and approaches. Soil Biol Biochem 67:192–211
Boddy L (2011) Fungal ecology: principles and mechanisms of colonization and competition by saprotrophic fungi. John Wiley & Sons, New York
Bonkowski M, Griffiths BS, Ritz K (2019) Food choices of soil animals: preferences, feedbacks, and consequences. Front Plant Sci 10:1427
Bradford MA, Jones TH, Bardgett RD (2002) The use of microbial indicators in soil and sediment pollution monitoring. Water Air Soil Pollut 2:63–94
Brady NC, Weil RR (2016) The nature and properties of soils. Pearson, Columbus, OH
Chaparro JM, Badri DV, Vivanco JM (2013) Rhizosphere microbiome assemblage is affected by plant development. ISME J 8:790–803

Compant S, Duffy B, Nowak J, Clément C, Barka EA (2005) Use of plant growth-promoting bacteria for biocontrol of plant diseases: principles, mechanisms of action, and future prospects. Appl Environ Microbiol 71(9):4951–4959

Cotrufo MF, Wallenstein MD, Boot CM, Denef K, Paul E (2013) The microbial efficiency-matrix stabilization (MEMS) framework integrates plant litter decomposition with soil organic matter stabilization: do labile plant inputs form stable soil organic matter? Glob Chang Biol 19(4):988–995

Crowther TW, Maynard DS, Leff JW et al (2014) Predicting the responsiveness of soil biodiversity to deforestation: a cross-biome study. Glob Chang Biol 20:2983–2994. https://doi.org/10.1111/gcb.12565

Dassen S, Cortois R, Martens H, de Hollander M, Kowalchuk GA, van der Putten WH, De Deyn GB (2017) Differential responses of soil bacteria, fungi, archaea and protists to plant species richness and plant functional group identity. Mol Ecol 26(15):4085–4098. https://doi.org/10.1111/mec.14175

Delgado-Baquerizo M, Oliverio AM, Brewer TE, Benavent-González A, Eldridge DJ, Bardgett RD et al (2016) A global atlas of the dominant bacteria found in soil. Science 352(6290):1104–1104

Delgado-Baquerizo M, Oliverio AM, Brewer TE, Benavent-González A, Eldridge DJ, Bardgett RD, Maestre FT, Singh BK, Fierer N (2018) A global atlas of the dominant bacteria found in soil. Science (New York, NY) 359(6373):320–325. https://doi.org/10.1126/science.aap9516

Delgado-Baquerizo M, Eldridge DJ, Liu YR, Sokoya B, Wang JT, Hu HW, He JZ, Bastida F, Moreno JL, Bamigboye AR, Blanco-Pastor JL, Cano-Díaz C, Illán JG, Makhalanyane TP, Siebe C, Trivedi P, Zaady E, Verma JP, Wang L et al (2021) Global homogenization of the structure and function in the soil microbiome of urban greenspaces. Sci Adv 7(28):eabg5809. https://doi.org/10.1126/sciadv.abg5809

Dini-Andreote F, Van Elsas JD (2019) The soil microbiome—an overview. In: Van Elsas JD, Trevors JT, Rosado AS, Nannipieri P (eds) Modern soil microbiology, 3rd edn. CRC Press, Baca Raton, CA, pp 37–48

Dubey A, Malla MA, Khan F, Chowdhary K, Yadav S, Kumar A, Sharma S, Khare PK, Khan ML (2019) Soil microbiome: a key player for conservation of soil health under changing climate. Biodivers Conserv 28(8–9):2405–2429. https://doi.org/10.1007/s10531-019-01760-5

Fierer N (2017) Embracing the unknown: disentangling the complexities of the soil microbiome. Nat Rev Microbiol 15:579–590. https://doi.org/10.1038/nrmicro.2017.87

Fierer N, Jackson RB (2006) The diversity and biogeography of soil bacterial communities. Proc Natl Acad Sci 103(3):626–631. https://doi.org/10.1073/pnas.0507535103

Fierer N, Ladau J, Clemente JC, Leff JW, Owens SM, Pollard KS et al (2012) Reconstructing the microbial diversity and function of pre-agricultural tallgrass prairie soils in the United States. Science 342(6158):621–624. https://doi.org/10.1126/science.1243768

Geisen S (2021) The future of (soil) microbiome studies: current limitations, integration, and perspectives. MSystems 6(4):e0061321. https://doi.org/10.1128/msystems.00613-21

Glick BR (2012) Plant growth-promoting bacteria: mechanisms and applications. Scientifica (Cairo) 2012:963401. https://doi.org/10.6064/2012/963401

González G, Dignac MF, Feder F, Rusch A, Richert C, Herrmann AM, Haider MZ, Rumpel C (2019) Earthworm casts enhance carbon and nitrogen mineralization: evidence from a microcosm experiment. Soil Biol Biochem 138:107568

Griffiths BS, Rombke J, Schmelz RM, Scheffczyk A, Faber JH, Bloem J, Bakonyi G, Mäder P (2016) Soil microbial community structure and function in a global change experiment. Plant Soil 401:221–234

Gupta A, Singh UB, Sahu PK, Paul S, Kumar A, Malviya D, Singh S, Kuppusamy P, Singh P, Paul D, Rai JP, Singh HV, Manna MC, Crusberg TC, Kumar A, Saxena AK (2022) Linking soil microbial diversity to modern agriculture practices: a review. Int J Environ Res Public Health 19(5):3141. https://doi.org/10.3390/ijerph19053141

Hartman K, van der Heijden MGA, Wittwer RA, Banerjee S, Walser JC, Schlaeppi K (2017) Cropping practices manipulate abundance patterns of root and soil microbiome members paving the way to smart farming. Microbiome 5(1):1–17. https://doi.org/10.1186/s40168-017-0267-5
Hartmann M, Brunner I, Hagedorn F, Bardgett RD, Stierli B, Herzog C, Chen X, Zingg A, Graf Pannatier E (2018) A decade of irrigation transforms the soil microbiome of a semi-arid pine forest. Mol Ecol 27:4067–4082
Hermans SM, Lear G, Case BS, Buckley HL (2023) The soil microbiome: an essential, but neglected, component of regenerative agroecosystems. iScience 26(2):106028. https://doi.org/10.1016/j.isci.2023.106028
Herridge DF, Peoples MB, Boddey RM (2008) Global inputs of biological nitrogen fixation in agricultural systems. Plant Soil 311(1–2):1–18
Herrmann AM, Witter E, Kühn O (2008) Effect of different land use on the composition of soil organic matter in particle-size separates in Nepal. Soil Sci Soc Am J 72(5):1365–1372
Hu L, Robert CAM, Cadot S, Zhang X, Ye M, Li B, Manzo D, Chervet N, Steinger T, van der Heijden MGA, Schlaeppi K, Erb M (2018) Root exudate metabolites drive plant-soil feedbacks on growth and defense by shaping the rhizosphere microbiota. Nat Commun 9(1):2738. https://doi.org/10.1038/s41467-018-05122-7
Islam W, Noman A, Naveed H, Huang Z, Chen HYH (2020) Role of environmental factors in shaping the soil microbiome. Environ Sci Pollut Res 27(33):41225–41247. https://doi.org/10.1007/s11356-020-10471-2
Jacoby R, Peukert M, Succurro A, Koprivova A, Kopriva S (2017) The role of soil microorganisms in plant mineral nutrition-current knowledge and future directions. Front Plant Sci 8:1617. https://doi.org/10.3389/fpls.2017.01617
Jin K, Shen J, Ashton RW, White RP, Dodd IC, Parry MAJ, Whalley WR (2010) The effect of plant growth promoting rhizobacteria on the growth and root system architecture of wheat grown under a range of ammonium concentrations. Ann Appl Biol 157:193–206
Kaiser C, Meyer H, Biasi C, Rusalimova O, Barsukov P, Richter A (2015) Driving forces of soil bacterial community structure, diversity, and function in temperate grasslands and forests. Sci Rep 5:17056
Kraut-Cohen J, Zolti A, Shaltiel-Harpaz L, Argaman E, Rabinovich R, Green SJ, Minz D (2020) Effects of tillage practices on soil microbiome and agricultural parameters. Sci Total Environ 705:135791
Kuzyakov Y, Razavi B (2019) Rhizosphere size and shape: temporal dynamics and spatial stationarity. Soil Biol Biochem 135:343–360. https://doi.org/10.1016/j.soilbio.2019.04.008
Lal R (2004) Soil carbon sequestration to mitigate climate change. Geoderma 335:10–19
Leake J, Johnson D, Donnelly D, Muckle G, Boddy L, Read D (2004) Networks of power and influence: the role of mycorrhizal mycelium in controlling plant communities and agroecosystem functioning. Can J Bot 82(8):1016–1045. https://doi.org/10.1139/b04-060
Lehmann J, Kleber M (2015) The contentious nature of soil organic matter. Nature 528(7580):60–68. https://doi.org/10.1038/nature16069
Li Y, Wang C, Wang T, Liu Y, Jia S, Gao Y, Liu S (2020) Effects of different fertilizer treatments on rhizosphere soil microbiome composition and functions. Land 9(9):329. https://doi.org/10.3390/land9090329
Lynch MDJ, Neufeld JD (2015) Ecology and exploration of the rare biosphere. Nat Rev Microbiol 13(4):217–229. https://doi.org/10.1038/nrmicro3400
Mattarozzi M, Di Zinno J, Montanini B, Manfredi M, Marengo E, Fornasier F, Ferrarini A, Careri M, Visioli G (2020) Biostimulants applied to maize seeds modulate the enzymatic activity and metaproteome of the rhizosphere. Appl Soil Ecol 148:103480. https://doi.org/10.1016/j.apsoil.2019.103480
Mau RL, Liu CM, Aziz M, Schwartz E, Dijkstra P, Marks JC et al (2015) Linking soil bacterial diversity and vegetation community structure through co-occurrence modeling. ISME J 9(3):849–856. https://doi.org/10.1038/ismej.2014.184

Mendes R, Garbeva P, Raaijmakers MJ (2013) The rhizosphere microbiome: significance of plant beneficial, plant pathogenic, and human pathogenic microorganisms. FEMS Microbiol Rev 37(5):634–663. https://doi.org/10.1111/1574-6976.12028

Mitter B, Pfaffenbichler N, Flavell R, Compant S, Antonielli L, Petric A, Berninger T, Naveed M, Sheibani-Tezerji R, von Maltzahn G, Sessitsch A (2017) A new approach to modify plant microbiomes and traits by introducing beneficial bacteria at flowering into progeny seeds. Front Microbiol 8:11. https://doi.org/10.3389/fmicb.2017.00011

Nannipieri P (2020) Soil is still an unknown biological system. Appl Sci 10(11):3717. https://doi.org/10.3390/app10113717

Nannipieri P, Ascher J, Ceccherini MT, Landi L, Pietramellara G, Renella G (2003) Microbial diversity and soil functions. Eur J Soil Sci 54(4):655–670. https://doi.org/10.1046/j.1351-0754.2003.0556.x

Nannipieri P, Ascher J, Ceccherini MT, Landi L, Pietramellara G, Renella G (2017) Microbial diversity and soil functions. Eur J Soil Sci 68(1):12–26. https://doi.org/10.1111/ejss.4_12398

Nannipieri P, Penton CR, Purahong W, Schloter M, van Elsas JD (2019) Recommendations for soil microbiome analyses. Biol Fertil Soils 55(8):765–766. https://doi.org/10.1007/s00374-019-01409-z

Narveer N, Vyas A, Kumar H, Putatunda C (2014) In vitro phosphate solubilization by Bacillus sp. NPSBS 3.2.2 obtained from the cotton plant rhizosphere. Biosci Biotechnol Res Asia 11(2):401–406. https://doi.org/10.13005/bbra/1288

Nesme J, Achouak W, Agathos SN, Bailey M, Baldrian P, Brunel D, Frostegård S, Heulin T, Jansson JK, Jurkevitch E, Kruus KL, Kowalchuk GA, Lagares A, Lappin-Scott HM, Lemanceau P, Le Paslier D, Mandic-Mulec I, Murrell JC, Myrold DD et al (2016) Back to the future of soil metagenomics. Front Microbiol 7:73. https://doi.org/10.3389/fmicb.2016.000733

Omotayo OP, Babalola OO (2021) Resident rhizosphere microbiome's ecological dynamics and conservation: towards achieving the envisioned sustainable development goals, a review. Int Soil Water Conserv Res 9(1):127–142. https://doi.org/10.1016/j.iswcr.2020.08.002

Orgiazzi A, Singh B, Wall D, Barrios E, Kandeler E, Moreira F, De Deyn G, Chotte J, Six J, Hedlund K, Briones M, Miko L, Johnson N, Ramirez K, Fierer N, Kaneko N, Lavelle P, Eggleton P, Lemanceau P, Bardgett R, Jeffery S, Fraser T, Behan Pelletier V, Van Der Putten W, Montanarella L, Jones A (2016) Global soil biodiversity Atlas. Publications Office of the European Union, Luxembourg

Osono T, Takeda H, Azuma JI (2009) Depolymerization of polychlorinated biphenyls by white-rot fungi: microscale evidence of the degradation mechanism. Appl Environ Microbiol 75(3):864–870

Peiffer JA, Spor A, Koren O, Jin Z, Tringe SG, Dangl JL, Buckler ES (2013) Diversity and heritability of the maize rhizosphere microbiome under field conditions. Proc Natl Acad Sci 110(16):6548–6553

Philippot L, Raaijmakers JM, Lemanceau P, van der Putten WH (2013) Going back to the roots: the microbial ecology of the rhizosphere. Nat Rev Microbiol 11(11):789–799

Ramirez KS, Leff JW, Barberán A, Bates ST, Betley J, Crowther TW, Kelly EF, Oldfield EE, Shaw EA, Steenbock C, Bradford MA, Wall DH, Fierer N (2014) Biogeographic patterns in below-ground diversity in New York City's Central Park are similar to those observed globally. Proc R Soc B Biol Sci 281(1795):20141988. https://doi.org/10.1098/rspb.2014.1988

Sánchez-López R, Garcia-Sanchez MJ, Hernandez-Apaolaza L, García-Sánchez F (2021) Biofertilizers: an environmentally friendly alternative to chemical fertilizers for sustainable agriculture. Agriculture 11(5):410

Santos LF, Olivares FL (2021) Plant microbiome structure and benefits for sustainable agriculture. Curr Plant Biol 26:100198. https://doi.org/10.1016/j.cpb.2021.100198

Schimel JP, Balser TC, Wallenstein M (2007) Microbial stress-response physiology and its implications for ecosystem function. Ecology 88(6):1386–1394. https://doi.org/10.1890/06-0219.1

Serna-Chavez HM, Fierer N, van Bodegom PM (2013) Global drivers and patterns of microbial abundance in soil. Glob Ecol Biogeogr 22(10):1162–1172. https://doi.org/10.1111/geb.12070

Singh BK, Bardgett RD, Smith P, Reay DS (2019) Microorganisms and climate change: terrestrial feedbacks and mitigation options. Nat Rev Microbiol 17(11):706–720. https://doi.org/10.1038/s41579-019-0256-x

Sinsabaugh RL, Lauber CL, Weintraub MN, Ahmed B, Allison SD, Crenshaw C et al (2016) Stoichiometry of soil enzyme activity at global scale. Ecol Lett 19(7):800–809. https://doi.org/10.1111/ele.12624

Smith SE, Read DJ (2008) Mycorrhizal symbiosis. Academic Press, Cambridge

Sokol NW, Slessarev E, Marschmann GL, Nicolas A, Blazewicz SJ, Brodie EL, Firestone MK, Foley MM, Hestrin R, Hungate BA, Koch BJ, Stone BW, Sullivan MB, Zablocki O, Trubl G, McFarlane K, Stuart R, Nuccio E, Weber P et al (2022) Life and death in the soil microbiome: how ecological processes influence biogeochemistry. Nat Rev Microbiol 20(7):415–430. https://doi.org/10.1038/s41579-022-00695-z

Starke R, Jehmlich N, Bastida F (2019) Using proteins to study how microbes contribute to soil ecosystem services: the current state and future perspectives of soil metaproteomics. J Proteome 198:50–58. https://doi.org/10.1016/j.jprot.2018.11.011

Stefan G, Cornelia B, Jörg R, Michael B (2014) Soil water availability strongly alters the community composition of soil protists. Pedobiologia 57:205–213

Sui Y, Gao X, Li H (2019) Effects of soil moisture on soil microbial communities: a review. J Resources Ecol 10(4):366–375. https://doi.org/10.5814/j.issn.1674-764x.2019.04.006

Tecon R, Or D (2017) Biophysical processes supporting the diversity of microbial life in soil. FEMS Microbiol Rev 41(5):599–623. https://doi.org/10.1093/femsre/fux039

Tisdall JM, Oades JM (1982) Organic matter and water-stable aggregates in soils. J Soil Sci 33(2):141–163. https://doi.org/10.1111/j.1365-2389.1982.tb01755.x

Ushio M, Wagai R, Balser TC, Kitayama K (2015) Variations in the soil microbial community composition of a tropical montane forest ecosystem: does tree species matter? Soil Biol Biochem 81:1–7. https://doi.org/10.1016/j.soilbio.2014.10.002

Vallino JJ, Stephan JG (2018) Mycorrhizae: a pathway for nutrient transfer in the soil-plant-atmosphere continuum. Crit Rev Plant Sci 37(1):1–23

Wang M, Garrido-Sanz D, Sansegundo-Lobato P, Redondo-Nieto M, Conlon R, Martin M, Mali R, Liu X, Dowling DN, Rivilla R, Germaine KJ (2021) Soil microbiome structure and function in ecopiles used to remediate petroleum-contaminated soil. Front Environ Sci 9:39. https://doi.org/10.3389/fenvs.2021.624070

Wardle DA, Bardgett RD, Klironomos JN, Setälä H, van der Putten WH, Wall DH (2012) Ecological linkages between aboveground and belowground biota. Science 337(6106):1603–1609. https://doi.org/10.1126/science.1224592

Williamson KE, Fuhrmann JJ, Wommack KE, Radosevich M (2017) Viruses in soil ecosystems: an unknown quantity within an unexplored territory. Annu Rev Virol 4(1):201–219. https://doi.org/10.1146/annurev-virology-101416-041639

Yurgel SN, Douglas GM, Dusault A, Percival D, Langille MGI (2018) Dissecting community structure in wild blueberry root and soil microbiome. Front Microbiol 9:1187. https://doi.org/10.3389/fmicb.2018.01187

Zhang H, Kim MS, Sun Y, Dowd SE, Shi H, Paré PW (2016) Soil bacteria confer plant salt tolerance by tissue-specific regulation of the sodium transporter HKT1. Mol Plant-Microbe Interact 29(11):819–829. https://doi.org/10.1094/MPMI-04-16-0085-R

Chapter 2
Tools and Techniques to Measure Soil Health

Khem Raj Meena, Sunita Meena, Rajendra Prasad Meena, Ashutosh Singh, and Arvind Kumar

2.1 Introduction

Soil health assessment and management is prime factor for sustainable development of agriculture. A good soil health is essential to achieve high productivity in terms of crop biomass production and/or produce quality (McBratney et al. 2012; Lehmann et al. 2020). Previously several studies have been conducted regarding the fitness of different soils to enhance the crop's yields and profit as well. But in recent years, this paradigm shifted toward soil quality instead of soil fitness to attract a global audience as it directly and indirectly impacts the agriculture yield, human health, and environmental quality. Soil is the keystone since it is the mother of all grains. Moreover, soil quality also plays a key role in water security, food security, and biodiversity protection (McBratney et al. 2014; Twomlow et al. 2008; Nelson et al. 2010). Soil health determines the vitality of living ecosystem and connects them with each other. A healthy soil reflects how well the soil can carry out its

K. R. Meena (✉)
Department of Biotechnology, School of Life Sciences, Central University of Rajasthan, Ajmer, Rajasthan, India

S. Meena
Department of Soil Sciences, Dr. Rajendra Prasad Central Agricultural University, Pusa, Samastipur, Bihar, India

R. P. Meena
Indian Agricultural Research Institute, New Delhi, India

A. Singh
Centre for Advanced Studies on Climate Change, Dr. Rajendra Prasad Central Agricultural University, Pusa, Samastipur, Bihar, India

A. Kumar
DBT-National Institute of Plant Genome Research, New Delhi, India

R. K. Bhatia, A. Walia (eds.), *Advancements in Microbial Biotechnology for Soil Health*, Microorganisms for Sustainability 50,
https://doi.org/10.1007/978-981-99-9482-3_2

environmental functions. Usually, healthy soils are represented by a sufficient supply of nutrients and appropriate organic contents to the growing plants, correct tilth and drainage, abundant presence of beneficial organisms over pathogens, absence of pathogens and hazardous chemicals, and high resistance to erosion and degradation. An improved understanding of soil heterogeneity at the farm-scale research can greatly help us in monitoring the changes in soil fertility and other associated properties over a time period. Research data obtained from 24 states showed that most of the farmers apply fertilizers more than the required limit without testing the soil health score or knowing the soil nutrients parameters (Dimkpa et al. 2017). This suggests that there is high need of awareness programs and economic and effective analytical tools/kits for soil testing so that farmer can apply and utilize them properly without much complications. To address this need, several laboratories developed numerous efficient tools and techniques for soil health analysis either both site and lab specific. Often these kits/tools which are available in market are based on colorimetry, spectroscopy, and sensor technology (Adamchuk et al. 2004). Despite the various rapid soil-testing options which are available in the world market, only a small number of tools are being used by the developing world (Dimkpa et al. 2017). In this chapter, our aim is to provide a brief analysis of simple and sensitive soil measures, tools, and techniques that can be employed for the easy management of soil health and basic soil fertility parameters to attain long-term agricultural productivity. The abundance of earthworms in the soil the characteristics simple visible biological marker of good soil. They are not only improve the soil physical texture, water retention and porosity but also greatly helps in the cycling and supplying of essential and important nutrients especially N, S, and P to the soils and plants (Stevenson 1986). Soil organic matter status is also an important characteristic that is influenced by the management practices such as tillage intensity, crop residue management, crop rotation, cropping intensity, and crop diversity (Fig. 2.1; Varvel 1994).

2.2 Methods for Assessment of Soil Health

2.2.1 Understanding of Soil Health

Farmers, as well as experienced crop growers, should have sufficient knowledge to estimate and evaluate the soil health and the quality of cropland. Typically, this ability is gained by the farmers through the years of experience of farming activities. Usually, skilled and educated farmers fractioned their cultivation land into the "good" and the "bad" categories depending on the soil health and crop productivities to rectify the measures for further improvement of "bad or poor" cropland (Guo 2021).

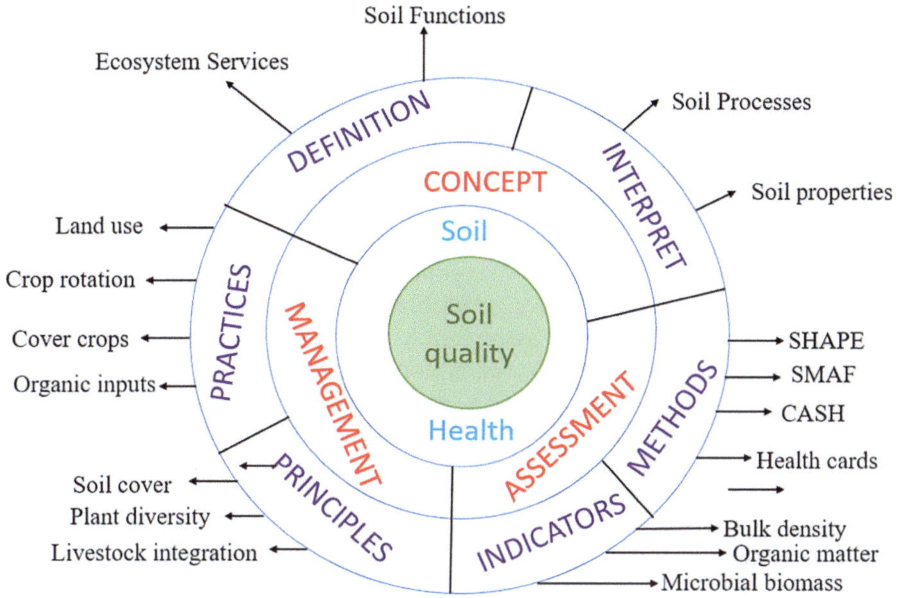

Fig. 2.1 Schematic presentation of focal areas for the management of soil. SHAPE (soil health assessment protocol and evaluation); SMA (soil management assessment framework); CASH (comprehensive assessment of soil health)

2.2.2 Soil Health Cards

A soil health card is a tool of field that is used to assess soil health. A soil health card contains a list of number of soil health indicators such as root growth and compaction, infiltration, soil odor, organic matter content, etc. and farmer's farming experiences along with the information of the local environment. Some of the indicators can be assessed directly without any need of the laboratory instruments. These health cards are very helpful in soil improvement and management and proper utilization of available land resources.

2.2.3 Comprehensive Assessment for Soil Health (CASH)

The CASH method was developed by Cornell University, USA, to examine the overall soil health from a group of parameters like physical, biological, and chemical properties collected from selected soil analysis. The obtained parameters are used to build the best recommendations for the management of that soil. The CASH consists of approximate 12 major soil health indicators which include soil surface hardness and subsurface hardness, active carbon, wet aggregate stability, soil organic matter, protein index, soil respiration, soil pH, extractable phosphorus, extractable potassium, water holding capacity, and micronutrients (Moebius-Clune et al. 2017).

2.3 Application of Tools/Techniques in the Measurement of Soil Health

Several techniques are available to determine the microbial morphological characteristics in the given soil sample, e.g., microbial respiration and biomass estimation. Many of them are being used to assess soil health. These include biogeochemical and physiological approaches, e.g., chloroform, fumigation extraction, substrate-induced respiration, the 1-day burst CO_2 test, and metabolic techniques (Costantini et al. 2016; Muñoz-Rojas et al. 2016; Griffiths et al. 2016). Methods like stable isotope probing, phospholipid fatty acid analysis (PLFA), and DNA probing have helped a lot to link soil biodiversity to soil processes. The microbial component of the soil represents approximately 0.1–0.3% of the total soil volume which is very less than the required. Since microbes facilitate 90% of the functions of a soil ecosystem, therefore this minimum microbial load must be available in a quality (Adhikari and Hartemink 2016). At present the characterization of the soil microbial community is increasingly being used to determine the environmental changes as an indicator of ecosystem recovery (Table 2.1).

Soil quality and soil function is an essential living system within the ecosystem and land boundaries for sustainable biological activities including to preservation of water quality, air environments, animals, plant, and humans (Doran and Safley 1997). Soil microorganisms respond rapidly against a hazardous environment if any. However, it is still unclear whether natural environmental factors play any role in damaging the soil microbiomes genotypic capacity or not in recovery of soil health after harsh conditions (Sparling 1997). Research on the soil microbiota resilience may act as an important field in study seasonal microorganism's fluctuation by using molecular approaches. Some of the general soil parameters are as follow.

2.3.1 Soil Organic Matter and Soil Organic Carbon

Soil organic matter (SOM) is significant parameter in the evaluation of soil fertility and its structure. SOM is heterogeneous that varies in particle size, carbon content, decomposition rate, and on turnover time, whereas changes in soil organic carbon (SOC) normally occur over a period of many years, and often it is difficult to identify small changes of SOC in short period of time. SOC is the main energy source for the growing soil microorganisms and often used as synonym of SOM. The SOM/SOC can be estimated by both colorimetric and volumetric methods. The SOM normally used to represent N availability index (the available nitrogen in the

Table 2.1 Soil health characteristics and their concerned soil health tests

Characteristics of soil health	Sub-characteristics of soil health	Soil health-concerned tests
Soil nutrients	Nutrient amount at the time of sampling	Soil nutrient test
	Soil ability to not losing nutrients to leaching/runoff	Cation interchange capacity
	Nutrient availability for the plant growth	pH
Soil organic matter	Organic matter	Total SOM (soil organic matter)
	Dead and living organic matter	Organic matter (active form)
		Particulate organic matter (POM)
Soil aggregates	Amount of stable aggregates	Aggregate stability
	Organisms play role in aggregate formation	Soil protein/glomalin
Diversity and growth of soil organisms	Growth (activity) of soil organisms	Soil respiration
	Decompositions rate by soil organisms	Nitrogen mineralization
	Populations with diversity of various soil organisms	Phospholipid Fatty Acid (PLFA)
Physical and chemical characteristics	Soil compaction	Bulk density
	Rainfall or irrigation ability of water to soak into the soil	Infiltration of the water
	Gritty or stickiness of a soil	Soil texture test

soil for plants) in the soil. Potassium dichromate ($K_2Cr_2O_7$), an oxidizing agent, is used for the estimation of oxidizable organic carbon. Walkely-Black (WB) method was once widely popular for determining the SOC (Yang et al. 2022). In WB method the chromic ions (proportional to the oxidized organic carbon) produced after addition of concentrated sulfuric acid (H_2SO_4) and K_2Cr_2O are measured. This method provides a universal factor for the correction to convert organic carbon to calculate correct quantified organic carbon present in the soil assuming ~77% recovery of soil organic carbon. The following formula can be used to calculate the SOM from given SOC.

$$\text{SOM} = \text{SOC} \times 1.72.$$

However, neither the Walkely-Black method, nor the combustion method (not described anywhere) are capable to differentiate the biologically active carbon from biologically inactive carbon. Therefore, mid-infrared spectroscopy came to fulfil this purpose, and it is widely accepted all over the world.

2.3.2 Soil pH

pH is an important characteristic to find out the acidity, alkalinity, or neutrality of a given soil. Any alteration in the soil's pH may greatly affect the nutrients like phosphorus, iron, molybdenum, manganese, and aluminum availability to the plants (Richard et al. 2020). Moreover, microbes are also very sensitive to pH of the surrounding medium; hence, soil pH may also alter the microbial diversity and their population. In acidic soils it is necessary to neutralize pH by application of lime before to start normal agricultural activities, and for this the pH measurement is the essential part for the calculation of lime requirement (LR). There are two methods to measure the soil pH in the laboratory:

1. Older method (pHw): In this one part of soil is suspended in five parts of water.
2. Newer method (pHc): In this one part of soil is suspended in five parts of 0.01 M $CaCl_2$ solution and pH measured.

For increasing the soil pH of an acidic soil, along with pH, soil organic matter, soil aluminum (for phytotoxicity), and soil texture should also be examined thoroughly, before adding the lime.

2.3.3 Estimation of Nitrogen Turnover

Bioavailable nitrogen (N) is also one of the key characteristics for evaluation of soil health. On the other hand, nitrogenous compounds like nitrate, nitrite, or N_2O are considered as pollutant which play a significant role in environmental pollution. Therefore, it is of great concern to understand the key processes involved in the nitrogen cycle in more details to define ways to cope up with these pollutants without affecting the agricultural productivity as well as environment. In nature, where at one side, two main nitrogen delivery processes like mineralization and nitrogen fixation are very common and well known for the enhancement of crop yield, on the other side, there are denitrification processes that cause significant loss of nitrogen from the bound pool. This has to be managed properly. Microbes are important for effective mineralization and mobilization of organically bound N (e.g., proteins) to plants (Ladd and Butler 1972). The microbially derived extracellular proteases (protein-degrading enzyme) are crucial for the hydrolysis or complete degradation of the complex macromolecular polypeptides into smaller molecules and make accessible for growing plants and other organisms. Research report indicates that differences in the proteases activities are not due to different microbial populations; instead, it is due to variation in the expression rates of the same community at a given time. Since the proteases are exo-enzymes and are not under control of microbial cells once released, therefore, the measurement of proteolytic activity cannot be counted as a sensible indicator to represent the actual microbial activity in a given soil sample. However, several attempts are made to correlate the ecophysiological conditions that cause an induction or repression of the proteases/peptidases expression in the habitat (Bach et al. 2001).

2.3.4 *Electrical Conductivity Measurement*

Electrical conductivity (EC) is the measurement of the ion's transportation in a solution between the cathode and anode. EC is generally considered as a measurement of the dissolved salts present in a given soil sample. The optimal EC levels in the soil ranged from 110 to 570 millisiemens per meter (mS/m). The excess EC may affect the plant growth. To estimate the EC of soil, collect soil sample from the 0 to 7 cm depth and dissolved in 1:5 ratio (soil:water) to make soil-water suspension, and filter to obtain homogenous solution and determine EC (Grace and Weier 2007). Some methods also recommend the soil:water ratio of 1:2.5, i.e., 10 g soil with 25 ml of distilled water to determine the electrical conductivity of the soil. The EC is the index which show the salinity of the soil and often measured by a standard EC meter.

2.3.5 *Water Repellence Test (Field Method)*

Water-repellent soils generally have tendency to repel water absorption after rainfall. In this case water failed to evenly infiltrate into soil until there is formation of a continuous water film on the soil particles, even though one can see the uneven wetting patterns. The water repellence is caused due to the formation of a long-chain series of polyethylene waxes that are attached to the sand grains (McDonald et al. 1990). Subsequently this leads to the reduction of crop and pasture production, and ultimately it convert soil into a bare soil. Field testing method is a simple and lab-based test; however, an abbreviated form is used.

2.3.6 *Salinity*

Salinity is critical in both arid and semiarid areas where excess transpiration results, and this results in the formation of salts in near-surface horizons. This can be measured by a simple field test using a handheld salinity meter, and it is very useful for conducting preliminary investigations, ongoing monitoring activities, and point sampling of selected areas. Electromagnetic mapping (EM) using specific instruments can be used to characterize and map spatial variability of soil and apparent salinity over the larger areas. This is a valuable tool for land planning and provides a rapid assessment of differences across a paddock. More precise water and soil laboratory tests can also be performed to confirm preliminary field testing where a possible salinity problem is suspected.

2.3.7 Available Phosphorus

Phosphorus is an important element that is required for healthy plant growth which plays several key functions such as photosynthesis, transformation of sugars and starches, energy transfer, nutrient movement within the plant, and transfer of genetic characteristics from one generation to the next. But its limit should not be crossed from an environmental hazardous limit (Sharpley et al. 1996). Farmer's management practices can influence available phosphorus through the fertilizers and animal manure applications at approximate neutral pH. The Bray Kurtz test's extractant (Dr. Bray and Dr. Kurtz in 1945) is a mixture of hydrochloric acid (HCl) and ammonium fluoride (NH_4F) solution which is generally recommended for testing of phosphorus in acidic and neutral soils ($pH \leq 7.0$) but not for the alkaline soils ($pH > 7.0$). In the Olsen method, sodium bicarbonate ($NaHCO_3$) extractant is used for the estimation of phosphorus in a given soil sample in which Ca-phosphates and phosphate are adsorbed on the surfaces of calcium and magnesium carbonates extracted from the soil with 0.5 M $NaHCO_3$ at pH 8.5. The hydroxide ion and carbonate ions in the $NaHCO_3$ solution controls the activity of Ca^{2+}, Al^{3+}, and Fe by calcium precipitation as carbonate and aluminum and iron as hydroxides. In this method, the dry soil (~5 g) is added with 100 ml of extracting solution and capped. Thereafter a half teaspoon of activated charcoal is added, and bottles are shaked for 30 min and filtered followed by mixing of approximate 3 ml of reagent to develop blue color (FAO 2021).

2.3.8 Microbial Carbon Biomass

It provides the data of the biological activities occurring within a soil sample. It reflects nutrient cycling processes which are essential to meet crop growth. It also influenced by management practices such as tillage intensity, crop type (annuals versus perennials), and crop residue management strategies. Microbial biomass measured generally by chloroform fumigation extraction method in which microorganisms are killed first by exposing fresh soil to the ethanol-free chloroform for 24 h followed by extraction of the carbon released from lysed microbial cells with salt solution. Analysis of the net increase in soluble carbon multiplied by a conversion factor for the determination of microbial total biomass carbon (Carter et al. 1999).

2.4 SMAF (Soil Management Assessment Framework)

The SMAF consists of three steps: indicator selection, indicator interpretation, and integration into a soil quality index (Andrews et al. 2004). The indicator selection step uses an expert system of decision rules to recommend indicators for inclusion in the assessment based on the user's stated management goals, location, and current practice. In the indicator interpretation step, observed indicator data is transformed into a unit less score based on clearly defined, site-specific relationships to the soil function. The soil functions of interest include crop productivity, physical stability, water and solute flow, nutrient cycling, contaminant filtering, buffering, and biodiversity. The indicator interpretation steps uses various factors (i.e., organic matter, climate, texture, region, slope, weathering, mineralogy, class, sampling time, crop, and analytical method) to adjust threshold values in the scoring curves that are used to assign a relative value of 0-1 for each type of data being collected.

2.5 Measurement of Indicator Tools

Soil health Indicator tools such as high rates of soil erosion, losses of organic matter, text of what the soil is being asked to do, inherent reductions in fertility, productivity, environmental, chemical and abiotic influences such a temperature, heavy metal contamination, degradation of air and precipitation. The target/optimum soil and water quality have sparked interest in the concept of not one standard for the USA or the other world. Instead, soil quality (SQ) and its assessment is a series of thresholds defined by the limiting factors that fully fulfil plant growth (Larson and Pierce 1991; Karlen et al. 2003). Although SQ has a variety of indicators that can be defined loosely, as those of soil definitions in the current literature properties and processes that have greatest sensitivity erasure (Doran and Safley 1997; Karlen et al. 2003). Some important fact to be emphasized is that SQ indicators should correlate well with soil functions (or ecosystem services). Indicator groups or the Minimum Data Sets (MDSs), used in the maintenance of biodiversity and habitat, are shown in Fig. 2.2 and Table 2.2.

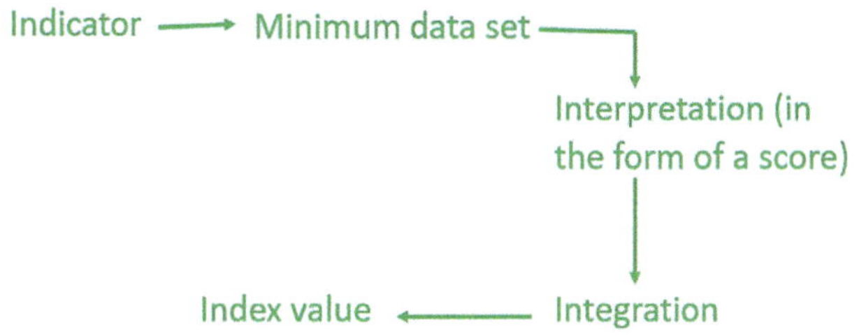

Fig. 2.2 Conceptual framework for the soil management assessment (Andrews 1998)

Table 2.2 Recommended indicators and methods (Ref…)

Soil process	Soil health indicators	Methods	Notes/remarks/reference
Cycling of organic matter and carbon sequestration	Soil organic carbon content	Dry combustion	Neslson and Sommers (1996)
		Wet oxidation	Gives numbers comparable to dry combustion but produces chemical wastes and is more labor intensive
Soil stability (infiltration)	Aggregation	macroaggregate stability (MAS)	Kemper and Rosenau (1986)
		NRCS wetted aggregation	Kemper and Rosenau (1986)
		Cornell sprinkle infiltrometer	Schindlbeck et al. (2016)
Microbial activity	Enzyme activity (EA)	β-glucosidase (BG)	Eivazi and Tabatabai (2002)
Carbon food source	Readily available carbon pool	Permanganate oxidizable C (POXC)	Weil et al. (2003)
		Particulate organic matter	–
Bioavailable nitrogen	Available organic nitrogen pool	Autoclaved citrate extractable (ACE) protein content	Schindlbeck et al. (2016)
Microbial diversity	Community structure	PLFA (phospholipid fatty acid)	Buyer and Sasser (2012)

The tests and tools listed in this section are laboratory methods. The application(s) of these tests to soil fertility investigations are reviewed and explained well earlier (Peverill et al. 1999).

2.6 Conclusion

Intensively, soil health studies practiced as a means of improving global, social, environmental, and economic sustainability. However, the soil health indicators and the assessment methods need further improvement and validation in terms of relevance, scientific validity, and economic and local adaptability. Importantly the soil health tool adoption and application can help in the rational use of fertilizer to make balance between soil health and crop productivity, and in saving money of the farmers and ranchers through accurate determination of plant's nutrient need. These improvements will ultimately benefit the environment by preventing the nutrient losses.

References

Adamchuk VI, Hummel JW, Morgan MT, Upadhyaya SK (2004) On-the-go soil sensors for precision agriculture. Comput Electron Agric 44:71–91

Adhikari K, Hartemink AE (2016) Linking soils to ecosystem services—a global review. Geoderma 262:101–111

Andrews SS (1998) Sustainable agriculture alternatives: Ecological and managerial implications of poultry litter management alternatives applied to agronomic soils. (Ph.D. diss). University of Georgia, Athens

Andrews SS, Karlen DL, Cambardella CA (2004) The soil management assessment framework: a quantitative evaluation using case studies. Soil Sci Soc Am J 68:1945–1962

Bach HJ, Hartmann A, Schloter M, Munch JC (2001) PCR primers and functional probes for amplification and detection of bacterial genes for extracellular peptidases in single strains and in soil. J Microbiol Methods 44:173–182

Buyer JS, Sasser M (2012) High throughput phospholipid fatty acid analysis of soils. Appl Soil Ecol 61:127–130

Carter MR, Gregorish EG, Angers DA, Beare MH, Sparling GP, Wardle DA, Voroney RP (1999) Interpretation of microbial biomass measurements for soil quality assessment in humid temperate regions. Can J Soil Sci 79:507–520

Costantini EAC, Branquinho C, Nunes A, Schwilch G, Stavi I, Valdecantos A, Zucca C (2016) Soil indicators 212 to assess the effectiveness of restoration strategies in dryland ecosystems. Solid Earth 7(397–213):414

Dimkpa C, Bindraban P, McLean JE, Gatere E, Singh U, Hellums D (2017) Methods for rapid testing of plant and soil nutrients. In: Sustainable agriculture reviews, vol 25. Springer, Cham, pp 1–43

Doran JW, Safley M (1997) Defining and assessing soil-health and sustainable productivity. In: Pankhurst C, Doube BM, Gupta V (eds) Biological indicators of soil-health. CAB International, Wallingford, pp 1–28. Food security, farming, and climate change to 2050: Scenarios, results, policy options

Eivazi F, Tabatabai MA (2002) Glucosidases and galactosidases in soils. Soil Biol Biochem 20:601–606

FAO (2021) Standard operating procedure for soil available phosphorus—Olsen method. FAO, Rome

Grace PR, Weier KL (2007) Soil-health assessment users guide 2007 version 1. Queensland University of Technology, Brisbane, Australia. 47pp

Griffiths BS, Römbke J, Schmelz RM, Scheffczyk A, Faber JH, Bloem J, Peres G, Cluzeau D, Chabbi A, Suhadolc M et al (2016) Selecting cost effective and policy-relevant biological indicators for European 293 monitoring of soil biodiversity and ecosystem function. Ecol Indic 69:213–223. Intl. Food Policy Res. Inst, 172: 20-28

Guo M (2021) Soil-health assessment and management: recent development in science and practice. Soil Syst 5:61

Karlen DL, Andrews SS, Doran JW, Wienhold BJ (2003) Soil quality: Humankind's foundation for survival. J Soil Water Conserv 58:171–179

Kemper WD, Rosenau RC (1986) Aggregate stability and size distribution. Methods Soil Anal: Part 1 Phys Mineral Methods 5:425–442

Ladd JN, Butler JHA (1972) Short-term assays of soil proteolytic enzyme activities using proteins and dipeptide derivates as substrates. Soil Biol Biochem 4:19–30

Larson WE, Pierce FJ (1991) Conservation and enhancement soil quality. In: Dumanski J et al (eds) Evaluation for sustainable land management in the developing world, Volume 2: Technical papers. Proceeding of the international workshop, Chiang Rai, Thailand. 15–21 Sept. 1991. International Board for Soil Research and Management, Bangkok, Thailand, pp 175–203

Lehmann J, Bossio DA, Kögel-Knabner I et al (2020) The concept and future prospects of soil-health. Nat Rev Earth Environ 1:544–553

McBratney A, Minasny B, Wheeler I, Malone BP (2012). Frameworks for digital soil assessment. In: Minasny B, Malone BP, McBratney AB (eds.), Digital soil assessment and beyond. Taylor & Francis Group, London, pp. 9–14

McBratney A, Field DJ, Koch A (2014) The dimensions of soil security. Geoderma 213:203–213

McDonald RC, Isbell RF, Speight JG, Walker J, Hopkins MS (1990) Australian soil and land survey, field handbook, 2nd edn. Inkata Press, Melbourne, Australia

Moebius-Clune BN, Moebiue-Clune DJ, Gugino BK, Idowu O, Schindelbeck R, Rostow A, van Es HM, Thies J, Shyler H, McBride M et al (2017) Comprehensive assessment of soil-health, vol 2017, 3rd edn. Cornell University, Ithaca, NY, pp 19–101

Muñoz-Rojas M, Erickson TE, Martini D, Dixon KW, Merritt DJ (2016) Soil physicochemical and 231 microbiological indicators of short, medium and long term post-fire recovery in semi-arid 232 ecosystems. Ecol Indic 63:14–22

Nelson GC, Rosegrant MW, Palazzo A, Gray I, Ingersoll C, Robertson R, Msangi S (2010) Food security, farming, and climate change to 2050: scenarios, results, policy options. Int Food Policy Res Inst 172:20–28

Neslson DW, Sommers LE (1996) Total carbon, organic carbon, and organic matter. In: Sparks DL et al (eds) Methods of soil analysis. Part 3, SSSA book series, Madison, WI, pp 961–1010

Peverill KI, Sparrow LA, Reuter DJ (eds) (1999) Soil analysis—an interpretation manual. CSIRO Publishing, Collingwood

Richard M, Damian B, Melissa C, Doug C, Gemma H, Mark I, Tim J, Bernard N, Darryl P and Abdur R (2020). Tools and systems for assessing soil-health. Victoria Resources Online

Schindlbeck JC, Kutterolf S, Freundt A, Alvarado GE, Wang KL, Straub SM, Hemming SR, Frische M, Woodhead JD (2016) Late Cenozoic tephrostratigraphy off shore the southern Central American Volcanic Arc: 1. Tephra ages and provenance. Geochem Geophys Geosyst 17(11):4641–4668. https://doi.org/10.1002/2016GC006503

Sharpley A, Daniel TC, Sims JT, Pote DM (1996) Determining environmentally sound soil phosphorus levels. J Soil Water Conserv 51:160–166

Sparling GP (1997) Soil microbial biomass, activity and nutrient cycling as indicators of soil-health. In: Pankhurst C, Doube BM, Gupta V (eds) Biological indicators of soil-health. CAB International, Wallingford, pp 97–119

Stevenson FJ (1986) Cycles of soil: carbon, nitrogen, phosphorus, sulfur, micronutrients. John Wiley and Sons, New York, NY, p 367

Twomlow S, Mugabe FT, Mwale M, Delve R, Nanja D, Carberry P, Howden M (2008) Building adaptive capacity to cope with increasing vulnerability due to climatic change in Africa—a new approach. Phys Chem Earth 33:780–787

Varvel GE (1994) Rotation and N fertilizer effects of changes in soil carbon and nitrogen. Agron J 86:319–325

Weil RR, Islam KR, Stine MA, Gruver JB, Samson Liebing SE (2003) Estimating active carbon for soil quality assessment: a simplified method for laboratory and field use. Am J Altern Agric 18:1–17

Yang S, Jansen B, Absalah S, Kalbitz K, Castro FO, Cammeraat EL (2022) Soil organic carbon content and mineralization controlled by the composition, origin and molecular diversity of organic matter: a study in tropical alpine grasslands. Soil Tillage Res 215:105203

Chapter 3
Bioengineered Microbes for Restoration of Soil Health

Shweta Sharma, Amit Kumar Kesharwani, and Aditya Kulshreshtha

3.1 Introduction

The global population is expanding staggeringly and is predicted to reach 9 billion by 2050, raising the food demand. Consequently, agricultural production needs to be raised at least 70–100% by 2050 (Tomlinson 2013). Although, agricultural practices like extensive cropping and exhaustive use of fertilizers and agrochemicals have tremendously increased crop yields. However, they adversely pollute the environment, leading to depleted natural resources, water quality, and soil fertility (Novotny et al. 2010). It also enhances greenhouse gas production, resulting in diverse stresses like salinity, drought, nutrient deficits, and soil erosion (Chataut et al. 2023; Wassie 2020). Moreover, leaching fertilizers and agrochemicals into water bodies has also polluted the aquatic environment, threatening aquatic life and human health (Tudi et al. 2021). Alternatively, sustainable agriculture has emerged as a proficient methodology to fulfil the demands of an expanding population with minimal environmental concerns. Moreover, these methods are also beneficial for nutrient recycling, preventing pests and diseases, and maintaining the sustainability of the global ecosystem (Singh et al. 2020).

The original version of this chapter has been revised. The correction to this chapter can be found at https://doi.org/10.1007/978-981-99-9482-3_16

S. Sharma
Department of Veterinary Pathology, Dr. GCN College of Veterinary & Animal Sciences, CSK Himachal Pradesh Agricultural University, Palampur, Himachal Pradesh, India

A. K. Kesharwani · A. Kulshreshtha (✉)
Division of Plant Pathology, ICAR-Indian Agricultural Research Institute, New Delhi, India

R. K. Bhatia, A. Walia (eds.), *Advancements in Microbial Biotechnology for Soil Health*, Microorganisms for Sustainability 50,
https://doi.org/10.1007/978-981-99-9482-3_3

Soil health is a vital component in sustainable agriculture and is defined by its functionality and ecological equilibrium to sustain plant health and productivity. A crucial factor affecting soil health is microorganisms' diversity in soil. It fosters agricultural sustainability by producing growth-promoting hormones, enhancing mineralization, nitrogen fixation, and mobilization of nutrients, making them available to plants (Chourasiya et al. 2017). A novel way to improve soil health is to use genetically modified bacteria. They can be exploited for xenobiotic degradation, biocontrol agents, biofertilizers, and managing various biotic and abiotic stress (Rafeeq et al. 2023). The most successful and long-lasting approach to restoring soil health can be achieved by utilizing the potential of the soil microorganisms (Saeed et al. 2021). This chapter discusses the prospects of bioengineered microbes in nutrient cycling, growth promotion, and remediation of pollutants to restore soil health and increase soil fertility. It also highlights the practical application of bioengineered microbes with challenges like their stability, efficiency, environmental safety concerns, gene exchange with native populations, and interactions with other species.

3.2 Engineered Microbes as Biofertilizers

During development, plants form intimate and continuous interactions with soil microbes to improve soil fertility. Engineering these microbes has emerged as a potential strategy for synthesizing metabolites and enzymes that improve nitrogen fixation, phosphorous solubilization, and phytohormone production, promoting plant growth and sustainability in agriculture (Table 3.1).

3.2.1 Nitrogen Fixation

Nitrogen, a building block for protein synthesis, is not only required as a precursor for synthesizing essential amino acids but also as a precursor for several metabolites essential for cellular metabolism and physiological maintenance. It is a limiting factor for soil fertility and is critical for crop productivity and plant growth (Moreau et al. 2019). In soil, nitrogen is available in both inorganic and organic forms. The organic form comprises soil biota, plant, and animal debris, which is not freely available and takes millions of years to convert into available forms, i.e., ammonium (NH_4^+) and nitrates (NO_3^-). On the other hand, the inorganic form comprises ammonia (NH_3), ammonium (NH_4^+), nitrite (NO_2^-), and nitrate (NO_3^-), which are insufficient to meet current agriculture goals (Andrews et al. 2013). In biological nitrogen fixation (BNF), nitrogen-fixing symbiotic bacteria transform atmospheric nitrogen (N_2) into a simple soluble and harmless form (NH_4^+ primarily). Exploiting BNF is the best way to enhance soil fertility. It can be accomplished by expressing nitrogenase into bacteria symbiotically associated with crops (Soumare et al. 2020).

Table 3.1 Application of genetically engineered microbes for biofertilization and bioremediation

Engineered microbe	Pathway/gene/peptide/ protein involved	Result/application	Reference
Biofertilization			
Mesorhizobium strains V-15b, ST-2, and PMI-6	*nifA* or *nodD* genes	Enhanced nodulation and bacterial symbiotic performance in chickpea	da-Silva et al. (2019)
Synechocystis sp. PCC 6803	35 *nif* genes from *Cyanothece* sp. ATCC 51142	Improved nitrogen fixation	Liu et al. (2018)
M. mediterraneum UPM-Ca36^{T}	*clpB* gene	Improved chickpea-rhizobia symbiosis	Paço et al. (2016)
Yarrowia lipolytica	A total of 14 genes from *A. thaliana* and *Fusarium fujikuroi*	Enhanced gibberellins production	Kildegaard et al. (2021)
E. coli MG1655	IAA biosynthesis pathway	Production of indole-3-acetic acid	Wu et al. (2021)
Bioremediation			
E. coli	Human MT2 and MT3 genes	Cadmium removal	Uckun et al. (2021)
E. coli	Metalloprotein from freshwater crab *Sinopotamon henanense*	Improved heavy metal binding affinities	Li et al. (2021)
Chlamydomonas reinhardtii	γ-glutamylcysteine synthetase	Cadmium bioremediation	Piña-Olavide et al. (2020)
Saccharomyces cerevisiae	A metallothionein gene (PtMT2b) from *Populus trichocarpa*	Cadmium bioremediation	De Oliveira et al. (2020)
E. coli BL21	Three lead binding proteins PbrR, PbrR691, PbrD, from the *Cupriavidus metallidurans* strain CH34	Lead bioremediation	Jia et al. (2020)
C. reinhardtii	*acr3* gene from *Pteris vittata*	Arsenic bioremediation	Ramírez-Rodríguez et al. (2019)
E. coli	Lead-binding peptide	Lead bioremediation	Nguyen et al. (2013)
E. coli	Pesticide-binding peptide	Fenitrothion bioremediation	Wang et al. (2020a, b)
Pseudomonas putida KT2440	Manganese peroxidase 2 (*mnp2*) gene of *Pleurotus ostreatus*	Diazinon bioremediation	Amiri et al. (2022)
Bacillus subtilis	Organophosphorus hydrolase and organophosphorus acid anhydrolase	Bioremediation of organophosphorus compounds	Song et al. (2019)
P. putida KT2440	P8 promoter-driven PNP degrading genes	Paranitrophenol (PNP) bioremediation	Huo et al. (2022)

(continued)

Table 3.1 (continued)

Engineered microbe	Pathway/gene/peptide/ protein involved	Result/application	Reference
P. putida KT2440	1,2-dichloroethane degradation pathway	1,2-dichloroethane bioremediation	Huo et al. (2023)
P. putida KT2440	Biodegradation pathway to mineralize TCP	1,2,3-trichloropropane (TCP) bioremediation	Gong et al. (2017)
P. putida KT2440	Naphthalene catabolic pathway	Naphthalene bioremediation	Fernández et al. (2012)
E. coli	Catechol 2,3-dioxygenase (C230)	Phenanthrene and pyrene bioremediation	Ahankoub et al. (2020)
E. coli DH5α	Alkane hydroxylase (alkB) of *P. putida* GPo1	Degradation of diesel oil	Luo et al. (2015)

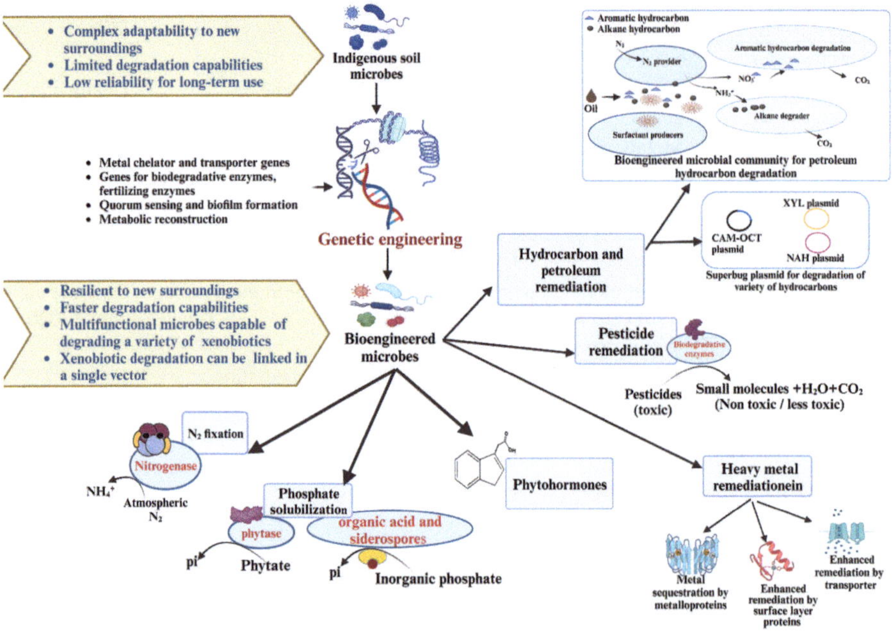

Fig. 3.1 The figure illustrates the genetic engineering of indigenous soil bacteria using various synthetic biology techniques. These bioengineered microbes enhance plant growth by providing phytohormones, fixing atmospheric nitrogen, and enhancing phosphorous solubilization. In addition, they can also be employed to degrade harmful environmental pollutants, including petroleum hydrocarbons, toxic pesticides, and heavy metals

Rhizobacteria are symbiotic nitrogen-fixing bacteria that reside in the rhizosphere of leguminous plants. They improve soil health, provide essential plant nutrients, produce growth regulators, transform organic matter, and degrade xenobiotics (Ali et al. 2017). Synthetic biology allowed the manipulation of microbes with nitrogenase activity to increase BNF (Fig. 3.1). The association of nitrogen-fixing microbes with cereal plants is limited; hence, they heavily depend on chemical fertilizers. In a natural adaptation, maize grown in nitrogen-deficient soil developed many aerial roots that secrete mucilage. It was enriched with diazotrophic bacteria, contributing to fixing atmospheric nitrogen (Van Deynze et al. 2018). Amy Wen and his coworkers developed a genome-edited nontransgenic strain of *Klebsiella variicola* for corn. Application of this strain showed a 122-fold increase in nitrogen fixation activity in a nutrient-rich environment (Wen et al. 2021). Likewise, in *Azotobacter vinelandii*, a *nifl* gene mutant strain showed constitutive *nif* genes expression and produced more ammonium. It promoted the growth of microalgae and cucumber plants without nitrogen fertilizers (Ambrosio et al. 2017). Similarly, *Pseudomonas protegens* Pf-5 was bioengineered with *nif* genes for constitutive expression of a functional nitrogenase. The recombinant bacteria increased ammonium concentration in soil and boosted productivity in *Arabidopsis*, alfalfa, tall fescue, and maize in nitrogen-deficient environments (Setten et al. 2013). Hydrogen-uptake (Hup) activity is closely associated with the BNF and can be exploited to engineer Rhizobacteria. *Hup* gene cluster from *Rhizobium leguminosarum* bv. *viciae* encoding hydrogenase activity was transferred to common bean-associated rhizobial strains lacking hydrogenase. Application of these strains enhanced the nodule efficiency and seed total nitrogen content in the bean plants (Torres et al. 2020). *P. protegens* CHA0, a root-colonizing bacterium with broad-spectrum biocontrol ability, was also engineered by deleting *retS* gene from the chromosome and integrating a 49-kb *Nif* nitrogen-fixing gene island from the *P.* stutzeri DSM4166. The mutant showed improved BNF and bactericidal activity and enhanced *Arabidopsis* growth in a nitrogen-deficient medium (Yu et al. 2019). In another study, using CRISPR/Cas9, the AZC_2928 gene of *Azorhizobium caulinodans* ORS571 was eliminated. The mutant bacterium displayed significantly improved chemotaxis and biofilm formation, which is crucial for improved nitrogen fixation (Wang et al. 2020a, b). These findings showed that bioengineered bacteria could improve nitrogen fixation and may replace harmful fertilizers.

3.2.2 Phosphorous Solubilization

Phosphorus is a crucial plant macronutrient for synthesizing nucleic acids, phospholipids, and ATP. It also regulates many metabolic processes, such as photosynthesis, cell division, and regulation of enzyme activities (Bechtaoui et al. 2021). Phosphorous is available in both inorganic and organic forms in the soil. Plants can only assimilate inorganic phosphorous through phosphate anions (PO_4^{3-}; Pi). In acidic soils, these phosphate anions combine with Al, Fe, and Mn, but in alkaline soils, they react with Ca or Mg, which are insoluble in water and unavailable to plants. Organic phosphates derived from soil biomass are diverse residues of plant, animal, and microorganisms and exist in the aggregated form of myo-inositol hexaphosphate in soil, chemically known as phytate or phytic acid (Rizwanuddin et al. 2023). Therefore, all the accumulated phosphorus in agricultural soil is insufficient to sustain agriculture and must be converted into soluble phosphate. The phosphorus-solubilizing bacteria (PSB) found in rhizospheric soil use diverse mechanisms to extract phosphorus from natural sources and make it available to plants (Kalayu 2019). Engineering PSB by expressing phosphatase and phytase enzymes or enhancing siderophores to dissolve inorganic phosphate is a promising strategy for getting phosphate from the soil and making it available to plants (Shulse et al. 2019).

Phosphorus is mainly found in soil as phytate, linked to different Ca, Fe, K, Mg, Mn, and Zn cations that are insoluble and inaccessible to plants (Liu et al. 2022b). Phytate must be dephosphorylated through phytase-mediated hydrolysis to add phosphorus to the soil (Rizwanuddin et al. 2023). The engineered microorganisms capable of hydrolyzing phytate could be an alternative to environmentally hazardous phosphorous fertilizers. Shulse et al. (2019) engineered root-associated *P. simiae*, *Ralstonia* sp., and *P. putida* strains that showed enhanced phytate hydrolysis. Further, inoculation of these strains also improved *A. thaliana* growth (Shulse et al. 2019). Many bacteria and fungi also secrete low-molecular-weight metal-chelating substances called siderophores. They can potentially solubilize phosphate by replacing phosphates in exchange for metal ions (Neilands 1984). In recent research, deleting a SreA transcription factor in *Talaromyces marneffei* fungi enhanced siderophore (coprogen B) production, which can be utilized for phosphorous solubilization in soil (Amsri et al. 2022). These findings demonstrate that genetically modified microbes can potentially solubilize phosphorous, which agriculture can use to increase plant growth (Fig. 3.1).

3.2.3 Phytohormones

Phytohormones are essential for plant growth and development as they regulate several physiological processes. They exist in minimal amounts inside plants and have been applied exogenously to increase plant yield and boost immunity to fight biotic and abiotic stresses (Zheng et al. 2021). Research on plant growth-promoting rhizobacteria (PGPR) has demonstrated their function in phytohormones synthesis and supported their usage as a key metabolic engineering target for improved soil health and sustainable agriculture (Egamberdieva et al. 2017). Several microorganisms have been engineered to create phytohormones with the potential to stimulate plant growth (Fig. 3.1).

Indole-3-acetic acid (IAA), a naturally occurring auxin, regulates cell division, differentiation, and apical plant dominance. *Cupriavidus pinatubonensis* JMP134 (a PGPR) was engineered to synthesize IAA under the control of a quorum-sensing signal. Application of this strain showed enhanced root growth in *A. thaliana* (Zúñiga et al. 2018). Similarly, abscisic acid (ABA), a widely used phytohormone in agriculture, has a role in several developmental processes in plants, including seed dormancy, cell elongation, and floral induction. *Yarrowia lipolytica*, an oleaginous yeast, was engineered by expressing genes involved in ABA biosynthesis and the mevalonate pathway and showed increased ABA production (Arnesen et al.

2022). Biogenic amines are formed by the amination and transamination of aldehydes and ketones or by the decarboxylation of amino acids. These nitrogenous compounds also serve as hormones in plants. Putrescine and cadaverine are generated by decarboxylation of ornithine and lysine, respectively. Cadaverine is a natural polyamine and regulates plant growth and development, seed germination, seedling growth cell signaling, and stress response (Jancewicz et al. 2015). *E. coli* was engineered to express a galactose utilization pathway to produce cadaverine from galactose (Kwak et al. 2017). Moreover, a genetically engineered microbial consortium consisting of *Corynebacterium glutamicum* cgl-FDK and *E. coli* BL-ABST-Spy was constructed to synthesize cadaverine from glycerol as a carbon source (Liu et al. 2022a).

3.3 Bioengineered Microbes to Improve Soil Health

Bioremediation is an emerging technology that utilizes microbes to consume, break down, or detoxify environmental pollutants. Restoring soil health and enhancing soil fertility can be achieved using rhizosphere bacteria to degrade and eliminate toxic contaminants. These rhizosphere bacteria secrete various biomolecules that degrade soil pollutants like heavy metals, pesticides, hydrocarbons, and insecticides and contribute to soil fertility (Saeed et al. 2021). However, indigenous rhizobacteria can only partially break down these pollutants. Bioengineered microbes, with their altered metabolic pathways, can be employed to speed up hypersecretions of various biomolecules, hence bioremediation. These microbes are engineered with sensor-based bio-affinity reporters to sense pollutants and remediate different compounds by manipulating the enzyme affinity and specificity or constructing a gene that modulates the regulatory pathway (Table 3.1) (Kumar et al. 2018). Furthermore, engineered microbes can be incorporated with many genes that potentiate the use of single microbes to degrade a wide range of xenobiotics (Pant et al. 2021). Now, it is possible to remediate heavy metals, pesticides, herbicides, and insecticides by synergistically combining biotechnology and genetic engineering with contemporary remediation approaches.

3.3.1 Heavy Metal Remediation

Heavy metals can pollute soil by both natural and anthropogenic means. Natural processes include weathering, metal corrosion, mineral erosion, and volcanic activities. Anthropogenic activities pollute soil by mining, smelting, electroplating, and industrial activity. The leading cause of heavy metal pollution is the discharge of industrial effluents containing hazardous heavy metals (Mokarram et al. 2020). Furthermore, many developing countries use water rich in harmful industrial effluents for agricultural irrigation. These heavy metals negatively interfere with many physiological and biochemical processes in plants, affecting their growth, development, and genetic stability (Li et al. 2020). Additionally, the biomagnification of heavy metals at the higher levels of the food chain results in serious ecological issues and endangers human and animal health (Ali and Khan 2019). Genetically modified microorganisms can efficiently remove toxic metals and decrease heavy metal pollution.

In a recent study, *E. coli* was genetically modified to express two human metallothioneins, MT2A and MT3. The recombinant *E. coli* was highly effective in removing hexavalent chromium from aqueous solutions (Akkurt et al. 2022). In another study, two surface proteins (Hpi and SlpA) of *Deinococcus radiodurans* were exploited to bioremediate cadmium and uranium. The *Synechococcus elongatus* smtA gene (encoding a metal-binding protein) was over-expressed independently or in fusion with Hpi or SlpA in *D. radiodurans*. The recombinant bacteria expressing Hpi or SlpA fusion protein removed cadmium more efficiently in an aqueous solution (Misra et al. 2021). Likewise, a plant cadmium (Cd) and zinc (Zn) transporter (AtHMA4) was overexpressed in microalgae *Chlamydomonas reinhardtii*. It showed a two- to threefold increase in Cd and Zn tolerance and may be used for the bioremediation of metal pollutants (Ibuot et al. 2020). Similarly, metallothionein genes, PsMT1 and PsMT2 from pea (*Pisum sativum* L.), were expressed in *R. leguminosarum* bv. *viciae*. These recombinant strains can form nodules on pea roots and have increased cadmium tolerance (Tsyganov et al. 2020). Even in very low concentrations, most heavy metals may cause toxic severe hazards to human health and the environment. Engineering microorganisms to break down heavy metals could be promising to fight against adverse consequences (Fig. 3.1).

3.3.2 *Pesticide Remediation*

Agriculture faces a considerable load due to annual increases in food consumption and exponential population growth. It primarily relies on the massive use of pesticides, which helps decrease pest infestation and crop loss to boost crop yield. Unregulated pesticide application causes their accumulation in soil, affecting fertility and long-term agricultural productivity. Furthermore, pesticide accumulation has many health and environmental hazards (Rani et al. 2021). Many microbes have been engineered to enhance the remediation of dangerous pesticides to make them less or nontoxic from polluted locations (Fig. 3.1).

Iprodione is a potentially carcinogenic pesticide extensively used to control fungal diseases in various crops. Gene-encoding iprodione hydrolase (IpaH) from *Pseudarthrobacter* sp. Y-5 was amplified and expressed in *Bacillus subtilis* WB800-ipaH strain that effectively removed iprodione residues in soil and proved a cost-effective methodology for the bioremediation of an iprodione-polluted environment (Zhang et al. 2023). Similarly, carbamate is a widely used pesticide in agriculture and forestry for controlling pests/bugs. Carboxylesterase is a critical enzyme for its degradation. The *Bacillus velezensis* carboxylesterase (carCby gene) possesses carbaryl hydrolysis activity. It was cloned and expressed on the outer membrane of *E. coli* cells. The whole-cell biocatalyst showed better stability and can be employed in carbamates-rich environments (Liu et al. 2022c). In another study, Lazarini-Martínez and coworkers transferred atrazine metabolizing genes from *Pseudomonas* sp. to many gram-negative bacteria. The modified bacteria showed enhanced mineralizing capabilities and can be employed for xenobiotic degradation (Lazarini-Martínez et al. 2019).

3.3.3 Petroleum Hydrocarbons Remediation

Soil contamination with hydrocarbon is a significant issue worldwide and the leading cause of soil pollution in some areas. Petroleum products are the most common pollutants that remain intact for a long time. Although it significantly impacts the marine environment, spillage during transportation has several negative repercussions on living things and the environment, depending on the spillage and exposure to other organisms. Bioengineered microbes or superbugs can be employed for petroleum remediation and engineered with plasmids with hydrocarbon-degrading genes (Fig. 3.1) (Singha and Pandey 2021). In 1979, Chakrabarty and his coworkers engineered *P. putida* by transferring different destructive plasmids capable of degrading various hydrocarbons. This superbug was the first genetically modified living organism to be granted a US patent after winning a legal battle (Pandey and Arora 2020). Similarly, for the biodegradation of diesel oil, the alkane hydroxylase (alkB) gene of *P. putida* GPo1 was expressed in *E. coli* DH5α. The recombinant *E. coli* increased diesel oil degradation at 24 h from 31% to 50% (Luo et al. 2015). In another study, *P. putida* KT2440 strain was modified by transforming NAH7 plasmid expressing the naphthalene catabolic pathway. The modified bacteria enhanced naphthalene mineralization (Fernández et al. 2012).

Developing engineered microbial communities has proven effective in degrading complex hydrocarbons compared to a single microorganism. Different strains are engineered with biodegradation pathways or related genes with desirable characteristics like multiple catalytic activities and cross-feeding that eliminate feedback inhibition and byproduct removal to improve degradation efficiencies in synthetic communities. Microbial communities can also be divided into "alkane-degrader and alkane-degrader" and "alkane-degrader and helper." In an alkane-degrader and alkane-degrader community, alkanes can be degraded by all members, although the kinds of alkanes that each member degrade are different (Wang et al. 2023). For instance, in an alkaline environment, coculture of bacteria *Acinetobacter baumannii* and fungi, *Talaromyces* sp., has been has been shown effective alkane-degrader and alkane-degrader community, with a total crude oil degradation rate of up to 80% (Zhang et al. 2021). In an "alkane-degrader and helper" community, one or some members do the alkane degradation while others help alkane-degrading members by providing surfactants or nitrogen, and thus both alkane-degrader and helper benefit from each other by cross-feeding (Fig. 3.1) (Wang et al. 2023).

3.4 Bioengineered Microbes: Challenges and Future Prospects

Advanced molecular biology techniques have revolutionized gene alternations, knockout, and gene transfer from one organism to another, contributing to the development of genetically engineered microorganisms (GEMs). Cutting-edge gene-editing methods have drawn much attention to engineering microorganisms. These tools provide simple, affordable, and reliable methods for successful gene expression and can promote plant growth and remove xenobiotics from the soil (Rafeeq et al. 2023; Saxena et al. 2019).

Bioremediation involving synthetic microbial communities is also effective in removing multiple xenobiotics. This method alters genes and metabolic pathways to enhance collaboration among microbial species, improving microbial consortia's remediation capacities in diverse polluted environments. Synthetic communities engineered with bacterial signaling are highly efficient while competing with other species in the natural environment. They develop multicellular behavior, allowing them to execute specialized tasks using quorum sensing and cell-to-cell communications (Sharma and Shukla 2020). However, the field application of GEMs is a critical step. It is associated with challenges since GEMs might not last in the natural environment for long and might not successfully combat high temperatures and pH. Further, the potential of GEMs is significantly impacted by unstable plasmid and recombinant DNA, mutations, ineffective gene expression, and horizontal gene transfers between GEMs and native bacteria. Moreover, some pollutant mixtures make microbial biodegradation challenging. These pollutant mixtures are harmful xenobiotics, poisonous, chemically inert, and do not break down well by GEMs in normal environmental conditions. Additionally, the application of GEMs in the native environment also raises questions about associated ecological risks and may affect indigenous soil microbiomes (Liu et al. 2019; Rebello et al. 2021). Hence, more research and field trials are needed to overcome strict regulatory requirements and legal concerns.

3.5 Conclusion

In recent years, synthetic biology and metabolic engineering advancements have developed many microbial strains and communities to remediate environmental pollutants. The microorganisms are modified by expressing different genes, pathways, and metabolic reconstruction. They can remediate various pollutants, including pesticides, heavy metals, and petroleum hydrocarbons. In addition, they can also enhance plant growth by providing phytohormones, fixing atmospheric nitrogen fixation, and phosphorous solubilization.

However, several limitations are associated with genetically modified organisms, including their stability and the possibility of genetic contamination to other microorganisms through horizontal gene transfer. Further, applying genetically modified organisms may also negatively affect human health and ecological balance. In this direction, the researchers need to have information on essential microbes, genes, and known and unknown pathways for bioremediation in a sustainable way. In addition, it is also equally important to promote genetic tools that do not harm the ecosystem. Thus, more research is needed to elucidate the risks and benefits of bioengineered microorganisms before releasing them into the environment.

References

Ahankoub M, Mardani G, Ghasemi-Dehkordi P, Mehri-Ghahfarrokhi A, Doosti A, Jami MS, Allahbakhshian-Farsani M, Saffari-Chaleshtori J, Rahimi-Madiseh M (2020) Biodecomposition of phenanthrene and pyrene by a genetically engineered *Escherichia coli*. Recent Pat Biotechnol 14:121–133. https://doi.org/10.2174/1872208314666200128103513

Akkurt Ş, Oğuz M, Alkan Uçkun A (2022) Bioreduction and bioremoval of hexavalent chromium by genetically engineered strains (*Escherichia coli* MT2A and *Escherichia coli* MT3). World J Microbiol Biotechnol 38:45. https://doi.org/10.1007/s11274-022-03235-2

Ali H, Khan E (2019) Trophic transfer, bioaccumulation, and biomagnification of non-essential hazardous heavy metals and metalloids in food chains/webs-Concepts and implications for wildlife and human health. Hum Ecol Risk Assess Int J 25:1353–1376. https://doi.org/10.1080/10807039.2018.1469398

Ali MA, Naveed M, Mustafa A, Abbas A (2017) The good, the bad, and the ugly of rhizosphere microbiome. In: Kumar V, Kumar M, Sharma S, Prasad R (eds) Probiotics and plant health. Springer, Singapore, pp 253–290. https://doi.org/10.1007/978-981-10-3473-2_11

Ambrosio R, Ortiz-Marquez JCF, Curatti L (2017) Metabolic engineering of a diazotrophic bacterium improves ammonium release and biofertilization of plants and microalgae. Metab Eng 40:59–68. https://doi.org/10.1016/j.ymben.2017.01.002

Amiri NA, Amiri FA, Faravardeh L, Eslami A, Ghasemi A, Rafiee M (2022) Enhancement of MBBR reactor efficiency using effective microorganism for treatment of wastewater containing diazinon by engineered *Pseudomonas putida* KT2440 with manganese peroxidase 2 gene. J Environ Manag 316:115293. https://doi.org/10.1016/j.jenvman.2022.115293

Amsri A, Srichairatanakool S, Teerawutgulrag A, Youngchim S, Pongpom M (2022) Genetic engineering of *Talaromyces marneffei* to enhance siderophore production and preliminary testing for medical application potential. J Fungi 8:1183. https://doi.org/10.3390/jof8111183

Andrews M, Raven JA, Lea PJ (2013) Do plants need nitrate? The mechanisms by which nitrogen form affects plants. Ann Appl Biol 163:174–199. https://doi.org/10.1111/aab.12045

Arnesen JA, Jacobsen IH, Dyekjær JD, Rago D, Kristensen M, Klitgaard AK, Randelovic M, Martinez JL, Borodina I (2022) Production of abscisic acid in the oleaginous yeast *Yarrowia lipolytica*. FEMS Yeast Res 22:foac015. https://doi.org/10.1093/femsyr/foac015

Bechtaoui N, Rabiu MK, Raklami A, Oufdou K, Hafidi M, Jemo M (2021) Phosphate-dependent regulation of growth and stresses management in plants. Front Plant Sci 12:679916. https://doi.org/10.3389/fpls.2021.679916

Chataut G, Bhatta B, Joshi D, Subedi K, Kafle K (2023) Greenhouse gases emission from agricultural soil: a review. J Agric Food Res 11:100533. https://doi.org/10.1016/j.jafr.2023.100533

Chourasiya D, Sharma MP, Maheshwari HS, Ramesh A, Sharma SK, Adhya TK (2017) Microbial diversity and soil health in tropical agroecosystems. In: Advances in soil microbiology: recent trends and future prospects: Volume 2: Soil-microbe-plant interaction. Springer, Singapore, pp 19–35 da-Silva JR, Menéndez E, Eliziário F, Mateos PF, Alexandre A, Oliveira S (2019) Heterologous expression of nifA or nodD genes improves chickpea-Mesorhizobium symbiotic performance. Plant Soil 436:607–621. https://doi.org/10.1007/s11104-019-03950-0

De Oliveira VH, Ullah I, Dunwell JM, Tibbett M (2020) Bioremediation potential of Cd by transgenic yeast expressing a metallothionein gene from *Populus trichocarpa*. Ecotoxicol Environ Saf 202:110917. https://doi.org/10.1016/j.ecoenv.2020.110917

Egamberdieva D, Wirth SJ, Alqarawi AA, Abd Allah EF, Hashem A (2017) Phytohormones and beneficial microbes: essential components for plants to balance stress and fitness. Front Microbiol 8:2104. https://doi.org/10.3389/fmicb.2017.02104

Fernández M, Niqui-Arroyo JL, Conde S, Ramos JL, Duque E (2012) Enhanced tolerance to naphthalene and enhanced rhizoremediation performance for *Pseudomonas putida* KT2440 via the NAH7 catabolic plasmid. Appl Environ Microbiol 78:5104–5110. https://doi.org/10.1128/AEM.00619-12

Gong T, Xu X, Che Y, Liu R, Gao W, Zhao F, Yu H, Liang J, Xu P, Song C, Yang C (2017) Combinatorial metabolic engineering of *Pseudomonas putida* KT2440 for efficient mineralization of 1, 2, 3-trichloropropane. Sci Rep 7:7064. https://doi.org/10.1038/s41598-017-07435-x

Huo K, Liu Y, Huang R, Zhang Y, Liu H, Che Y, Yang C (2022) Development of a novel promoter engineering-based strategy for creating an efficient para-nitrophenol-mineralizing bacterium. J Hazard Mater 424:127672. https://doi.org/10.1016/j.jhazmat.2021.127672

Huo K, Wang S, Zhao W, Guo H, Xiong W, Liu R, Yang C (2023) Creating an efficient 1, 2-dichloroethane-mineralizing bacterium by a combination of pathway engineering and promoter engineering. Sci Total Environ 878:163140. https://doi.org/10.1016/j.scitotenv.2023.163140

Ibuot A, Webster RE, Williams LE, Pittman JK (2020) Increased metal tolerance and bioaccumulation of zinc and cadmium in *Chlamydomonas reinhardtii* expressing a AtHMA4 C-terminal domain protein. Biotechnol Bioeng 117:2996–3005. https://doi.org/10.1002/bit.27476

Jancewicz AL, Gibbs NM, Masson PH (2015) Cadaverine's functional role in plant development and environmental response. Front Plant Sci 7:870. https://doi.org/10.3389/fpls.2016.00870

Jia X, Li Y, Xu T, Wu K (2020) Display of lead-binding proteins on Escherichia coli surface for lead bioremediation. Biotechnol Bioeng 117:3820–3834. https://doi.org/10.1002/bit.27525

Kalayu G (2019) Phosphate solubilizing microorganisms: promising approach as biofertilizers. Int J Agron:4917256. https://doi.org/10.1155/2019/4917256

Kildegaard KR, Arnesen JA, Adiego-Pérez B, Rago D, Kristensen M, Klitgaard AK, Hansen EH, Hansen J, Borodina I (2021) Tailored biosynthesis of gibberellin plant hormones in yeast. Metab Eng 66:1–11. https://doi.org/10.1016/j.ymben.2021.03.010

Kumar NM, Muthukumaran C, Sharmila G, Gurunathan B (2018) Genetically modified organisms and its impact on the enhancement of bioremediation. In: Varjani S, Agarwal A, Gnansounou E, Gurunathan B (eds) Bioremediation: applications for environmental protection and management. Energy, environment, and sustainability. Springer, Singapore, pp 53–76. https://doi.org/10.1007/978-981-10-7485-1_4

Kwak DH, Lim HG, Yang J, Seo SW, Jung GY (2017) Synthetic redesign of *Escherichia coli* for cadaverine production from galactose. Biotechnol Biofuels 10:20. https://doi.org/10.1186/s13068-017-0707-2

Lazarini-Martínez A, Pérez-Valdespino A, Martínez FH, Ordaz NR, Galíndez-Mayer J, Juárez-Ramírez C, Curiel-Quesada E (2019) Assembly of an atrazine catabolic operon and its introduction to Gram-negative hosts for robust and stable degradation of triazine herbicides. FEMS Microbiol Lett 366:fnz233. https://doi.org/10.1093/femsle/fnz233

Li C, Quan Q, Gan Y, Dong J, Fang J, Wang L, Liu J (2020) Effects of heavy metals on microbial communities in sediments and establishment of bioindicators based on microbial taxa and function for environmental monitoring and management. Sci Total Environ 749:141555. https://doi.org/10.1016/j.scitotenv.2020.141555

Li X, Ren Z, Crabbe MJC, Wang L, Ma W (2021) Genetic modifications of metallothionein enhance the tolerance and bioaccumulation of heavy metals in *Escherichia coli*. Ecotoxicol Environ Saf 222:112512. https://doi.org/10.1016/j.ecoenv.2021.112512

Liu D, Liberton M, Yu J, Pakrasi HB, Bhattacharyya-Pakrasi M (2018) Engineering nitrogen fixation activity in an oxygenic phototroph. MBio 9:e01029–e01018. https://doi.org/10.1128/mbio.01029-18

Liu L, Bilal M, Duan X, Iqbal HM (2019) Mitigation of environmental pollution by genetically engineered bacteria-current challenges and future perspectives. Sci Total Environ 667:444–454. https://doi.org/10.1016/j.scitotenv.2019.02.390

Liu S, Mi J, Song K, Qi H, Zhang L (2022a) Engineering synthetic microbial consortium for cadaverine biosynthesis from glycerol. Biotechnol Lett 44:1389–1400. https://doi.org/10.1007/s10529-022-03306-2

Liu X, Han R, Cao Y, Turner BL, Ma LQ (2022b) Enhancing phytate availability in soils and phytate-p acquisition by plants: a review. Environ Sci Technol 56:9196–9219. https://doi.org/10.1021/acs.est.2c00099

Liu Y, Wang X, Nong S, Bai Z, Han N, Wu Q, Huang Z, Ding J (2022c) Display of a novel carboxylesterase CarCby on *Escherichia coli* cell surface for carbaryl pesticide bioremediation. Microb Cell Factories 21:97. https://doi.org/10.1186/s12934-022-01821-5

Luo Q, He Y, Hou DY, Zhang JG, Shen XR (2015) GPo1 alkB gene expression for improvement of the degradation of diesel oil by a bacterial consortium. Braz J Microbiol 46:649–657. https://doi.org/10.1590/S1517-838246320120226

Misra CS, Sounderajan S, Apte SK (2021) Metal removal by metallothionein and an acid phosphatase PhoN, surface-displayed on the cells of the extremophile, *Deinococcus radiodurans*. J Hazard Mater 419:126477. https://doi.org/10.1016/j.jhazmat.2021.126477

Mokarram M, Saber A, Sheykhi V (2020) Effects of heavy metal contamination on river water quality due to release of industrial effluents. J Clean Prod 277:123380. https://doi.org/10.1016/j.jclepro.2020.123380

Moreau D, Bardgett RD, Finlay RD, Jones DL, Philippot L (2019) A plant perspective on nitrogen cycling in the rhizosphere. Funct Ecol 33:540–552. https://doi.org/10.1111/1365-2435.13303

Neilands JB (1984) Siderophores of bacteria and fungi. Microbiol Sci 1:9–14

Nguyen TT, Lee HR, Hong SH, Jang JR, Choe WS, Yoo IK (2013) Selective lead adsorption by recombinant *Escherichia coli* displaying a lead-binding peptide. Appl Biochem Biotechnol 169:1188–1196. https://doi.org/10.1007/s12010-012-0073-2

Novotny V, Wang X, Englande AJ, Bedoya D, Promakasikorn L, Tirado R (2010) Comparative assessment of pollution by the use of industrial agricultural fertilizers in four rapidly developing Asian countries. Environ Dev Sustain 12:491–509. https://doi.org/10.1007/s10668-009-9207-2

Paço A, Brígido C, Alexandre A, Mateos PF, Oliveira S (2016) The symbiotic performance of chickpea rhizobia can be improved by additional copies of the clpB chaperone gene. PLoS One 11:e0148221. https://doi.org/10.1371/journal.pone.0148221

Pandey P, Arora NK (2020) Prof. Ananda Mohan Chakrabarty: the Superbug superhero. Environ Sustain 3:333–335. https://doi.org/10.1007/s42398-020-00117-x

Pant G, Garlapati D, Agrawal U, Prasuna RG, Mathimani T, Pugazhendhi A (2021) Biological approaches practised using genetically engineered microbes for a sustainable environment: a review. J Hazard Mater 405:124631. https://doi.org/10.1016/j.jhazmat.2020.124631

Piña-Olavide R, Paz-Maldonado LM, Alfaro-De La Torre MC, García-Soto MJ, Ramírez-Rodríguez AE, Rosales-Mendoza S, Bañuelos-Hernández B, García De la-Cruz RF (2020) Increased removal of cadmium by *Chlamydomonas reinhardtii* modified with a synthetic gene for γ-glutamylcysteine synthetase. Int J Phytoremediation 22:1269–1277. https://doi.org/10.1080/15226514.2020.1765138

Rafeeq H, Afsheen N, Rafique S, Arshad A, Intisar M, Hussain A, Bilal M, Iqbal HM (2023) Genetically engineered microorganisms for environmental remediation. Chemosphere 310:136751. https://doi.org/10.1016/j.chemosphere.2022.136751

Ramírez-Rodríguez AE, Bañuelos-Hernández B, García-Soto MJ, Govea-Alonso DG, Rosales-Mendoza S, Alfaro de la Torre MC, Monreal-Escalante E, Paz-Maldonado LM (2019) Arsenic removal using *Chlamydomonas reinhardtii* modified with the gene acr3 and enhancement of its performance by decreasing phosphate in the growing media. Int J Phytoremediation 21:617–623. https://doi.org/10.1080/15226514.2018.1546274

Rani L, Thapa K, Kanojia N, Sharma N, Singh S, Grewal AS, Srivastav AL, Kaushal J (2021) An extensive review on the consequences of chemical pesticides on human health and environment. J Clean Prod 283:124657. https://doi.org/10.1016/j.jclepro.2020.124657

Rebello S, Nathan VK, Sindhu R, Binod P, Awasthi MK, Pandey A (2021) Bioengineered microbes for soil health restoration: present status and future. Bioengineered 12:12839–12853. https://doi.org/10.1080/21655979.2021.2004645

Rizwanuddin S, Kumar V, Singh P, Naik B, Mishra S, Chauhan M, Saris PE, Verma A, Kumar V (2023) Insight into phytase-producing microorganisms for phytate solubilization and soil sustainability. Front Microbiol 14:1127249. https://doi.org/10.3389/fmicb.2023.1127249

Saeed Q, Xiukang W, Haider FU, Kučerik J, Mumtaz MZ, Holatko J, Naseem M, Kintl A, Ejaz M, Naveed M, Brtnicky M (2021) Rhizosphere bacteria in plant growth promotion, biocontrol, and bioremediation of contaminated sites: a comprehensive review of effects and mechanisms. Int J Mol Sci 22:10529. https://doi.org/10.3390/ijms221910529

Saxena G, Kumar N, More N, Bharagava RN (2019) Metagenomics: a genomic tool for monitoring microbial communities during bioremediation of environmental pollutants. In: Microbes for sustainable development and bioremediation. CRC Press, pp 155–176

Setten L, Soto G, Mozzicafreddo M, Fox AR, Lisi C, Cuccioloni M, Angeletti M, Pagano E, Díaz-Paleo A, Ayub ND (2013) Engineering *Pseudomonas protegens* Pf-5 for nitrogen fixation and its application to improve plant growth under nitrogen-deficient conditions. PLoS One 8:e63666. https://doi.org/10.1371/journal.pone.0063666

Sharma B, Shukla P (2020) Designing synthetic microbial communities for effectual bioremediation: a review. Biocatal Biotransformation 38:405–414. https://doi.org/10.1080/10242422.2020.1813727

Shulse CN, Chovatia M, Agosto C, Wang G, Hamilton M, Deutsch S, Yoshikuni Y, Blow MJ (2019) Engineered root bacteria release plant-available phosphate from phytate. Appl Environ Microbiol 85:e01210–e01219. https://doi.org/10.1128/AEM.01210-19

Singh BK, Trivedi P, Egidi E, Macdonald CA, Delgado-Baquerizo M (2020) Crop microbiome and sustainable agriculture. Nat Rev Microbiol 18:601–602. https://doi.org/10.1038/s41579-020-00446-y

Singha LP, Pandey P (2021) Rhizosphere assisted bioengineering approaches for the mitigation of petroleum hydrocarbons contamination in soil. Crit Rev Biotechnol 41:749–766. https://doi.org/10.1080/07388551.2021.1888066

Song T et al (2019) Surface display of organophosphorus-degrading enzymes on the recombinant spore of Bacillus subtilis. Biochem Biophys Res Commun 510(1):13–19. https://doi.org/10.1016/j.bbrc.2018.12.077

Soumare A, Diedhiou AG, Thuita M, Hafidi M, Ouhdouch Y, Gopalakrishnan S, Kouisni L (2020) Exploiting biological nitrogen fixation: a route towards a sustainable agriculture. Plan Theory 9:1011. https://doi.org/10.3390/plants9081011

Tomlinson I (2013) Doubling food production to feed the 9 billion: a critical perspective on a key discourse of food security in the UK. J Rural Stud 29:81–90. https://doi.org/10.1016/j.jrurstud.2011.09.001

Torres AR, Brito B, Imperial J, Palacios JM, Ciampitti IA, Ruiz-Argüeso T, Hungria M (2020) Hydrogen-uptake genes improve symbiotic efficiency in common beans (*Phaseolus vulgaris* L.). Antonie Van Leeuwenhoek 113:687–696. https://doi.org/10.1007/s10482-019-01381-6

Tsyganov VE, Tsyganova AV, Gorshkov AP, Seliverstova EV, Kim VE, Chizhevskaya EP, Belimov AA, Serova TA, Ivanova KA, Kulaeva OA, Kusakin PG, Kitaeva AB, Tikhonovich IA (2020) Efficacy of a plant-microbe system: *Pisum sativum* (L.) cadmium-tolerant mutant and *Rhizobium leguminosarum* strains, expressing pea metallothionein genes PsMT1 and

PsMT2, for cadmium phytoremediation. Front Microbiol 11:491191. https://doi.org/10.3389/fmicb.2020.00015

Tudi M, Daniel Ruan H, Wang L, Lyu J, Sadler R, Connell D, Chu C, Phung DT (2021) Agriculture development, pesticide application and its impact on the environment. Int J Environ Res Public Health 18:1112. https://doi.org/10.3390/ijerph18031112

Uçkun AA, Uçkun M, Akkurt Ş (2021) Efficiency of *Escherichia coli* Jm109 and genetical engineering strains (*E. coli* MT2, *E. coli* MT3) in cadmium removal from aqueous solutions. Environ Technol Innov 24:102024. https://doi.org/10.1016/j.eti.2021.102024

Van Deynze A, Zamora P, Delaux PM, Heitmann C, Jayaraman D, Rajasekar S, Graham D, Maeda J, Gibson D, Schwartz KD, Berry AM (2018) Nitrogen fixation in a landrace of maize is supported by a mucilage-associated diazotrophic microbiota. PLoS Biol 16:e2006352. https://doi.org/10.1371/journal.pbio.2006352

Wang M, Ding M, Yuan Y (2023) Bioengineering for the microbial degradation of petroleum hydrocarbon contaminants. Bioengineering 10:347. https://doi.org/10.3390/bioengineering10030347

Wang X, Lv S, Liu T, Wei J, Qu S, Lu Y, Zhang J, Oo S, Zhang B, Pan X, Liu H (2020a) CRISPR/Cas9 genome editing shows the important role of AZC_2928 gene in nitrogen-fixing bacteria of plants. Funct Integr Genomics 20:657–668. https://doi.org/10.1007/s10142-020-00739-8

Wang Y, Maruthamuthu MK, Jeong J, Yoo IK, Kim TW, Hong SH (2020b) Development of fenitrothion adsorbing recombinant *Escherichia coli* by cell surface display of pesticide-binding peptide. J Biotechnol 322:90–95. https://doi.org/10.1016/j.jbiotec.2020.07.014

Wassie SB (2020) Natural resource degradation tendencies in Ethiopia: a review. Environ Syst Res 9:33. https://doi.org/10.1186/s40068-020-00194-1

Wen A, Havens KL, Bloch SE, Shah N, Higgins DA, Davis-Richardson AG, Sharon J, Rezaei F, Mohiti-Asli M, Johnson A, Abud G (2021) Enabling biological nitrogen fixation for cereal crops in fertilized fields. ACS Synth Biol 10:3264–3277. https://doi.org/10.1021/acssynbio.1c00049

Wu H, Yang J, Shen P, Li Q, Wu W, Jiang X, Qin L, Huang J, Cao X, Qi F (2021) High-level production of indole-3-acetic acid in the metabolically engineered *Escherichia coli*. J Agric Food Chem 69:1916–1924. https://doi.org/10.1021/acs.jafc.0c08141

Yu F, Jing X, Li X, Wang H, Chen H, Zhong L, Yin J, Pan D, Yin Y, Fu J, Xia L (2019) Recombineering *Pseudomonas protegens* CHA0: an innovative approach that improves nitrogen fixation with impressive bactericidal potency. Microbiol Res 218:58–65. https://doi.org/10.1016/j.micres.2018.09.009

Zhang M, Li Q, Bai X, Gao S, Zhu Q, Ye B, Zhou Y, Qiu J, Yan X, Hong Q (2023) Construction of an unmarked genetically engineered strain *Bacillus subtilis* WB800-ipaH capable of degrading iprodione and its pilot application. Int Biodeterior Biodegradation 176:105527. https://doi.org/10.1016/j.ibiod.2022.105527

Zhang X, Kong D, Liu X, Xie H, Lou X, Zeng C (2021) Combined microbial degradation of crude oil under alkaline conditions by Acinetobacter baumannii and Talaromyces sp. Chemosphere 273:129666. https://doi.org/10.1016/j.chemosphere.2021.129666

Zheng Y, Wang X, Cui X, Wang K, Wang Y, He Y (2021) Phytohormones regulate the abiotic stress: an overview of physiological, biochemical, and molecular responses in horticultural crops. Front Plant Sci 13:1095363. https://doi.org/10.3389/fpls.2022.1095363

Zúñiga A, Fuente FDL, Federici F, Lionne C, Bônnet J, De Lorenzo V, González B (2018) An engineered device for indoleacetic acid production under quorum sensing signals enables *Cupriavidus pinatubonensis* JMP134 to stimulate plant growth. ACS Synth Biol 7:1519–1527. https://doi.org/10.1021/acssynbio.8b00002

Chapter 4
Microbial Inoculum Improvement for Soil Health

Khushneet Kaur Sekhon and Keshani

4.1 Introduction

A necessary aspect of the terrestrial ecosystem, soil, is a complex and heterogeneous matrix. It is a vital resource not just for agriculture but also for maintaining various biological processes including the upkeep of biogeochemical cycles and decomposition. Despite the fact that the area inhabited by microbes is just 5% of the overall space (Fierer et al. 2012), the soil microbiome, which consists of bacteria, fungi, protozoa, and viruses, maintain virile soil by performing functions like that of nutrient uptake, decomposition of the organic matter, and disease suppression (Müller et al. 2016). The role of microorganisms within soil health is hence essential for maintaining the soil health and fertility. By enhancing the population of beneficial microorganisms in soil, such as bacteria and fungi, a more balanced and diverse soil ecosystem that supports healthy plant growth and enhances soil fertility can be created.

Microbial inoculum refers to the intentional introduction of population of microorganisms into specific environment, such as soil, for beneficial or enhanced outputs like improving soil fertility (Babalola and Glick 2012). The use of microbial inoculums in soil can improve the structure, enhance nutrient availability, minimize the need for synthetic fertilizers, and mitigate the negative impacts of certain soil-borne diseases. Microbial inoculum improvement therefore is the primary focus of increasing the diversity and quantity of beneficial microorganisms in soil. This practice gained significant recognition in recent years due to the believed benefits in maintaining soil health and improving crop productivity.

K. K. Sekhon · Keshani (✉)
Department of Microbiology, Punjab Agricultural University, Ludhiana, Punjab, India
e-mail: keshani@pau.edu

R. K. Bhatia, A. Walia (eds.), *Advancements in Microbial Biotechnology for Soil Health*, Microorganisms for Sustainability 50,
https://doi.org/10.1007/978-981-99-9482-3_4

In modern agricultural methods, the nutritional content of agriculture products and the health of customers are adversely impacted by the heavy reliance on agro-chemicals, including fertilizers, herbicides, and pesticides. These chemical inputs' use encourages the buildup of hazardous substances in the soil. A number of synthetic fertilizers, including HCl and sulfuric radicals, include acid radicals that make the soil more acidic and are thus harmful to the health of the soil and plants. The overconsumption and careless usage of these chemicals led to contaminated food, weed and disease resistance, and undesirable environmental effects (Önder et al. 2011), all of which have a substantial detrimental influence on soil and human health (Ojo 2016).

Some examples of microbial inoculum that are beneficial are certain bacteria that can fix nitrogen from the air and make it available to plants (*Rhizobium, Azospirillum*), while others can suppress plant diseases by competing with pathogenic microorganisms for nutrients and space. Similarly, fungi can help to break down the organic matter and make nutrients available to the plants, while protozoa can control populations of other soil microorganisms and help to recycle nutrients (Kundan et al. 2015).

There are various types of microbial inoculum available, each with its own unique benefits. Some common types include compost tea, mycorrhizal fungi, and rhizobia. Compost tea is made by steeping compost in water and allowing it to ferment, creating a concentrated solution of microorganisms that can be applied to the soil. Symbiotic connections between mycorrhizal fungus and plant roots help to absorb nutrients from the soil and makes the nutrients accessible to the plant bacteria (Rani et al. 2017). Similarly, *Rhizobia* live in the root nodules of leguminous plants, where they fix nitrogen from the air and make it available to the plants.

In addition to these specific types of inoculum, there are also general-purpose microbial inoculants available in the market that contain a mixture of different microorganisms for more comprehensive soil health improvement. These products are often a mixture of bacteria, fungi, and other microorganisms that work together to perform a variety of functions in the soil.

The use of microbial inoculum is a promising approach for improving soil health, as it is sustainable and can help to decrease the requirement of chemical fertilizers and pesticides. It is important, however, to carefully select the right inoculum for the relevant needs and follow the manufacturer's instructions for application, as incorrect use can have unintended consequences. Microbial inoculum can be a valuable tool for improving soil health, but it should be used with care and in conjunction with other sustainable soil management practices. However, despite of the presence of so many inocula in the market, there is still a room of improvement of microbial inoculants to enhance the soil health and fertility more effectively.

4.2 Microbial Inoculants and Soil Health

Soil health refers to the capacity of a soil ecosystem to support biological, chemical, and physical processes that sustain plant and animal life. A healthy soil is characterized by a balanced and diverse community of microorganisms, a stable and fertile soil structure, and a sufficient supply of nutrients and water (Lehmann et al. 2020). Soil health is essential for the productivity, resilience, sustainability of agriculture, forestry, and natural ecosystems, as it supports crop growth, carbon sequestration, water filtration, and nutrient cycling. Maintaining soil health requires minimizing soil disturbance, increasing organic matter inputs, minimizing chemical inputs, and promoting biodiversity (Karlen et al. 2019).

Soil health degradation can occur due to variety of both natural and human-induced factors. Chemical pollution is observed to be the most significant factor in reducing soil fertility (Suyal et al. 2016). Synthetic fertilizers and pesticides cause soil ecosystems to become contaminated, resulting in decreased soil biodiversity, decreased fertility, and finally poor soil health. This is due to the inhibition of organic matter decomposition and changed nutrient cycling (Bisht and Chauhan 2020). As a result of the use of standard modern agriculture, continuous nutrient flow and the control of soil health become increasingly difficult.

The use of improved and efficient microbial inoculum as an alternative to agro-pesticides therefore can provide several benefits to improved soil health and crop productivity. Some observed benefits of artificial inoculants include protection against pathogens, as biocontrol agents and as substitute for chemical pesticides (Alori et al. 2017).

The homogenous combination of a chosen beneficial strain and a carrier that can stabilize and protect the strain while it is being transported and stored is known as the microbial inoculant formation or bioformulation. Good bioformulation needs to be efficient, nonpolluting, easily biodegradable, have a high capacity for water retention, and have a long enough shelf life (Sahu et al. 2018).

4.2.1 Benefits of Microbial Inoculants

The use of microorganism inoculants might help to speed up plant development and boost agricultural production. A sizable number of microbial (bacterial) species, often found in the rhizosphere of plants, have the capacity to positively influence soil fertility and plant development. The usage of microbial inoculants as an antagonistic measure against disease causing bacteria and fungi has been reported to increase growth and yield of crops like sweet potato, radish, sugar beet, and potato crops by 1–2% (Tariq et al. 2022). Moreover, promising findings are emerging by investigating the applications of microorganisms commercially. Nevertheless, additional microbial-plant interactions are still required for field applications to be more successful.

4.2.1.1 Nitrogen

The biogeochemical cycling process of atmospheric nitrogen into the biosphere, which is a crucial source of nitrogen for plant production, begins with nitrogen fixation. Bacteria in particular play a key role in N cycling. Since bacteria carry the essential enzyme nitrogenase, which specifically converts atmospheric nitrogen to ammonia, they have been observed to fix atmospheric nitrogen only in symbiotic or symbiotic relationships (Lindström and Mousavi 2020). Both legume and nonlegume plants engage in a process called symbiotic nitrogen fixation in the soil. Rhizobial inoculation of legumes has the power to improve the soil's nitrogen status. Around 40,000–48,000 million kilograms of nitrogen were reported to be fixed yearly by the legume-rhizobium relationship, as opposed to 98,000 million kilograms of nitrogen fertilizer each year (Mendoza-Suárez et al. 2021). The capacity of rhizobial inoculants to fix large amounts of nitrogen into the soil has the capacity to minimize the cost of industrial nitrogen fertilizers, lowering farmers' input costs. Nitrogen fixation by a properly noduled legume is a critical and necessary component of sustainable agriculture. Therefore, usage of rhizobial inoculum to accomplish efficient nitrogen fixation in soil necessitates the compatibility of the rhizobial inoculants and legume as well as their environmental adaptability (Liu et al. 2020).

4.2.1.2 HCN Production

When microorganisms colonize plant root surfaces, they can act as biocontrol agents. Cyanide is an extremely poisonous chemical generated by certain microorganisms that works as a metabolic inhibitor and biocontrol agent. Cyanide is biosynthesized, excreted, and metabolized by bacteria, algae, and fungus (Sehrawat et al. 2022). Host plants are largely unaffected by the deleterious impact of cyanide during HCN generation by related microorganisms; hence, certain rhizobia can operate as biological anti-weed agents. One of the primary benefits of HCN production in the soil is its ability to suppress plant pathogens. HCN is a potent antimicrobial agent that can inhibit the growth of many plant pathogens, including bacteria, fungi, and nematodes (Dimkić et al. 2022). By reducing the populations of these pathogens, HCN-producing microbes can help to protect plants from disease and promote overall plant health.

4.2.1.3 Phosphate Solubilization Microbes

They play a major role in the natural phosphorus cycle, occurring through the repeated redox reaction of phosphorus molecules. P-solubilizing bacteria boost plant P absorption, resulting in increased agricultural output. Many bacterial strains have been found to be effective phosphate solubilizers (Qarni et al. 2021), including *Bacillus*, *Pseudomonas*, *Rhizobium*, and others. The formation of organic acids is the mechanism for solubilization of mineral phosphate, and phosphatases have an important role in the mineralization of organic phosphorous in soil. Phosphate-solubilizing microorganisms (PSM) provide an alternate path for ecologically conscious agriculture to meet plant phosphorus needs. In addition to their part in improving plant growth, phosphate solubilization microbes can also have a positive impact on soil health (Tian et al. 2021). By breaking down insoluble phosphorus compounds, they release other nutrients that are bound to the phosphorus, such as nitrogen and potassium, making them available to plants as well. This can improve the overall nutrient content of soil and promote the development of beneficial microorganisms. Phosphate solubilization microbes can also reduce the need for synthetic fertilizers that might have unfavorable impacts on soil fertility.

4.2.1.4 Siderophore Production

Iron is a vital factor for all living things. When iron is scarce, microorganisms such as bacteria generate low-molecular-weight molecules known as siderophores in order to compete for ferric ion (Srivastava 2023). Siderophores are molecules produced by certain soil bacteria and fungi that are involved in the uptake of iron, a vital micronutrient for development of plants. Siderophores can help enhance soil health by promoting the growth of beneficial bacteria and fungus. This is because siderophores produced by one microbe can be used by other microbes in the soil to obtain iron, allowing them to thrive as well as contribute to the overall ecosystemic health of soil. Siderophore production can also help to reduce the populations of harmful microbes in the soil by limiting their access to iron, which is crucial for their growth as well as survival (Daniel et al. 2022). Furthermore, some siderophores can chelate other micronutrients such as copper and zinc, making these nutrients more available to plants. This can help to improve the overall nutritional content of soil and promote better plant growth.

4.2.2 Challenges of Using Microbial Inoculants

While the use of microbial inoculants can provide significant benefits for soil health and plant growth, there are also some challenges associated with their use. Some of the challenges of using microbial inoculants include the following:

1. Environmental variability: The effectiveness of microbial inoculants can be affected by environmental factors such as temperature, moisture, and soil type. Variability in environmental conditions can affect the survival, activity, and persistence of the inoculated microorganisms, leading to inconsistent results (Shuaib et al. 2021).
2. Compatibility with other inputs: Microbial inoculants may not be compatible with other factors such as fertilizers, insecticides, and herbicides. Certain syn-

thetic compounds can have negative effects on inoculated microorganisms' survival and function, reducing their effectiveness (Lopes et al. 2021).
3. Quality control: The quality and consistency of microbial inoculants can vary depending on the manufacturing process and storage conditions (Rojas-Sánchez et al. 2022). Poor-quality inoculants may contain low numbers of viable microorganisms or contaminants that can have negative effects on soil health.
4. Cost: Microbial inoculants can be expensive compared to other soil inputs such as fertilizers and pesticides. The cost of production and application can be a limiting factor for their adoption by small-scale farmers and low-income countries (Naamala and Smith 2020).
5. Lack of knowledge: The use of microbial inoculants requires a thorough grasp of soil microbiology, crop ecology, and application techniques. Lack of knowledge and training can lead to incorrect application methods and poor results.

Despite these challenges, the use of microbial inoculum has the potential to provide significant benefits for soil health and plant growth. Therefore, ongoing research and development efforts are being focused toward the improvement of the quality, effectiveness, and affordability of microbial inoculants and to promote their adoption by farmers and land managers. Some examples of commercially used microbial inoculants are given in the table below (Table 4.1).

4.3 Techniques for Inoculum Improvement

Approaches based on "omics" have provided the means to address the intricate global biological systems that support distinct microbial activities. By combining knowledge from many plant-microbe species, these technical advancements also sped up the usage of genome-scale tools in applied and developing model microbes and supported provisional research. Our understanding of microbial metabolism has been greatly improved by the computer-aided software annotation and comparative

Table 4.1 Commercial microbial inoculants

Sr. no.	Name of microbial inoculant	Use	Main manufacture	References
01.	Rhizobium inoculant	Nitrogen-fixing symbiotic bacteria	BioWorks Inc., USA	Kebede et al. (2020)
02.	Plant growth-promoting bacteria (PGPB) inoculant	Commercial PGPB inoculants are used to improve plant growth and yield, particularly under stress conditions	BiotaMax, Netherlands	Glick (2012)
03.	Actinomycetes inoculant	Commercial actinomycetes improve soil fertility	Agro Biosciences, USA	Taechowisan et al. (2003)
04.	Compost tea	Promote the growth of beneficial microorganisms like	Earthfort, USA	Islam et al. (2016)

assessment of DNA sequences, which provides microbiologists information and usable data about gene activity, structures, biological, metabolic, and regulatory systems and pathways as well as evolution of microbial genomes (Chelliah et al. 2022). When all these concurrent analyses of soil conditions and plant-microbe interaction are concerned, combining methods for molecular profiling has only commenced. Profiling technology can simultaneously explore many genes, transcripts, proteins, or metabolites, which benefits soil health and plant-microbe interactions. There are several techniques that can be used to improve the quality and effectiveness of microbial inoculum. Some of these techniques include the following:

4.3.1 Isolation and Identification Methods

The first criterion of microbe isolation is that it is able to be cultivated at the laboratory scale. The methodology of isolation necessitates the knowledge of ideal temperature, oxygen, and nutrition requirements. There are two methods for isolating organisms: (1) agar plate streaking and (2) the pour plate technique. Agar plate streaking entails consecutive dilutions until recognized isolation of colonies is observed. This is a quick qualitative approach that is widely used to isolate discrete colonies. The four directions or quadrant streak is mainly performed. During pour plate method, sample should be diluted properly before pouring onto the plate. Individual colonies form from the separated cells. These microbes are then characterized using standard procedures followed by physiological and biochemical testing.

4.3.1.1 FAME Analysis

This approach is used to identify and characterize diverse microorganisms. The FAME analysis uses fatty acid grouping to identify microbial species at the species level and also describes community makeup (Schutter and Fuhrmann 2001).

4.3.1.2 BIOLOG or Physiological Profiling

This is based on the usage of changeable solitary carbon source. It comprises a 96-well microtiter plate with separate plates for gram-positive and gram-negative microorganisms. In 95 wells, each plate has a variety of carbon sources, while the final well is devoid of substrate and is the control. This approach was initially designed for clinical microbes, but now is frequently used for identifying soil microbial communities (Stanley et al. 2016).

4.3.1.3 Genomic Profiling

The process involves several steps. First, soil samples are collected and processed to extract DNA from the microbial communities. Next, the DNA is sequenced using high-throughput sequencing technologies, such as Illumina sequencing (Nkongolo and Narendrula-Kotha 2020) or nanopore sequencing (Burton et al. 2020). The resulting data is then analyzed using bioinformatics tools to identify the various types of microbes present in the sample.

Genomic profiling can provide detailed information about the communities of microbes in the soil, including the diversity of species, the abundance of specific microorganisms, and their functional capabilities. This information can be used to better understand the role of microbes in soil health, nutrient cycling, and other important soil processes.

4.3.2 Formulation Techniques

Additional supplements, such as nutrients, growth factors, or other microbes, can be added to the inoculum to increase its efficiency and improve the development and survival of the inoculated microorganisms.

The formulation of bioinoculants is a homogenous combination of advantageous strain and carrier providing the strain stability as well as shielding throughout transit and storage (Sarbani and Yahaya 2022). The carrier is a live latent microbe's vehicle that offers protection and a supporting habitat for the microbial population. Good bioformulation needs to be efficient, nontoxic, easily biodegradable, have a high-water absorption and retention capacity, and have a long storage life (Chaudhary et al. 2020). The formulation technique enhances biofertilizer performance and shelf life. Occasionally, the formulation comprises cell protectants including desirable microbial organisms which improve the efficacy of spores under unfavorable conditions for longer duration of time (Singh et al. 2021). Various formulations are produced based on the effectiveness and rate of survival.

4.3.2.1 Solid Carrier formulation

The favorable strain is combined with a solid transporter in this formulation. It is used to shift strain from the lab to the field. This mixture mostly employs peat, powder, and granules. It offers a safe and nourishing habitat for microorganisms that create microcolonies. Peat should be adaptable, easy to sanitize, nontoxic, and have a high-water absorption capacity (Amenaghawon et al. 2021). Following drying, the peat is sifted through 250 m sieves and mixed with the appropriate strain. Curing is the process of incubating peat at a specified temperature for bacterial growth. Peat is mostly utilized for ecto-mycorrhizal and arbuscular mycorrhizal fungi (AMF) inoculants (Aini et al. 2019). In glucose medium, ecto-mycorrhiza grows swiftly and produces sporophore, which is utilized for field inoculation. Before planting, infected peat is applied to the surface of the seeds using mechanical tumble machines.

The biggest disadvantage of utilizing peat is its variable composition and quality. Polluting chemicals sometimes can be produced from peat during the sterilizing procedure, which might impact microbial survival and proliferation (Malusá et al. 2012). Grains have been utilized in place of peat since they offer better benefits. Granules, such as peat, are covered by microorganisms' live cells and are composed of calcite, granite, and silicate granules. They are far superior to peat and offer several advantages. These grains are easy and manageable, cleaner, and more convenient to move and store. Inoculants are applied on the furrow near the seed exterior during granule formation to facilitate lateral-root interactions. According to several studies, granule formulation outperforms other inoculants in respect to complete biomass, nitrogen fixing and buildup, and nodule count (Zaidi et al. 2017). Granular inoculants have a significant benefit in stress circumstances like moisture stress, soil pH, and wet soil because of their strong nitrogen fixation as well as nodule forming potential.

In addition to peat, a carrier made of charcoal, i.e., biochar, can be used as a reliable carrier for inoculant formation. Biochar increases survivability of bioinoculants and is environmentally benign and free of hazardous components. The major advantage of employing it is that may be kept without being sterilized due to the low water content of charcoal. It is devoid of wax elements and has no blending issues, providing efficient plant health assurance for crop fields.

4.3.2.2 Formulation via Liquid Carrier

Liquid formulation is a preparation of microbes containing efficient bacteria capable of solubilizing, fixing, or mobilizing vital nutrients through biological processes. In the liquid formulation, microbes that mobilize phosphate and potassium, nitrogen-fixing microbes (NFM), and various other microorganisms are utilized. Researchers have recently focused on the uptake, degradation, and dissolution of micronutrients through activities of manganese, sulfur, and zinc solubilizer microorganisms. Because of its advantageous benefits, this formulation is the most recent and promising technique above other carrier-type formulation. The carrier-based composition has limited storage duration of 2–3 months and therefore is not able be maintained through the crop cycle.

To solve the problem, a liquid formulation is utilized, with the ability to extend the survivability of chosen strain through the crop season. In contrast to carrier-based formulations, where bulk sterilization is not possible, this formulation allows for adequate sterilization so that contamination may be managed. It increases the strain's viability and survival by maintaining a high moisture capacity for longer periods of time. It has a desired strain and various preservatives to extend shelf life during harsh conditions to 19–25 months (Chandra et al. 2018). *Azospirillum* is a free-living, micro-aerophilic rhizobacterium that promotes plant development and is mostly utilized to create liquid formulations. Other phosphate-mobilizing bacteria from the genera *Aspergillus*, *Bacillus*, *Penicillium*, and *Pseudomonas* are also utilized in addition to *Azospirillum*. For this formulation, an *Azospirillum* population (1.66×10^8 CFU/ml) in trehalose (16 mM) and a phosphate solubilizer bacteria (3.66×10^8 CFU/ml) in PVP (3%) are utilized (Surendra Gopal and Baby 2016). With the maximum microbial population (10^8 CFU/ml) at room temperature, PVP (3%) and trehalose (16 mM) were shown to be extremely acceptable bioinoculants for extending the storage time of phosphate-solubilizing bacteria (PSB) and *Azospirillum* sp., for a period of10 months.

4.3.2.3 Formulation via Metabolites

The microbial metabolites in metabolite formulations assist in supplying bioregulators, vital nutrients, and defense against phytopathogens. Strains from the genera *Mesorhizobium*, *Rhizobium*, *Pseudomonas*, and *Trichoderma* and certain mycorrhizal fungus are employed for metabolite formation. Metabolite bioformulation has recently been created to address the shortcomings of cell-based formulation (Aamir et al. 2020). According to reports, under stressful conditions, the rhizobial strain's flavonoids biomolecules boost nitrogen fixation and nodulation. Leguminous host plants that relate to rhizobial organisms release lipochitooligosaccharides. In rhizobial-deficient areas, lipochitooligosachharide was said to aid in symbiosis.

Increased growth was seen in crop yields of both nonleguminous and leguminous plants in harsh soil, when PGPR treated with l-methionine (a precursor to ethylene), which serves as the main metabolite, was utilized. Nutrients like starch effluent, amino acids, and molasses have reportedly been employed as beneficial additives. These additions can lengthen the shelf life and increase the capacity of a good strain to survive in a tough soil environment (Chaudhary et al. 2020). According to some reports, PSB (phosphate-solubilizing bacteria) also secretes biosurfactants that have antibacterial, emulsifying, wetting, antiviral, and insecticidal properties (Ijaz et al. 2019). Liquid bioinoculants with biosurfactants are mostly sprayed on the plant's aerial portions.

The high cost and mass manufacture of this formulation are its main disadvantages. The plant growth-promoting (PGP) strain is combined with carriers and additives in future metabolite formulation, which boosts microbial activity and hence promotes soil health. In this formulation, primary and secondary metabolites can also be found as attractants, adjuvants, and stimulants.

4.3.2.4 Polymeric Formulation

Alginate beads are typically utilized in polymeric formulations. It is a naturally occurring polymeric compound consisting of d-mannuronic acid and l-glucuronic acid, both of which come from the fungus *Macrocystis pyrifera* (brown algae). According to studies, this is also made from the macro-algae *Sargassum sinicola* (Chaudhary et al. 2020). Recent suggestions propose utilization of several bacterial isolates, including *Azotobacter* and *Pseudomonas*, for these formulations. Alginate beads come in two sizes, micro- and macrobeads, with diameters between 2–3 mm and 50–200 m, respectively (colony forming unit per gram). This mixture improves the stability of chemotaxis bacteria, plasmid, and mushroom growth in the host cell. AMF is employed in the methods for alginate matrix strapping.

For some time now, many methods for polymeric formulation exploiting latent cell encapsulation in gel matrixes for diverse purposes were established. The gel matrix has the power to extend storage life of desirable isolate under difficult stress (biotic as well as abiotic) circumstances. First, during formulation, this method was applied to the *Pseudomonas fluorescens*, *A. brasiliense*, and *Aspergillus* (filamentous fungal) isolates (Chaudhary et al. 2020). The multistep and challenging process of creating alginate beads takes place at room temperature (Cassán et al. 2020). Alginate is said to be naturally nontoxic, biodegradable, and released gradually in the field.

4.3.2.5 Bioencapsulation and Co-inoculation

Living microbial cells are trapped in polymeric matrix components during capsulation process to increase the survival of germs. Immobilization is the bioencapsulation of favorable strain in a polymeric type of array. More than a single kind of microbe is employed and used in environmental and agricultural technological sphere during the co-immobilization process. It is frequently utilized for many different applications to achieve a protective structure in the food, pharmaceutical, agricultural, and other sectors.

Microbial inoculants are bioencapsulated in three steps: The first step entails absorbing beneficial components in the matrix (solid or liquid), the next stage involves misting a solvent onto solid components to create dispersion, and the final stage involves stabilizing through chemical-physical processes (gelatin) and polymerization. To store, protect against dehydration, and handle cultures, bioencapsulation has the tendency to sustain the viability and stability of these cells (Mancera-López et al. 2023). Furthermore, bioencapsulation shields bacteria from biological and abiotic stressors.

The pace at which the encapsulated matrix degrades is directly correlated with the biological activities of the soil microflora. For usage as biocontrol agents or agents that promote plant development, microbial agents are encapsulated. To create a suitable habitat for microbial cells, encapsuled cells are kept at ambient temperature for an extended length of time as dried capsules. An immobilized microbe-synthetic matrix is created during fermentation, and this formulation is employed for a variety of processes, including the production of amino and organic acids, various vitamins, enzymes, and bioremediation (Chaudhary et al. 2020). The primary function of the encapsulated microbial cell formulation is to give temporary refuge in the challenging soil environment during rhizosphere colonization.

The encapsulation process's main drawbacks are its high cost and demanding bead creation procedures. By eliminating inhibitory products and inducing a physical or biochemical response, co-inoculation works synergistically to improve the performance of other bacteria. Co-inoculating enhances the uptake and sorption of nitrogen, phosphorus, and other nutrients. The main adverse consequence of co-inoculation is the varied feeding requirements of various bacteria under in vitro conditions (Muthuraja and Muthukumar 2022). The combined inoculation of *Rhizobium* with PGP rhizobacteria has demonstrated potential in boosting nodulation and prevention of pathogenic diseases (Swarnalakshmi et al. 2020).

By using these techniques, it is possible to produce high-quality microbial inoculum that can effectively improve soil health and productivity. However, it is crucial to carefully analyze the specific soil and crop conditions, as well as other inputs and management practices, to ensure optimal results.

4.3.3 Characterization and Application of Microbial Inoculants

(a) As biofertilizers

Utilization of microbe inoculants as biological fertilizers for improving soil fertility and to promote higher yield of crops (Yadav and Sarkar 2019). Biofertilizers are live microorganisms that are incorporated in the soil for enhanced availability of nutrients, stimulating root growth, and improving soil structure. There are several types of microbial inoculants used as biofertilizers, including the following:

1. Nitrogen-fixing bacteria: These bacteria convert atmospheric nitrogen into a form that plants can use. Examples of bacteria fixing nitrogen include *Rhizobium*, *Azospirillum*, and *Azotobacter*.
2. Phosphate-solubilizing bacteria: These bacteria release phosphorus from insoluble compounds in the soil, making it available to plants. Examples of phosphate-solubilizing bacteria include *Bacillus* and *Pseudomonas*.
3. Mycorrhizal fungi: These fungi form a symbiotic association with plant roots, increasing their uptake nutrient and tolerance to environmental stress. Examples of mycorrhizal fungi include *Glomus* and *Rhizophagus*.
4. PGP bacteria: Plant growth hormones, enzymes, and other beneficial compounds that enhance growth and tolerance of stress are produced by PGP bacteria. Examples of plant growth-promoting bacteria include *Bacillus*, *Pseudomonas*, and *Azospirillum*.

Usage of microbial inoculants as such can provide several benefits for healthier soil and plant growth. By enhancing nutrient availability, microbial inoculants can reduce the need for synthetic fertilizers, thereby reducing environmental pollution and costs.

(b) As Biocontrol Agents.

Microbial inoculants can also be utilized as biocontrol agents to suppress crop pathogens as well as pests, reducing the need for synthetic pesticides.

Biocontrol agents are living microorganisms that can antagonize, parasitize, or compete with plant pathogens and pests, thereby reducing their population and damage (Weng et al. 2022). There are several types of microbial inoculants that can be used as biocontrol agents, including the following:

1. *Trichoderma* spp.: These fungi can colonize plant roots and soil, producing enzymes that degrade plant pathogens and induce systemic resistance in plants.
2. *Bacillus thuringiensis* (Bt): This bacterium produces a toxin that specifically targets certain insect pests, making it an effective biopesticide.
3. *Pseudomonas* spp.: These bacteria produce metabolites that limit growth and spread of plant pathogens, enhance plant development, and promote systemic resistance.
4. Entomopathogenic fungi: These fungi can infect and kill insect pests, reducing their population and damage to crops, thus reducing the use of chemical pesticides.

The use of microbial inoculants as such provides various benefits for soil health and ecosystem sustainability. By reducing the use of synthetic pesticides, microbial inoculants can reduce environmental pollution, improve food safety, and promote biodiversity. They can also increase resilience to environmental stressors.

Additionally, microbial inoculants can improve soil structure, increase water-holding capacity, and promote biodiversity, leading to more resilient and sustainable agroecosystems. However, the effectiveness of microbial inoculants as biofertilizers can be influenced by factors such as environmental conditions, soil type, and application methods. Therefore, it is important to carefully select and apply the appropriate microbial inoculant for the specific crop and soil conditions.

The method of inoculum application, such as seed coating, soil drenching, or foliar spraying, can affect the distribution and survival of microorganisms in the soil. Choosing an appropriate application method can help ensure optimal colonization of the microorganisms (Fig. 4.1).

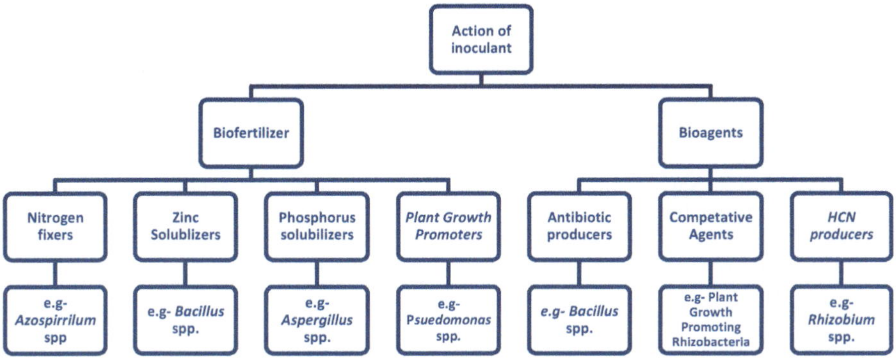

Fig. 4.1 Application of microbial inoculants

4.4 Conclusion

Microbial inoculum improvement is an effective approach to enhance fertility and diversity of favorable microbes in the soil. These microorganisms can enhance nutrient availability, control soil-borne diseases, improve soil structure and permeability, and ultimately raise crop productivity. The use of microbial inoculants can also reduce the dependency on chemical fertilizers and pesticides, making it an environmentally sustainable alternative. However, it is important to choose the right type of microbial inoculant for a specific crop and soil type, as well as to ensure proper storage, application, and management. Soil biodiversity is under threat from various factors like intensive agriculture, land-use change, and climate change, and this can have negative impacts on soil fertility and ecosystem services. By enhancing the diversity of microbes in the soil, microbial inoculation can help to counteract these negative effects and promote the restoration and maintenance of healthy and diverse soil ecosystems.

Microbial inoculum improvement techniques can enhance the effectiveness and efficiency of these products by optimizing the selection, isolation, production, and formulation of beneficial microorganisms. Commercial microbial inoculants are available in the market for different applications and purposes, and it is important to carefully select and apply the product, pertaining to the specific requirements of the soil and crop. Overall, microbial inoculum improvement is an important aspect of sustainable agriculture and ecosystem management, and ongoing research and innovation in this field can lead to further advancements in soil microbiology and plant-microbe interactions. Therefore, further research and development of microbial inoculants and their application methods are needed to fully harness the potential benefits of this approach for soil health and agricultural sustainability.

References

Aamir M, Rai KK, Zehra A, Dubey MK, Kumar S, Shukla V, Upadhyay RS (2020) Microbial bioformulation-based plant biostimulants: a plausible approach toward next generation of sustainable agriculture. In: Microbial endophytes. Woodhead Publishing, pp 195–225

Aini N, Yamika WSD, Ulum B (2019) Effect of nutrient concentration, PGPR and AMF on plant growth, yield, and nutrient uptake of hydroponic lettuce. Int J Agric Biol 21:175–183

Alori ET, Glick BR, Babalola OO (2017) Microbial phosphorus solubilization and its potential for use in sustainable agriculture. Front Microbiol 8:971

Amenaghawon AN, Anyalewechi CL, Kusuma HS (2021) Fabrication approaches for biofertilizers. Biofertil Study Impact 75:491–515

Babalola OO, Glick BR (2012) The use of microbial inoculants in African agriculture: current practice and future prospects. J Food Agric Environ 10(3&4):540–549

Bisht N, Chauhan PS (2020) Excessive and disproportionate use of chemicals cause soil contamination and nutritional stress. In: Larramendy ML, Soloneski S (eds) Soil contamination—threats and sustainable solutions. IntechOpen, London, UK, pp 1–10

Burton AS, Stahl SE, John KK, Jain M, Juul S, Turner DJ et al (2020) Off earth identification of bacterial populations using 16S rDNA nanopore sequencing. Genes 11(1):76

Cassán F, Coniglio A, López G, Molina R, Nievas S, de Carlan CLN et al (2020) Everything you must know about *Azospirillum* and its impact on agriculture and beyond. Biol Fertil Soils 56:461–479

Chandra D, Barh A, Sharma IP (2018) Plant growth promoting bacteria: a gateway to sustainable agriculture. In: Microbial biotechnology in environmental monitoring and cleanup. IGI Global, pp 318–338

Chaudhary T, Dixit M, Gera R, Shukla AK, Prakash A, Gupta G, Shukla P (2020) Techniques for improving formulations of bioinoculants. 3 Biotech 10:1–9

Chelliah R, Banan-MwineDaliri E, Khan I, Wei S, Elahi F, Yeon SJ et al (2022) A review on the application of bioinformatics tools in food microbiome studies. Brief Bioinform 23(2):bbac007

Daniel AI, Fadaka AO, Gokul A, Bakare OO, Aina O, Fisher S et al (2022) Biofertilizer: the future of food security and food safety. Microorganisms 10(6):1220

Dimkić I, Janakiev T, Petrović M, Degrassi G, Fira D (2022) Plant-associated Bacillus and Pseudomonas antimicrobial activities in plant disease suppression via biological control mechanisms—a review. Physiol Mol Plant Pathol 117:101754

Fierer N, Lauber CL, Ramirez KS, Zaneveld J, Bradford MA, Knight R (2012) Comparative metagenomic, phylogenetic and physiological analyses of soil microbial communities across nitrogen gradients. ISME J 6(5):1007–1017

Glick BR (2012) Plant growth-promoting bacteria: mechanisms and applications. Scientifica 2012:963401. https://doi.org/10.6064/2012/963401

Ijaz M, Ali Q, Ashraf S, Kamran M, Rehman A (2019) Development of future bioformulations for sustainable agriculture. In: Microbiome in plant health and disease: challenges and opportunities. Springer, Berlin/Heidelberg, pp 421–446

Islam MK, Yaseen T, Traversa A, Kheder MB, Brunetti G, Cocozza C (2016) Effects of the main extraction parameters on chemical and microbial characteristics of compost tea. Waste Manag 52:62–68

Karlen DL, Veum KS, Sudduth KA, Obrycki JF, Nunes MR (2019) Soil health assessment: past accomplishments, current activities, and future opportunities. Soil Tillage Res 195:104365

Kebede E, Amsalu B, Argaw A, Tamiru S (2020) Eco-physiological and physiological characterization of cowpea nodulating native rhizobia isolated from major production areas of Ethiopia. Cogent Biol 6(1):1875672

Kundan R, Pant G, Jadon N, Agrawal PK (2015) Plant growth promoting rhizobacteria: mechanism and current prospective. J Fertil Pestic 6(2):9

Lehmann J, Bossio DA, Kögel-Knabner I, Rillig MC (2020) The concept and future prospects of soil health. Nat Rev Earth Environ 1(10):544–553

Lindström K, Mousavi SA (2020) Effectiveness of nitrogen fixation in rhizobia. Microb Biotechnol 13(5):1314–1335

Liu J, Yu X, Qin Q, Dinkins RD, Zhu H (2020) The impacts of domestication and breeding on nitrogen fixation symbiosis in legumes. Front Genet 11:00973

Lopes MJDS, Dias-Filho MB, Gurgel ESC (2021) Successful plant growth-promoting microbes: inoculation methods and abiotic factors. Front Sustain Food Syst 5:606454

Malusá E, Sas-Paszt L, Ciesielska JJTSWJ (2012) Technologies for beneficial microorganisms inocula used as biofertilizers. Sci World J 2012:491206. https://doi.org/10.1100/2012/491206

Mancera-López ME, Barrera-Cortés J, Mendoza-Serna R, Ariza-Castolo A, Santillan R (2023) Polymeric encapsulate of *Streptomyces* mycelium resistant to dehydration with air flow at room temperature. Polymers 15(1):207

Mendoza-Suárez M, Andersen SU, Poole PS, Sánchez-Cañizares C (2021) Competition, nodule occupancy, and persistence of inoculant strains: key factors in the Rhizobium-legume symbioses. Front Plant Sci 12:690567

Müller DB, Vogel C, Bai Y, Vorholt JA (2016) The plant microbiota: systems-level insights and perspectives. Ann Rev Genet 50:211–234

Muthuraja R, Muthukumar T (2022) Co-inoculation of halotolerant potassium solubilizing Bacillus licheniformis and Aspergillus violaceofuscus improves tomato growth and potassium uptake in different soil types under salinity. Chemosphere 294:133718
Naamala J, Smith DL (2020) Relevance of plant growth promoting microorganisms and their derived compounds, in the face of climate change. Agronomy 10(8):1179
Nkongolo KK, Narendrula-Kotha R (2020) Advances in monitoring soil microbial community dynamic and function. J Appl Genet 61(2):249–263
Ojo J (2016) Pesticides use and health in Nigeria. IFE J Sci 18(4):981–991
Önder M, Ceyhan E, Kahraman A (2011) Effects of agricultural practices on environment. Biol Environ Chem 24:28–32
Qarni A, Billah M, Hussain K, Shah SH, Ahmed W, Alam S et al (2021) Isolation and characterization of phosphate solubilizing microbes from rock phosphate mines and their potential effect for sustainable agriculture. Sustainability 13(4):2151
Rani A, Singh R, Kumar P, Shukla G (2017) Pros and cons of fungicides: an overview. Int J Eng Sci Res Technol 6(1):112–117
Rojas-Sánchez B, Guzmán-Guzmán P, Morales-Cedeño LR, Orozco-Mosqueda MDC, Saucedo-Martínez BC, Sánchez-Yáñez JM et al (2022) Bioencapsulation of microbial inoculants: mechanisms, formulation types and application techniques. Appl Biosci 1(2):198–220
Sahu PK, Gupta A, Singh M, Mehrotra P, Brahmaprakash GP (2018) Bioformulation and fluid bed drying: a new approach towards an improved biofertilizer formulation. In: Sengar R, Singh A (eds) Eco-friendly agro-biological techniques for enhancing crop productivity. Springer Singapore, pp 47–62. https://doi.org/10.1007/978-981-10-6934-5_3
Sarbani NMM, Yahaya N (2022) Advanced development of bio-fertilizer formulations using microorganisms as inoculant for sustainable agriculture and environment—a review. Malaysian J Sci Health Technol 8(1):92–101
Schutter ME, Fuhrmann JJ (2001) Soil microbial community responses to fly ash amendment as revealed by analyses of whole soils and bacterial isolates. Soil Biol Biochem 33(14):1947–1958
Sehrawat A, Sindhu SS, Glick BR (2022) Hydrogen cyanide production by soil bacteria: biological control of pests and promotion of plant growth in sustainable agriculture. Pedosphere 32(1):15–38
Shuaib M, Azam N, Bahadur S, Romman M, Yu Q, Xuexiu C (2021) Variation and succession of microbial communities under the conditions of persistent heavy metal and their survival mechanism. Microb Pathog 150:104713
Singh M, Bhasin S, Madan N, Suyal DC, Soni R, Singh D (2021) Bioinoculants for agricultural sustainability. In: Microbiological activity for soil and plant health management. Springer, Singapore, pp 629–641. https://doi.org/10.1007/978-981-16-2922-8_25
Srivastava N (2023) Siderophore production in iron uptake and plant biofortification. In: Plant microbiome for plant productivity and sustainable agriculture. Springer Nature Singapore, Singapore, pp 313–329
Stanley J, Preetha G, Stanley J, Preetha G (2016) Pesticide toxicity to microorganisms: exposure, toxicity and risk assessment methodologies. In: Pesticide toxicity to non-target organisms: exposure, toxicity and risk assessment methodologies. Springer, Netherlands, pp 351–410
Surendra Gopal K, Baby A (2016) Enhanced shelf life of Azospirillum and PSB through addition of chemical additives in liquid formulations. Int J Sci Environ Technol 5(4):2023–2029
Suyal DC, Soni R, Sai S, Goel R (2016) Microbial inoculants in sustainable agricultural productivity. Res Pers 1:311–318
Swarnalakshmi K, Yadav V, Tyagi D, Dhar DW, Kannepalli A, Kumar S (2020) Significance of plant growth promoting rhizobacteria in grain legumes: growth promotion and crop production. Plan Theory 9(11):1596
Taechowisan T, Peberdy JF, Lumyong S (2003) Isolation of endophytic actinomycetes from selected plants and their antifungal activity. World J Microbiol Biotechnol 19:381–385
Tariq JA, Ahmed B, Abro MA, Ismai M, Memon RM (2022) Antifungal and antibacterial activity of rhizobacterial microflora. Agric Sci J 4(2):35–42

Tian J, Ge F, Zhang D, Deng S, Liu X (2021) Roles of phosphate solubilizing microorganisms from managing soil phosphorus deficiency to mediating biogeochemical P cycle. Biology 10(2):158

Weng W, Yan J, Zhou M, Yao X, Gao A, Ma C et al (2022) Roles of arbuscular mycorrhizal fungi as a biocontrol agent in the control of plant diseases. Microorganisms 10(7):1266

Yadav KK, Sarkar S (2019) Biofertilizers, impact on soil fertility and crop productivity under sustainable agriculture. Environ Ecol 37(1):89–93

Zaidi A, Khan MS, Saif S, Rizvi A, Ahmed B, Shahid M (2017) Role of nitrogen-fixing plant growth-promoting rhizobacteria in sustainable production of vegetables: current perspective. Microb Strategies Vegetable Prod:49–79. https://doi.org/10.1007/978-3-319-54401-4_3

Chapter 5
Microbial Metabolites and Their Role to Improve Soil Health

Bharti Minhas, Poonam Katoch, Arun Singha, Naveen Minhas, and Neelam Kaushik

5.1 Introduction

Soil is a biologically active, porous medium that covers the earth's surface. Soil is formed from the weathering of rocks, consisting of particles of various sizes that are surrounded by gases and water that vary greatly in the composition and amount. There is a continuous exchange of molecules and ions between all three matter states (solid, liquid, and gases), and this exchange is carried out by physical, chemical, and biological processes. A unique balance is maintained in the ecosystem by these processes, which contribute to soil health (Doran and Parkin 1994). Soil provides the medium for the development of roots, and plants depend on soil to fulfil their requirements for all nutrients and water.

Historically, the word soil health referred to the status of soil in terms of agricultural fertility or productivity (Singer and Ewing 2000). Later on, soil health stated as "the ability of soil to act as an extremely important living system to support the development of living organisms, improve or preserve the traits of air and water, and also assist animal and plant welfare within the environment and the boundaries of land use" (Doran and Zeiss 2000). Soils are an integral part of terrestrial

B. Minhas (✉) · P. Katoch · A. Singha
Department of Microbiology, Himachal Pradesh University, Summer Hill, Shimla, Himachal Pradesh, India

N. Minhas
Intermediate Reference Laboratory for Tuberculosis, Govt. TB Sanatorium Dharampur, Solan, Himachal Pradesh, India

N. Kaushik
Department of Biotechnology, Himachal Pradesh University, Summer Hill, Shimla, Himachal Pradesh, India

R. K. Bhatia, A. Walia (eds.), *Advancements in Microbial Biotechnology for Soil Health*, Microorganisms for Sustainability 50,
https://doi.org/10.1007/978-981-99-9482-3_5

ecosystems, and it becomes necessary to maintain their stability from various natural and man-made disruptions, including drought, deforestation, climate change, insect infestations, etc. (Gregorich and Carter 1997). Interactions between plant and microbe are the prominent areas of research within the agriculture ecosystem.

Soil has its own physical, chemical, and biological properties. Earlier, researchers emphasized the importance of the physical and chemical properties of soil only. However, there are various reports on maintaining soil fertility through soil biodiversity. The interdependency of the soil's properties (physical, biological, and chemical) is well recognized nowadays (Fitter 2005). Physical properties like the quantity of soil organic matter (SOM) directly related to the type of microbial flora in the soil, which varies with soil health and its depth. Soil fertility and water retention capacity are increased by SOM. Microbial biomass, enzyme activities, soil respiration, and the diversity of microbes within the soil indicate soil health, of which microbial diversity is considered an excellent parameter of soil health (Abbott and Murphy 2003).

Further, soil fertility could be measured in terms of its components, i.e., biological fertility, chemical fertility, and physical fertility. These terms are of general significance and aren't defined in specific units. The biological fertility of the soil is widely used in agricultural production systems. Biological fertility is an equally important aspect of healthy soil. But the benefits provided by soil organisms have been largely ignored with the addition of indigenous nutrient supplies or synthetic fertilizers. This approach to modern agriculture leads to serious contamination of the environment by pesticides and nutrients, including nitrogen, phosphorus, and even trace elements (Pahalvi et al. 2021a, b).

Microorganisms play an essential function in improving soil health. The natural compounds obtained from these microorganisms act as a primary reservoir of nutrients to enhance plant growth. Microorganisms have a chief role in nutrient recycling in the environment and increasing crop productivity (Singh et al. 2017). Soil contains millions of microorganisms, e.g., algae, fungi, protozoa, viruses, and various bacteria (nitrogen fixation bacteria, azobacteria, actinomycetes, and rhizobacteria), which increase the fertility of soil and production of crop (Shah et al. 2021). They perform various activities, which include phosphorus solubilization, nitrogen fixation, suppression of insect pests and pathogens, improvement of plant stress tolerance, decomposition, and soil aggregation. Beneficial microbial metabolites are potential replacements for harmful pesticides and chemical fertilizers. Utilization of excessive chemical fertilizers, deforestation, and soil erosion due to rains or winds have resulted in desertification, which further causes the loss of millions of tonnes of fertile top soil and indirectly silts up rivers and lakes (Cardoso et al. 2013).

Various human activities contributed to desertification, the loss of biodiversity, and the deprivation of nutrients as well as organic matter from the soil. It has become imperative to achieve better soil health with increasing efforts in reforestation and the recovery of degraded areas. This aim could be attained by the utilization of organic amendments, the reintroduction of plants, microorganisms, and soil fauna. An overview of various challenges to soil health and their impact on plant health has been illustrated in Fig. 5.1.

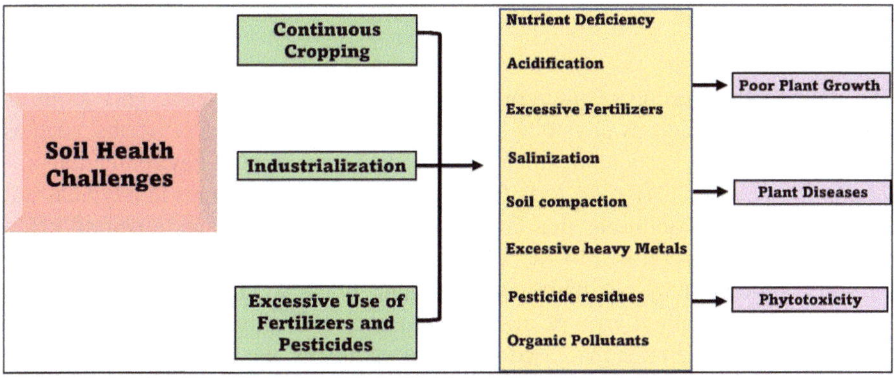

Fig. 5.1 Overview of soil health challenges and their impact on plant health

The health of the soil directly relies on healthy food to which public health is linked. Therefore, there is a requirement for regular assessment and evaluation of biological, physical, and chemical soil fertility to obtain good soil health. Microorganisms are the main nutrient cycling agents and have also been associated with plants, either through positive or negative interactions. A better equilibrium of beneficial microorganisms in the soil results in improved crop yield at a minimum cost and further helps in minimization of the use of pesticides or synthetic fertilizers, favoring high sustainability. This chapter primarily focuses on the roles of microorganisms and their metabolites as potential sources of nutrients in maintaining soil health. A wide variety of microorganisms are associated with enhancing the fertility of soil and healthy crop production.

5.2 Soil as Vibrant Habitat for a Huge Diversity of Microbes

Soil is a complex and heterogeneous component of terrestrial ecosystems, which serve as a diverse habitat for microbes on earth (Dischinger et al. 2009). Microbial communities present in the soil are an active component of ecosystems. Soil microbes spend all or part of their lives in the soil environment and live together by interacting with each other. These interactions lead to improved soil fertility and productivity. Microorganisms include diversified organisms like bacteria, fungi, algae, protozoa, viruses, and certain animals such as rotifers (Bhattarai et al. 2015). Akande and Adekayode (2019) reported that bacteria, being the most numerous type of microbe, constitute a significant part of the soil biomass, followed by fungi.

5.2.1 *Bacteria*

The most leading group of microorganisms in the soil is bacteria. This group is responsible for numerous physiological activities occurring in ecosystems, such as cycling of nitrogen, carbon, metal, and the biodecomposition of environmental pollutants (Deng et al. 2018). Bacteria form spores and microcolonies in a typical soil environment. Soil conditions that favor bacterial growth are water, a neutral pH, a relatively high temperature (30–40 °C), and abundant organic matter. They improve the soil composition by facilitating vital nutrients to the soil habitat, releasing plant hormones to boost plant growth, and regulating or inhibiting plant pathogens. The availability of nutrients is increased by microorganisms, e.g., nitrogen is produced by fixing atmospheric nitrogen. Nitrogen has an crucial role in improving plant growth by helping in the synthesis of various enzymes, chlorophyll, proteins, nucleic acids, etc. (Shah et al. 2021). Sustainable agricultural production systems depend heavily on the process of biological nitrogen fixation. Bacteria involved in different types of nitrogen fixation are *Rhizobium*, *Cyanobacteria*, *Anabaena*, *Nostoc*, *Beijerinckia*, *Azobacter*, and *Azospirillum* species.

5.2.2 *Fungi*

Fungi predominate in the soil as total biomass due to their larger size. Fungi are present almost in every environment and can adapt to a wide temperature and pH range (Frac et al. 2018). However, lower pH or slightly acidic soils are suitable habitats for most of the fungi (Lavelle and Spain 2002). Under dry conditions, fungi continue to survive and grow even at low moisture conditions. Fungi have the ability to synthesize and release a wide range of extracellular enzymes because of which they have the capability to decompose various kinds of organic matter, which thereby helps in the regulation of nutrient balance. Saprophytic fungi, also known as decomposers, degrade lignin and cellulose in the soil. Zygomycetes and sugar fungi degrade the simple sugars, and other fungi breakdown the more recalcitrant organic residues that are high in cellulose, hemicelluloses, or lignin. After this decomposition, some of the byproducts may turn into dark organic matter called humus (Ingham 2009). Fungi have the ability to store toxic metals like copper, cadmium, lead, mercury, and zinc in their fruiting bodies. However, this absorbance affects the growth and reproduction of fungi. They also act as an essential player in nitrogen fixation, phytohormone production, disease suppression, and protection against drought (Frac et al. 2018). Inspite, there are some pathogenic fungi including *Pythium*, *Rhizoctonia*, and *Verticillium*, that cause huge economic losses in agriculture every year.

5.2.3 *Algae*

Algae may range in size from 1 μm to over 50 m. Algae are ubiquitously distributed in different habitats, e.g., soils, lakes, rivers, oceans, or extreme environments (Song et al. 2022). Algae increase soil organic matter, enhance phosphorus availability and its uptake by excreting various organic acids, enhance nitrogen availability through biological nitrogen fixation, release bioactive extracellular substances that enhance plant development and growth, and produce biocontrol agents that act on plant pathogens. They also produce exopolysaccharides that improve soil structure (Ibraheem and Abdel-Raouf 2007).

Microalgae such as diatoms, green algae, and cyanobacteria are used as soil biofertilizers. They are usually richer in nitrogen, phosphorus, potassium, and other trace elements (Sasaki et al. 2020; Song et al. 2022). These elements promote crop growth and improve soil fertility. Algae can be used for reduction of erosion as well as waste water treatment due to their ability to act as surface-binding agents.

5.2.4 *Protozoa*

Protozoa are ubiquitously present in almost all habitats, including extreme environments. Some soils have high load of protozoa that are mainly active in the rhizosphere region, like bacteria. Soil protozoa have a broad range of functions; they act as consumers of bacteria, primary producers, predators of protozoa, fungal feeders, and also as parasites of plants, metazoans, and protists (Geisen et al. 2018). Protozoa provide nutrients to plants and other living organisms present in soil through the process of mineralization. Protozoa eat bacteria to meet their carbon and nitrogen requirements. But the amount of carbon provided by bacteria to protozoa exceeds the amount of carbon required. They release the excess nitrogen in the form of ammonium (NH^{4+}) near the rhizosphere. Bacteria and other living forms present in the soil habitat rapidly take up most of the ammonium released; only some of this is used up by the plant. Protozoa regulate bacterial populations by feeding on them and, in turn, increasing decomposition rates and soil aggregation (Raouf et al. 2012). Protozoa act as biocontrol agents and suppress many diseases by competing with pathogens or feeding on pathogens. Encysted protozoa, along with spores or cysts of plant-growth-promoting rhizobacteria (PGPR), act as potential biofertilizer.

5.2.5 *Viruses*

Soil viruses are the largest genetic diversity repository on the Earth and may likely contain numerous unique viral species. Several studies support the idea that the majority of viruses found in soil are bacteriophages (viruses that infect bacteria).

The occurrence and abundance of viruses in soil are attributed to various factors. These include pH, temperature, altitude, type of soil, quantity of organic matter in soil, and its water content. The soil's pH is the main factor of the viral community in agricultural soil as it affects the attachment of viruses to soil particles (Bi et al. 2021). Viruses kill microorganisms *through* viral infection and lysis, releasing nutrients like nitrogen, carbon, sulphur, and phosphorus. These released nutrients were further absorbed by other microorganisms and plants (Kuzyakov and Mason-Jones 2018). Viruses may also acquire specific host genes, transfer them, and cause gene acquisition, which may alter the robustness of soil microbes. Soil viruses functions as an important player in biogeochemical cycles, affecting the concentration and availability of gases and nutrients in the environment, contributing to food webs linked to organic matter decomposition, and hence increasing agricultural productivity. Various studies via metagenomics analysis showed that viral communities present in the soil were significantly distinct from those in other environments (e.g., human fecal samples, marine sediment, and seawater environments), and scanty data is available on the studies of soil viruses (Cao et al. 2022).

5.3 Soil Health

Earlier, the terms "soil quality" and "soil health" were regarded to be interchangeable. In 1971, Mausel coined the term "soil quality," defining it as "the ability of soils to yield corn, wheat, and soybeans under conditions of high-level management" (Bunemann et al. 2018). The deliberation over soil quality versus soil health arose immediately in the 1990s after the notion of soil quality was criticized. The term soil quality describes a soil's inherent qualities, impacted by natural factors and soil formation processes. However, soil health refers to changes in soil characteristics brought on over time by human management and soil usage. Soil health would capture the ecological attributes of the soil in contrast to soil quality. These attributes have implications beyond their quality, composition, or ability to produce a certain crop. Earlier, agricultural production was the primary focus of most definitions of soil quality, but Doran and Parkin in 1994 focused on environmental and ecosystem health issues related to soil quality. They defined it as "soil capacity to function within ecosystem boundaries in order to support plants, animal productivity, and soil biodiversity." Healthy soil supports human health by improving or maintaining the quality of the water and air. Soil degradation has become a major environmental concern for soil, which may cause a partial or whole loss of its productive capacity. Soil health is deteriorated mainly by industrialization, the indiscriminate use of chemical fertilizers, and excessive flooding (Allen et al. 2011).

5.3.1 Soil Health Indicators

Soil has its own dynamic and inherent properties, and these properties are determinants of soil function. Soil health evaluation does not depend on any single soil property or on a universal set of indicators but varies with every unique location (Riley 2001). Soil indicators chosen to evaluate soil health consist of a complex group of measurable physical, biological, and chemical characteristics that are directly related to soil functions. However, these categories are not always well defined since a soil property or indicator might have an impact on multiple soil functions or categories. A link between the soil health indicators and beneficial soil functions has previously been suggested (Parisi et al. 2005). Physical indicators of soil health include soil texture, aggregation, bulk density, and available water capacity. Chemical indicators include soil pH (6.0–7.5 for general crop productivity), soil nitrate, electrical conductivity, and other mineral nutrients. However, the release of carbon dioxide (CO_2), organic matter, organic carbon, soil microorganisms and macroorganisms, etc. are biological indicators of the soil (Doran and Parkin 1996).

5.3.2 Properties of Healthy Soil

5.3.2.1 Adequate Soil Tilth

Soil tilth is the physical state of the soil as it is related to plant development. A soil with good tilth is crumbly, dark with organic matter, and well structured. Also have good conditions for seed germination, root proliferation, water infiltration, and aeration, which benefit both the crop and the environment. A soil with good tilth has no large and hard clods. The different pore sizes of good tilth soil have been represented in Fig. 5.2.

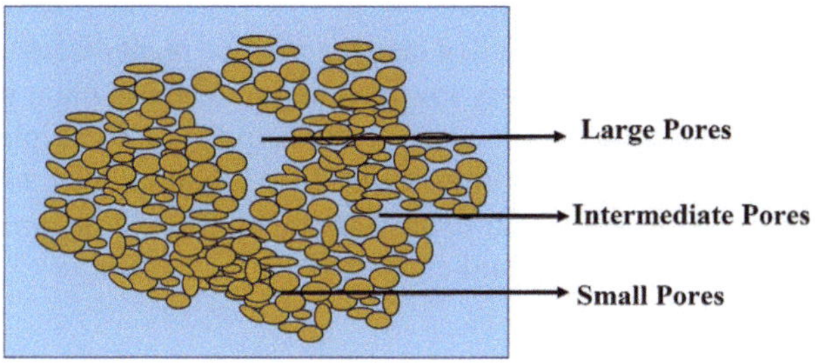

Fig. 5.2 Good tilth soils have a different range of pore sizes

5.3.2.2 Sufficient Depth

The depth of the soil is directly related to its capacity for nutrients and water, which strongly impact the genera of plants that can grow there. Deeper soils may contain extra water and nutrients for plants relative to shallow soils. Plants growing in deeper soils are mechanically stronger than those growing in shallow soils (Usharani et al. 2019).

5.3.2.3 Good Water Storage and Drainage

Healthy soils have high total porosity and large to small pore sizes distributed over a wide range to absorb soil water. These large pores supply water to the medium pores, and the medium pores further conduct water to the small pores for later use. This stored water is used by plants during dry spells (Mandal et al. 2021).

5.3.2.4 Adequate Supply of Nutrients and Minerals

The natural cycling of nutrients maintains the essential nutrients required for plant growth in the soil. The nutrient cycling occurred mainly through the breakdown of biomass. This cycle incorporates physical, chemical, and biological processes. There are 17 essential nutrients needed for plant growth. A deficiency of any one of these nutrients in the soil leads to low crop yield. It is therefore necessary to maintain an adequate and balanced nutrient supply for optimal plant growth. An excess of nutrients, especially nitrogen and phosphorus, can lead to leaching, which contributes to surface and groundwater pollution as well as toxicity to microbial communities and plants (Parikh and James 2012).

5.3.2.5 Reduced Population of Insect Pests and Plant Pathogens

Pests and plant pathogens are capable of causing various plant diseases that result in crop damage. There should be a very low population of these plant pathogens and pests to increase agricultural productivity. The activity of these plant pathogens could be reduced by direct competition for nutrients, habitat, etc. with soil pathogens. Moreover, the defense system of healthy plants should be strong enough to resist various kinds of pests (Usharani et al. 2019).

5.3.2.6 Presence of Abundant Beneficial Microorganisms

The presence of a diverse variety of beneficial organisms is a characteristic feature of healthy soil. A number of microorganisms aid in plant thriving or survival. The functioning of soil depends on the organisms associated with it. They aid in the putrefaction of organic matter, nutrient cycling, maintaining soil structure, and biological control of plant pests. They also produce antibiotics that halt the spread of diseases, or they may sometimes contribute to the phenomenon of acquired resistance. There are some bacteria that aid in plant development by producing siderophores (molecules that chelate iron) that limit the availability of this crucial element to pathogens and other microflora (Usharani et al. 2019).

5.3.2.7 Less Weed Pressure

Weed pressure is a major barrier to the healthy production of crops. Weeds compete with crops to meet the essential requirements needed for plant growth, e.g., water and nutrients. Less nutrients are consequently available to the crop. In addition to this, weeds can hinder solar energy, interfere with stand setup and harvest operations, and also harbor pathogens and pests that can spread disease to the crop of interest (Usharani et al. 2019).

5.3.2.8 Free of Chemicals and Toxins

Healthy soils should be free or devoid of harmful chemicals and toxins. Potentially harmful chemicals have been shown to disturb the biological balance of the soil. The disturbance of this balance is likely caused by the disruption of physiological processes, the degradation of soil microorganism cellular membranes, and protein denaturation. This imbalance is brought on by an excess of potentially harmful chemicals and toxins (Bansal 2018).

5.3.2.9 Resistant to Degradation

A healthy soil with a wide population of living organisms is less vulnerable to erosion (through water or wind) or other unfavorable events such as excessive rainfall, extreme drought, vehicle compaction, disease outbreaks, etc. Soil degradation causes the loss of important soil fertility by affecting its physical, chemical, and biological properties.

5.3.2.10 Resilience when Unfavorable Conditions Occur

A healthy biological component of the soil is essential for soil resiliency because it allows the soil to recover from stress. The living component of soil, often known as "soil biology or biota," is crucial to numerous soil functions and for maintaining the overall soil quality. The main functions include the following:

1. Provision of nutrients and their cycling.

2. Prevention of pests and pathogens.
3. Growth factor production.
4. Availability of water.

5.3.2.11 Common Soil Constraints

Any soil feature that limits plant development and reduces agricultural productivity is referred to as a soil constraint. These could either be physical, chemical, or biological and may restrict farm sustainability, environmental quality, and agricultural output. Only management practices can be changed to address these issues.

5.4 The Interconnection of Plants with Soil Microbes

Soil organisms interact with plants in many distinct ways and can be characterized into different types of interactions like neutral, competitive, exploitative, commensal, and mutualistic. The majority of interaction studies in modern plant science have focused mainly on pathogenic effects, such as reducing abiotic stress conditions or herbivory and infection. The study of the beneficial ecological interactions that aid in plant growth, however, has also long been people's area of curiosity. It was well known in the nineteenth century that some microorganisms (e.g., symbiotic nitrogen-fixing bacteria and mycorrhizal fungi) play essential roles in plant growth by enhancing mineral nutrition. In the 1980s, numerous bacterial strains (e.g., *Pseudomonas*, *Azospirillum*, etc.) were identified to exhibit effects that aid in promoting plant growth (Teintze et al. 1981; Lin et al. 1983). Since 2000s, the focus of research has switched from examining specific bacterial strains to employing metagenomics to explore the diversity and abundance of the root microbiota. According to the results of sequencing research, the rhizospheric habitat is a hub of ecological richness (Jacoby et al. 2017).

The rhizosphere region is a narrow area of soil around a plant root that is inhibited by a large number of microorganisms. Rhizosphere microorganisms, which are regarded as the second set of plant genomes, are essential for promoting plant growth, nutrient uptake, nutrient acquisition, and the control of fungal plant diseases. The health of soil and crops is dependent on the microflora available, particularly in the rhizosphere region (Shah et al. 2021).

When microbes such as bacteria, fungus, algae, and actinomycetes interact with plants in the soil, they can produce auxins, salicylates, gibberellins, cytokinins, and abscisic acid in plant tissues. Phytohormones are produced by various microorganisms associated with roots and induce tolerance in plants to diverse abiotic stresses (Andreote et al. 2014; Egamberdieva et al. 2017).

5.5 Role of Microbes in Improving Soil Health

Numerous soil microorganisms contribute to improving soil fertility and crop yield (Gougoulias et al. 2014). The soil's properties are influenced by its pH, redox potential, organic content, and biomass density (Pandey et al. 2020). The microscopic culture's composition in the soil's structure is also significantly impacted by these factors. A healthy soil must include a wide range of soil microorganisms that are active. They are directly correlated with crop production and sustainable agriculture. Microbes can improve soil health by aiding in the breakdown of organic matter, releasing nutrients essential for plant growth, strengthening soil structure, and fighting with harmful microbes. They can also coexist peacefully with plants as mycorrhizal fungi, which facilitate a better uptake of nutrients from the soil. Overall, a diverse population of microorganisms in the soil can lead to healthier or more productive plants (Lombard et al. 2011; Glick 2018).

Soil microorganisms improve plant growth through different mechanisms. The mechanisms include the creation of metabolites, the transformation of atmospheric nitrogen into ammonia, the generation of plant growth regulators, and also providing resistance against pathogens. Rhizospheric soil's microbial diversity supports soil productivity and health. Arbuscular fungi and plant growth-promoting rhizobacteria (PGPR) develop an association with plants and release various metabolites in the soil. Hence, they significantly encourage plant development and contribute toward sustainable agriculture (Rasul et al. 1998). Rhizosphere microorganisms can dissolve insoluble calcium phosphate. *Micrococcus*, *Pseudomonas*, *Achromobacter*, *Agrobacterium*, *Bacillus*, *Burkholderia*, *Rhizobium*, *Flavobacterium*, *Erwinia*, and *Aerobacter* are bacterial species that can solubilize phosphorus. The rhizosphere region frequently contains a significantly greater concentration of phosphate-solubilizing bacteria (Cunningham and Kuiack 1992; Goldstein 1995; Khalid 2012). Figure 5.3 depicts the role of microorganisms in enhancing soil health.

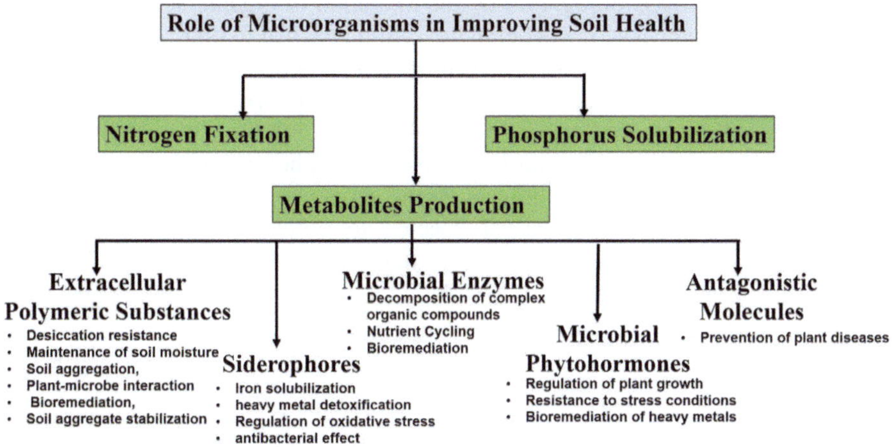

Fig. 5.3 Role of microorganisms and their metabolites in improving soil health

5.5.1 Phosphate Solubilization

Phosphate solubilization is the process by which bacteria, fungi, and other microorganisms change the insoluble phosphate found in soil into its plant absorbable solubilized form. These phosphate-solubilizing microbes produce enzymes that change phosphates into a form that plants can take more easily. The second essential element for plant growth and development is phosphorus, after nitrogen. The biogeochemical cycling of phosphorus in natural and farming environments is fundamentally influenced by soil-dwelling microorganisms like bacteria, fungi, and actinomycetes. Organic and inorganic phosphates are sources of phosphorus in the soil. Most insoluble mineral complexes that include inorganic phosphorus are found in soil. Some of these arise due to utilization of chemical fertilizers. However, organic matter is a significant source of immobilized phosphorus and contributes between 20 and 80% of the phosphorus in soil (Richardson 1994).

For plants, phosphate, being one of the most crucial macronutrients, is added to the soil as phosphate fertilizer. However, soluble inorganic phosphate used in the form of chemical fertilizer quickly immobilizes and will no longer be available to plants (Malboobi et al. 2009). At the moment, regulation of soil phosphorus is mostly done to maximize crop yield and reduce phosphorus loss from soils. The major area of research in agriculture has now focused on the utilization of phosphate-solubilizing bacteria as soil inoculums to increase plant growth and yield. Additionally, through phosphate-heavy metal precipitation, phosphate-phosphorus compounds can immobilize heavy metals, particularly lead (Pb), in the contaminated environment because the majority of phosphorus compounds are not readily soluble in soils. To repair polluted soil, phosphate-solubilizing microbes have the ability to improve phosphate-induced metal immobilization. Due to the problem of eutrophication, there is a restriction on the concentration of phosphate that can be introduced to the environment.

Bacteria are more efficient than fungi at solubilizing phosphorus. The bacterial solubilization activity is brought on by the liberation of organic acids (oxalic and citric acids) that contribute in chelating the cation attached to phosphate and transform it into a soluble form that plants can use (Chen et al. 2006). Microorganisms that solubilize phosphate are crucial for providing phosphorus to plants. The majority of the phosphorus in the Earth is still interlocked in different insoluble forms and is not accessible to plants. The indiscriminate use of phosphorus fertilizer reduces soil quality and has detrimental effects on the economy and ecology. There is an inherent ability for some bacterial species to solubilize phosphorus. Such phosphorus-solubilizing bacterial species, when used as an inoculant, concurrently increase crop yield and phosphorus availability to plants.

5.5.2 *Nitrogen Fixation*

Nitrogen is a crucial element of various plant molecules, such as DNA/RNA, amino acids as well as proteins. Hence the nitrogen fixation process is crucial for the growth and well-being of plants. The transformation of atmospheric nitrogen (N_2) into plant and other living organisms utilizable forms such as ammonia (NH_3) or nitrate is a process known as nitrogen fixation (NO^{3-}). It is possible to artificially induce nitrogen fixation by using nitrogen-fixing plants or fertilizers, or it can occur naturally in soil due to the actions of specific microorganisms like bacteria and cyanobacteria. Some agricultural techniques, like crop rotation and the blending of organic materials into the soil, can also help with nitrogen fixation.

Nitrogen, a crucial nutrient, is not readily accessible to plants in the form of gaseous nitrogen (N_2) because of the tremendous energy needed to hydrolyze the triple bonds between the two atoms (Cain et al. 2011). Rhizobacteria transform nitrogen gas (N_2) into ammonia (NH_3), which is then made available to the host plant to support and encourage plant development. In return, the host plant creates amino acids for the bacteria to use, and further, they do not need to assimilate ammonia (Willey et al. 2011). Leg hemoglobin, a hemoglobin protein generated within the nodules, provides oxygen to the rhizobacteria because they need it to perform their metabolic processes (Cain et al. 2011). Plant growth-promoting rhizobacteria (PGPR) with an ability to fix N_2 through nitrogenase activity and stimulate the development of a specific host plant include *Azoarcus* sp., *Beijerinckia* sp., *Rhizobium* sp., *Klebsiella pneumoniae*, and *Pantoea agglomerans* (Antoun et al. 1998; Hurek et al. 1994; Baldani et al. 1997; Riggs et al. 2001). The most studied PGPR are the rhizobia (*Azorhizobium*, *Rhizobium*, *Allorhizobium*, *Sinorhizobium*, *Bradyrhizobium*, and *Mesorhizobium*) species for their capacity to fix N_2 in their leguminous hosts. The first commercially available microbial inoculants for improvement of legume crops were established in the 1890s.

In soil, diazotrophic microorganisms, which include bacteria like *Azotobacter* and Archaea, can fix nitrogen naturally. Some nitrogen-fixing bacteria coexist harmoniously with certain plant species, particularly legumes (Zahran 1999). In the case of nitrogen fixation on rice roots, looser nonsymbiotic connections between plants and diazotrophs are frequently called as associative. Some termites and fungi work together to fix nitrogen (Sapountzis et al. 2016).

5.5.3 *Production of Metabolites*

Microbial metabolites are intermediates in microbial metabolism. Primary metabolites and secondary metabolites are two kinds of metabolites produced by microorganisms during their growth. Primary metabolites are those molecules that are needed for the growth of microorganisms and are used as biotransformation agents (Singh et al. 2022). The organic molecules including nucleotides, amino acids, and fermentation end products, viz., ethanol, lactic acid, and other organic acids constitute the major primary metabolites of microorganisms. Secondary metabolites include organic compounds which are not directly related to the growth and development of microorganisms and are synthesized in the stationary phase of growth. Antibiotics, toxins, bacteriocines, pigments, etc. are some examples of secondary metabolites and are majorly utilized in healthcare activities (Demain 1999; Singh et al. 2017). These have both ecological and biogeochemical effects as well as potential in various industries, including agriculture, food, and chemical industries. Soil microorganisms like bacteria, protozoa, and fungi are essential for the synthesis of metabolites because they can degrade organic matter and can liberate nutrients that plants may ingest. The overall health of the plants can be affected by a variety of metabolites that these microbes can produce. Additionally, the soil's physical and chemical properties also have a significant impact on the production of metabolites.

5.5.4 *Biofertilizers*

Beneficial microbes increase agricultural production and reduce disease incidence, which may be beneficial for the overall health of the environment. The consortium of these beneficial microbes is known as biological fertilizer, or biofertilizer. These biofertilizers enhances the number of primary nutrients available to the recipient plant by colonizing the rhizosphere, or plant interior to encourage growth (Vessey 2003). The use of chemical herbicides and fertilizers can be eliminated as a result of biofertilizers. The microflora in the biofertilizers rebuild the soil's organic matter and nutrient cycle.

Moreover, biofertilizers are free of any chemicals that could harm the soil's living organisms. Bio-fertilizers are more economical than chemical fertilizers and are organic, eco-friendly agricultural inputs. Long-established biofertilizers include *Rhizobium*, *Azospirillum*, *Azotobacter*, and blue-green algae (BGA) (Khalid 2012).

5.6 Role of Microbial Metabolites in Soil Improvement

Microorganisms are ubiquitous in nature and are invisible players that contribute toward the sustainability of all life forms on the Earth. Microbial metabolism is the chief source of nutrients and helps in the recovery of various micronutrients as well as macronutrients in the environment. In the soil ecosystem, microorganisms secrete a range of metabolites that mediate various ecological interactions such as competition, communication, and beneficial interaction with other living forms and the environment (Charlop-Powers et al. 2014). Thus, the availability of a plethora of microbes and their metabolites in the soil assesses the quality of that soil (Furtak and Gajda 2018).

Microbial metabolites, including extra-polymeric substances, siderophores, microbial enzymes, phytohormones, and antagonistic molecules, are some important metabolites that act as an important player in soil and plant health. Some important microbial metabolites have been described in the following section.

5.6.1 Extracellular Polymeric Substances

Microbial extra-polymeric substances or exopolysaccharides (EPS) are mixed polymers synthesized by diverse microbial organization (algae, fungi, and bacteria). These are high-molecular-weight polymers, majorly consisting of polysaccharides, proteins, and nucleic acids. Microbial EPS has been classified as soluble EPS (S-EPS) and bound EPS (B-EPS). S-EPS is found outside the cell wall of bacteria and is available in the form of soluble macromolecules, slime, and colloids, whereas B-EPS is available in the form of capsules, sheaths, and condensed gel and is present very closely to the bacterial cell. Further, two forms of B-EPS have been identified: loosely bound EPS and tightly bound EPS. Exopolysaccharides are polyionic in nature, either due to the presence of the pyruvate group, i.e., ketal-linked, or uronic acid. The EPS has a polyionic nature due to the association of inorganic components such as phosphates and sulphates. However, some EPS have also been found to be neutral (Naseem et al. 2018). Microbial EPS is important in performing various microbial metabolic and physiological activities, viz., it functions as an electron acceptor, a source of energy, a nutrient source, and also acts as a protection barrier against various environmental stress conditions. Other processes such as adhesion, cellular communication, exchange of genetic information, carbon storage, entrapment of nutrients, etc. have also been performed by microbial EPS (Sukanya et al. 2022).

Exopolysaccharides have a prominent influence on improving soil as well as soil health because they help with the maintenance of soil moisture content, desiccation resistance, soil aggregation and stabilization, plant-microbial interaction, surface adhesion, and bioremediation, which thereby helps in restoring soil structure and, thus soil health. In water-deficient stress conditions, EPS-synthesizing bacteria accumulate higher amounts of organic molecules such as proline, free amino acids, and carbohydrates (Naseem et al. 2018). Proline has been reported to decrease osmotic potential in plants and contribute in maintaining high turgor pressure, which is required for a plant's metabolic as well as physiological activities. Thus, the presence of proline in the area inoculated with EPS-producing bacteria indicates the ability of EPS to combat water deficits and environmental stress (Naseem et al. 2018). Seeding of soil with EPS-producing bacteria also results in numerous favorable effects on salt-affected soil (Ashraf et al. 2005). Beneficial attributes of EPS-producing bacteria in the soil have been given below:

- Improvements in basic soil features such as soil aggregation, aeration, permeability, water holding capacity, bulk density, etc.

- Increase in nutrient recycling in the root zone.
- Improvement in soil fertility by providing plant growth hormones and essential nutrients.

Exopolysaccharides can interact with metallic pollutants including Cd (II), Cu (II), and Pb (II) (Fig. 5.4) because of their huge surface area. This interaction leads to the biotransformation of metallic pollutants, which helps in biosorption (Kumawat et al. 2021). Biosorption of metallic pollutants has been reported in EPS extracted from some *Bacillus* species isolates (Shameer 2016). Further, EPS of *Bacillus cereus* KMS3-1 has also been identified to biotransform toxic pollutants such as Cd (II), Cu (II), and Pb (II) into nontoxic inorganic compounds CdS, CuS, and PbO, respectively, highlighting the ability of EPS to enhance the quality of soil by reducing the metallic pollutants in the soil (Mathivanan et al. 2021). Figure 5.4 illustrates the interaction of exopolysaccharides with heavy metals.

Soil moisture content is an essential factor that determines the quality of the soil. In a study by Roberson and colleagues, amendment of sandy soil with extracted EPS of the *Pseudomonas* strain leads to an improvement in the water holding capacity of that soil, which reveals the ability of EPS to hold water and maintain soil moisture under water-deficient conditions (Roberson and Firestone 1992). Three mechanisms have been proposed by which the microbial EPS can retain soil moisture. Mechanisms involved are (1) direct holding of moisture content by EPS within its polymeric matrix; (2) promotion of soil aggregation, which leads to reduction in soil pore size, resulting in the retention of excess water; and (3) modification of the water repellency of soil, which leads to the formation of more hydrophobic micropores that inhibit water evaporation. Water deficiency is the major factor that reduces soil quality and limits crop yields. Further insights in the field of the application of microbial EPS in different soil stress conditions can help in improving the soil quality.

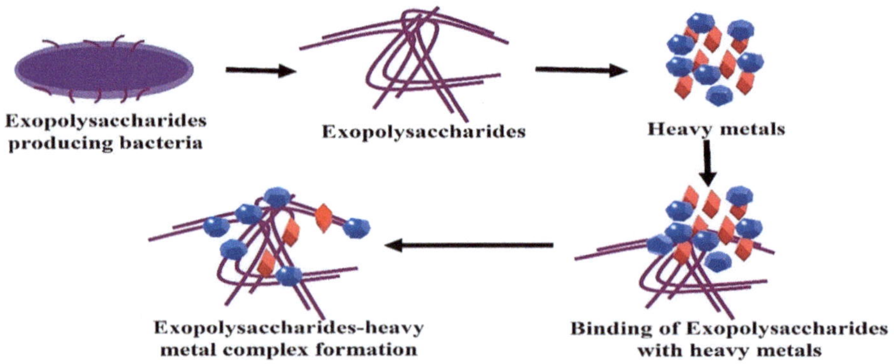

Fig. 5.4 Mechanism of exopolysaccharides interaction with heavy metals

5.6.2 Siderophores

Various factors such as rapid industrialization, urbanization, fast agricultural practices, invasion of alien species, overexploitation of natural resources, poor governance management strategies, etc. result in the degradation of more than 33% of the land. These factors resulted in the deterioration of soil health with respect to its nutrients, moisture, pH, etc. (Singh et al. 2022). Nutrient discrepancy is a central parameter for soil fertility. Among nutrients, iron is an important component needed for the survival of all living forms present on Earth because of its requirement for various cellular activities, including DNA metabolism, protein and fatty acid synthesis, and other enzymatic and metabolic reactions (Ahmed and Holmstrom 2014). Iron is one among the most abundant elements present on the Earth's crust; however, aerobic and alkaline conditions in the soil cause the oxidation of soluble iron (Fe^{2+}) into its insoluble form (Fe^{3+}), which cannot be taken up by autotrophs directly. Crop and soil management are the two main strategies through which iron is made available to plants. The delivery of inorganic iron ($FeSO_4$) or synthetic or semisynthetic Fe chelates to foliar and roots are the major practices involved in crop management. In soil management, the addition of inorganic salts (NO^{3-} or NH^{4+}) to the soil is an important method for making iron available to the crops. The incorporation of inorganic salts leads to a modification in soil pH, which can enhance the solubility of iron for plant uptake. The other method for soil management is the addition of synthetic chelates (e.g., Fe-EDTA) to the soil. The abovementioned practices are cheaper and more effective; however, their overuse may also have a negative environmental impact (Radzki et al. 2013). Moreover, advances in microbiological applications have led to the utilization of biological iron chelators referred to as siderophores, which are liberated by microorganisms under iron-starved environmental conditions.

Siderophores are low-molecular-weight iron-chelation organic molecules secreted by microorganisms, which are required for chelation of ferric ions from inorganic and organic molecules present in the soil and hence make them available for the growth of microbes and plants. Different types of siderophores with diverse chemical structures are released by microorganisms. In spite of chelation, siderophores can also interact with other metals including zinc, sulphur, vanadium, etc. Other functions of siderophores, such as sequestration of heavy metal toxins, transport of boron, cellular signaling, regulation of oxidative stress, and antibacterial effects, have also been reported (Ahmed and Holmstrom 2014). Cui and colleagues reported solubilization of phosphate by *Streptomyces* sp. siderophores in iron-rich soil, which further emphasizes the important role of siderophores in increasing soil fertility (Cui et al. 2022). In recent years, siderophores are being explored for their applications in the bioremediation of heavy metals, inhibition of phytopathogens, and promotion of plant growth.

Soil bioremediation is an essential factor in the maintenance of a healthy ecosystem in the soil. Various human activities, like rapidly growing industries, improper disposal of metal waste, utilization of pesticides and chemical fertilizers, spillage of petrochemicals, etc., cause pollution of soil with heavy metals and metalloids (Zhang et al. 2010). The capacity of microbial siderophores to bind with metals enhances their utilization in the bioremediation of heavy metals found in the soil. The removal of arsenic from contaminated soil by siderophores secreted by *Pseudomonas azotoformans* has been reported by Nair and colleagues (2007). According to Diels et al. (2009), soil inoculation with citrate siderophore-producing *Cupriavidus metallidurans* led to a 16-fold reduction in soil heavy metals. The potent role of siderophores in the mobilization of metals from mine waste has also been reported. Siderophores from *Pseudomonas fluorescens* showed the mobilization of several metals from waste material from a uranium mine. Thus, siderophores secreted by microorganisms can have the potential to be utilized as potential bioremediation agents for management and improvement of soil health (Saha et al. 2016a, b).

5.6.3 Microbial Enzymes

The determination of enzymes available in the soil is a measure for the assessment of soil quality in any area. Soil enzymes function as an essential factor in biogeochemical cycles, decompositions, and in the maintenance of qualitative microbial survival in the soil (D'Alo et al. 2021). Enzymes such as hydrolases and glucosidases are associated with the breakdown of organic matter, whereas other enzymes including amidases, ureases, arylsulfatases, and phosphatases are known to be involved in nutrient mineralization (Jamir et al. 2019). All these enzymes are required for maintaining soil fertility and are thus used as essential indicators of soil management. The enzymes of the soil serve as sensitive probes of changes in environmental conditions and pollution levels (Jamir et al. 2019). These enzymes are secreted by plants and other life forms found in the soil; however, microbes, including bacteria, fungi, and algae are the major source of soil enzymes (Bayranvand et al. 2021).

The action of enzymes is influenced by numerous extrinsic environmental factors, including temperature, pH, moisture content, the presence of substrate, etc. Substrate is the limiting factor for the catalytic activity of the enzymes. Plant cell wall material acts as a major substrate for various degrading enzymes. Plant cell wall contain cellulose, hemicellulose, pectin, lignin, and various proteins. However, the structural complexity and heterogenicity of these polymers require specific enzymes for their degradation (Sobucki et al. 2021). Enzymes secreted by microorganisms play a central function in the degradation of these complex organic molecules for the production of fertile soil. A consortium of enzymes secreted by microorganisms is required for the sequential decomposition of all complex molecules available in the soil (Wallenstein and Burns 2011).

Oxidative enzymes and hydrolytic enzymes are two major categories of enzymes found in the soil and act as central players in the degradation of organic molecules and nutrient cycling. Phenol oxidases and peroxidases are examples of oxidative enzymes, whereas hydrolytic enzymes include α-glucosidase, cellobiohydrolase, leucine aminopeptidase, phosphomonoesterase, etc. (Sinsabaugh 2010; Uwituze et al. 2022). The list of microbial enzymes involved in nutrient cycling in wheat and maize-season soil is provided in Table 5.1.

5.6.4 Phytohormones Released by Microorganism

Phytohormones are the regulatory proteins needed by plants for their proper growth. Various phytohormones have been identified in the supernatant of microbial cultures (fungi and bacteria) isolated from soil. Important phytohormones secreted by microorganisms and their related functions have been given in Fig. 5.5. Several species of *Bacillus* secrete phytohormones, namely, auxins, gibberellins, abscisic acids, and cytokinins (Poveda and Gonzalez-Andres 2021). In addition, some bacteria and fungi can also secrete other phytohormones, including ethylene, salicylic acid, and jasmonic acid (Pozo et al. 2015; Egamberdieva et al. 2017). The rhizobacterial strains of various microbial species, viz., *Pseudomonas stutzeri* MTP40, *Pseudomonas putida* MTP50, and *Stenotrophomonas maltophilia* MTP42 were identified to secrete auxin, cytokinin, and gibberellic acid, and their function in plant protection and growth enhancement was reported by previous studies (Patel and Saraf 2017). All these hormones were described to regulate plant growth and stimulate plant signaling under biotic and abiotic stress conditions (Egamberdieva et al. 2017). Important microbial phytohormones along with their functions in plant growth regulation and development are given in Fig. 5.5. In vitro synthesis of phytohormones secreted by microorganisms can be utilized as bioformulations for improving soil and, thus, plant health as well.

The profile of isolated phytohormone producing rhizobacteria varies based on the form of plant root and soil. Auxin or IAA producing *Mycobacterium* species were isolated from the roots of orchids (Tsavkelova et al. 2007), while bacterial species including *Azotobacter*, *Mycoplana*, *Cellulomonas*, *Azospirillum*, and *Rahnella*

Table 5.1 Enzyme involved in nutrient cycling in wheat and maize-season soil (Ullah et al. 2019)

Nutrient cycle	Enzyme
Nitrogen cycle	α-glucosidase
	N-acetyl-β-glucosaminidase
	L-leucine aminopeptidase
Carbon cycle	β-cellobiosidase
	β-glucosidase
	β-xylosidase
Phosphorus cycle	Phosphatase
Sulphur cycle	Sulfatase
Phenolic compounds oxidation	Phenol oxidase
	Peroxidase

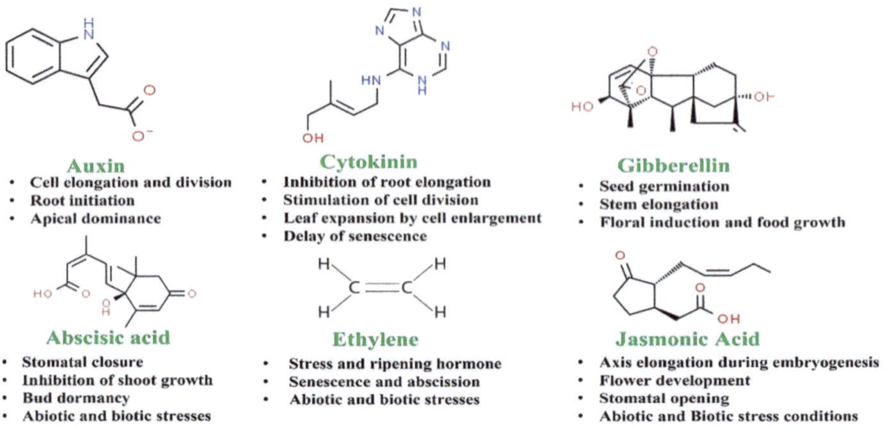

Fig. 5.5 Important role of microbial phytohormones in regulation of plant growth

were found to be present in the wheat rhizosphere (Egamberdiyeva and Höflich 2003). A highly efficient auxin producing strain of *Ignatzschineria* species was isolated from food processing waste, which can be potentially used for increasing soil fertility (Park et al. 2021). Under unsuitable condition of abiotic stress, the phytohormone synthesis in the plant decreases, which leads to reduced plant growth due to deteriorated soil quality; however, this compliance can be compensated with the help of phytohormone-producing soil microorganisms. Phytohormones also play a crucial function in the phytoremediation of heavy metals. Improved maize growth has been observed in the cadmium polluted soil in the presence of indole acetic acid-producing rhizobacteria isolated from the bitter gourd rhizoids, highlighting the potential role of auxin in phytoremediation of pollutants including heavy metals (Ahmad et al. 2016).

Under the stress of high salt concentrations in the soil, plants reduce their ability to produce phytohormones. However, the potential of bacterial strains to survive under a variety of stress conditions provides phytohormones to the plants, which aid in healthy plant development.

5.6.5 Antagonistic Molecules

Soil is a hub of diverse living forms that, in addition to beneficial microorganisms, also contain detrimental organisms such as pests, predators, competitors, parasites, and other potentially harmful organisms that affect plant growth and development (Kinkel et al. 2011). Soil microbiota, including some fungi and bacteria, can produce some antagonistic molecules with toxic effects on the growth of detrimental organisms. Filamentous actinomycetes, *Bacillus* species, and *Pseudomonas* species are the major classes of microbes which have been identified to produce a number of antibiotics with the potential to protect the soil and plants from various detrimental infections (Samaddar et al. 2021).

Application of chemical fertilizers has reduced the number of plant diseases in the past few decades; however, the employment of chemical fertilizers results in the accretion of various chemicals in the soil that can cause various adverse effects such as contamination of the environment due to the accumulation of harmful chemicals, the emergence of resistant plant pathogenic microorganisms, etc. (Wu et al. 2016). Molinatto and colleagues reported synthesis of various antimicrobial metabolites against phytopathogens by *Bacillus amyloliquefaciens* subsp. *Plantarum* which also stimulates the production of systemic resistance in plants (Molinatto et al. 2016). An appreciable reduction in soil-borne pathogens and enhancement in beneficial microflora were observed when wheat rhizosphere was treated with *Paenibacillus jamilae* HS-26 strain obtained from cucumber rhizosphere soil, revealing their potential to act as biocontrol agents (Wang et al. 2019). Various mechanisms of action are used by microorganisms to function as biocontrol agents (Kohl et al. 2019). These are as follows:

1. Induction of plant resistance against diseases through plant-microbe interactions.
2. Competition for nutrient availability between beneficial soil and pathogens.
3. Hyperparasitism.
4. Antibiosis by antimicrobial molecules.

Broad-spectrum antimicrobial metabolite production has been described by various bacterial genera such as *Agrobacterium*, *Serratia*, *Bacillus*, *Stenotrophomonas*, *Pseudomonas*, *Streptomyces*, etc. Lipopolysaccharides like iturin and surfactin have been explored as potential antimicrobial metabolites produced by *Bacillus* species (Ongena and Jacques 2008). In *Pseudomonas* species, pyrrolnitrin and phenazine have been reported to have an antimicrobial effect (Raaijmakers and Mazzola 2012).

5.7 Major Microbial Groups as Metabolite Producers

5.7.1 *Arbuscular Mycorrhiza*

The beneficial interactions of microorganisms with plants are crucial for the health, development, and yield of plants as well as soil (De Vries et al. 2020). Arbuscular mycorrhizal fungi (AMF) are one among the essential groups of microbes that form symbiotic interactions with about 70% of vascular plants (Brundrett and Tedersoo 2018). Symbiotic association in the form of branched structures called arbuscules has been characterized in arbuscular mycorrhiza. These arbuscules assist in the transport of required nutrients from the fungal cell to plant cells and the movement of carbon compounds from plant cells to fungal cells. Recognition of signal molecules released by plants and fungi is the initial step in the establishment of symbiotic associations. Lipochitooligosaccharides, or Myc factors, are released by fungi, and signaling molecules such as strigolactones are released by plants. Following recognition of signal molecules, arbuscular mycorrhiza forms a hyphopodium on the root epidermidis, colonizes the root cortex and spreads the hyphae in the surrounding soil as extra radical mycelium, which contributes in increased access of water and minerals to plants from the soil (Bowles et al. 2016). Association of AMF with medicinal plants has also been reviewed by Zhao and colleagues to increase industrially important secondary metabolites of plants (Zhao et al. 2022).

Soil structure is a central factor that determines the soil's strength and health. Extraradical mycelium formed by arbuscular mycorrhizae helps in holding the soil particles, thus participating in the soil aggregation process. A glycoprotein (glomalin) is secreted by extraradical mycelium and is associated with maintenance of soil aggregates due to its hydrophobic nature. Glomalin acts as glue, which transforms soil microaggregates into macroaggregates (Zou et al. 2016). These soil macroaggregates help in the improvement of the gas interchange ability of the soil, water and mineral retention, water infiltration, reduction of surface runoff, reduction in soil erosion, and reduction in nutrient and organic matter losses. Further, after a particular period of time, mycelia renews itself, and degraded mycelia helps in increasing the soil's organic content (Thirkell et al. 2017). The quantification of glomalin in the soil is utilized as a measure of soil health. Glomalin quantification is done using the Bradford method through the detection of glomalin-related soil proteins (GRSP). Determination of GRSP in the soil predicts other factors such as carbon content, root exudates, and enzyme activity in the soil, which are required for improved soil health (Gupta 2020).

Nutrient recycling and the presence of nutrients in readily available forms are important for soil to be fertile and healthy. AMF plays a significant function in the nitrogen, carbon, and phosphorus assimilation in plants. In addition, AMF also participates in the solubilization of insoluble phosphorus into its soluble form, thus improving soil fertility (Etesami et al. 2021). Previous reports suggest that AMF recruits phosphate-solubilizing bacteria in the soil, which helps in phosphate solubilization. Various processes such as chelation, acidification, exchange reactions, and organic acids production by AMF are other proposed mechanisms by which AMF can convert insoluble form of phosphorus into its soluble form that can be utilized by the plants (Behera et al. 2014). The synthetic form of alkaline phosphatase also helps in phosphorus solubilization (Liu et al. 2013). Arbuscular mycorrhizae secrete proteinaceous substances into the soil, which improves the soil's organic content. Increased water holding ability and drought resistance have been reported in soil containing arbuscular mycorrhiza. AMFs interact with a plethora of soil microorganisms and improve soil fertility. Positive, neutral, and negative interactions between microorganisms and AMFs directly and indirectly influence soil health (Fall et al. 2022).

5.7.2 Plant Growth-Promoting Rhizobacteria

Plant growth-promoting rhizobacteria (PGPR) are soil microorganisms that are associated with plant roots and improve the health of the soil. PGPR exists as a symbiotic associations with plant roots or as free-living microbes. PGPR helps in sustaining soil fertility as it solubilizes various nutrients such as potassium, iron, zinc, phosphorus, etc. using diverse strategies including the production of enzymes, siderophores, organic acids, etc. (Saha et al. 2016a, b; Hodge 2017). An arsenal of plant growth-stimulating and antimicrobial compounds is secreted by PGPR for the maintenance of plant and soil health. Inoculation of wheat seeds with *Pseudomonas moraviensis* and *Bacillus cereus* with L-tryptophan improved the soil's health by increasing organic matter, nutrient bioavailability, proline content, and antioxidant activities in the soil (Ul Hassan and Bano 2015). Further, PGPRs also showed the capacity of plants to survive hypersaline stress conditions in the soil, highlighting their importance in the survival of plants under unsuitable stress conditions (Upadhyay and Singh 2015).

Various mechanisms by which PGPRs improve soil health include the following:

- Biological nitrogen fixation.
- Phosphate solubilization, zinc solubilization, and potassium solubilization.
- Synthesis of phytohormones and siderophores.
- Production of volatile organic compounds such as *N,N*-dimethylhexadecylamine, dimethyl disulfide, 2,3-butanediol, acetoin, and hydrogen cyanide helps in plant growth by promoting nutrient uptake and preventing the growth of phytopathogens.
- Hydrolytic enzyme production.

Further, PGPRs act as an important alternative in bioremediation for the deterioration of hazardous materials from the soil. A crucial role of PGPRs in the biogeochemical cycling of heavy metals such as cadmium and lead has been demonstrated in previous studies, highlighting their role in the cleaning up of contaminated soil. Excessive use of pesticides in the soil has a hazardous impact on the soil as well as plant health. The long shelf life and nonbiodegradable nature of these pollutants lead to the accumulation of these pesticides in the food chain, leading to severe health problems in humans. However, utilization of PGPRs in biotransformation, biomineralization, bioaccumulation, biodegradation, bioremediation, and cometabolism are effective approaches to the degradation of these pesticides in soil (Khatoon et al. 2020).

5.8 Conclusion

Crop health is significantly affected by the quality of the soil. As a result, farmers have focused on enhancing soil quality with chemical fertilizers and pesticides for years. Chemical fertilizers, however, have a number of drawbacks, including health issues and environmental contamination. Consumer demand for organic food has evolved as a result of concerns about food safety and the desire to buy more healthful products, which has prompted farmers to use sustainable agricultural methods. The ultimate objective of ecological soil management is to develop an underground environment that is healthy, has a good soil structure, has a variety of thriving soil organisms, and has enough nutrients to support high crop yields. As a result, plants are given the best possible environment for growth. A variety of advantageous microbes can be employed to raise crop quality and yield. Sustainable agriculture relies heavily on beneficial microbes since they may promote plant development and combat diseases in an environmentally responsible manner. Bacteria, viruses, fungi, actinomycetes, protozoa, and nematodes make up the soil's beneficial microorganisms. They generate metabolites and plant growth regulators that significantly influence the development and growth of plants. Soil microbes break down organic waste and recycle soil nutrients in their ecological habitat. The tolerance of plants to plant diseases is significantly influenced by soil microorganisms. They nourish crops, improve soil quality, and ultimately increase crop yields. Good soil health is linked with sustainable agriculture. For more sustainable agricultural systems, it is imperative to focus research on improving plant-soil microbial interactions.

References

Abbott LK, Murphy DV (eds) (2003) Soil biological fertility: a key to sustainable land use in agriculture. Springer Science and Business Media

Ahmad I, Akhtar MJ, Asghar HN, Ghafoor U, Shahid M (2016) Differential effects of plant growth-promoting rhizobacteria on maize growth and cadmium uptake. J Plant Growth Regul 35:303–315

Ahmed E, Holmstrom SJ (2014) Siderophores in environmental research: roles and applications. Microb Biotechnol 7(3):196–208

Akande GM, Adekayode FO, Adekayode FO (2019) Identification of soil microbial population under different land use. Trop Plant Res 6(1):90–100

Allen DE, Singh BP, Dalal RC (2011) Soil health indicators under climate change: a review of current knowledge. In: Singh B, Cowie A, Chan K (eds) Soil health and climate change. Springer, Berlin, pp 25–45

Andreote FD, Gumiere T, Durrer A (2014) Exploring interactions of plant microbiomes. Sci Agric 71:528–539

Antoun H, Beauchamp CJ, Goussard N, Chabot R, Lalande R (1998) Potential of Rhizobium and Bradyrhizobium species as plant growth promoting rhizobacteria on non-legumes: effect on radishes (Raphanussativus L.). Mol Microb Ecol Soil 204(1):57–67. results from an FAO/IAEA coordinated research programme

Ashraf M, Hasnain S, Hussain F (2005) Exo-polysaccharides (EPS) producing biofilm bacteria in improving physico-chemical characteristics of the salt-affected soils. In: Proceedings of the first international conference on environmentally sustainable development v, pp 1–3

Baldani J, Caruso L, Baldani VL, Goi SR, Döbereiner J (1997) Recent advances in BNF with non-legume plants. Soil Biol Biochem 29(5-6):911–922

Bansal OP (2018) The influence of potentially toxic elements on soil biological and chemical properties, metals in soil—contamination and remediation. IntechOpen, Rijeka

Bayranvand M, Akbarinia M, SalehiJouzani G, Gharechahi J, Baldrian P (2021) Distribution of soil extracellular enzymatic, microbial, and biological functions in the C and N-cycle pathways along a Forest altitudinal gradient. Front Microbiol 12:660603

Behera BC, Singdevsachan SK, Mishra RR, Dutta SK, Thatoi HN (2014) Diversity, mechanism and biotechnology of phosphate solubilising microorganism in mangrove—a review. Biocatal Agric Biotechnol 3(2):97–110

Bhattarai A, Bhattarai B, Pandey S (2015) Variation of soil microbial population in different soil horizons. J Microbiol Exp 2(2):00044

Bi L, Yu DT, Du S, Zhang LM, Zhang LY, Wu CF, Xiong C, Han LL, He JZ (2021) Diversity and potential biogeochemical impacts of viruses in bulk and rhizosphere soils. Environ Microbiol 23(2):588–599

Bowles TM, Barrios-Masias FH, Carlisle EA, Cavagnaro TR, Jackson LE (2016) Effects of arbuscular mycorrhizae on tomato yield, nutrient uptake, water relations, and soil carbon dynamics under deficit irrigation in field conditions. Sci Total Environ 566:1223–1234

Brundrett MC, Tedersoo L (2018) Evolutionary history of mycorrhizal symbioses and global host plant diversity. New Phytol 220(4):1108–1115

Bunemann EK, Bongiorno G, Bai Z, Creamer RE, De Deyn G, De Goede R, Fleskens L, Geissen V, Kuyper TW, Mader P, Pulleman M (2018) Soil quality—a critical review. Soil Biol Biochem 120:105–125

Cain ML, Bowman L, William D, Sally D (2011) Chapter 16: Change in communities. Ecology:359–362

Cao MM, Liu SY, Bi L, Chen SJ, Wu HY, Ge Y, Han B, Zhang LM, He JZ, Han LL (2022) Distribution characteristics of soil viruses under different precipitation gradients on the Qinghai-Tibet plateau. Front Microbiol 13:848305

Cardoso EJBN, Vasconcellos RLF, Bini D, Miyauchi MYH, Santos CAD, Alves PRL, Paula AMD, Nakatani AS, Pereira JDM, Nogueira MA (2013) Soil health: looking for suitable indicators. What should be considered to assess the effects of use and management on soil health. Sci Agric 70:274–289

Charlop-Powers Z, Owen JG, Reddy BVB, Ternei MA, Brady SF (2014) Chemical-biogeographic survey of secondary metabolism in soil. Proc Natl Acad Sci 111(10):3757–3762

Chen YP, Rekha PD, Arun AB, Shen FT, Lai WA, Young CC (2006) Phosphate solubilizing bacteria from subtropical soil and their tricalcium phosphate solubilizing abilities. Appl Soil Ecol 34(1):33–41

Cui K, Xu T, Chen J, Yang H, Liu X, Zhuo R, Peng Y, Tang W, Wang R, Chen L, Zhang X (2022) Siderophores, a potential phosphate solubilizer from the endophyte Streptomyces sp. CoT10, improved phosphorus mobilization for host plant growth and rhizosphere modulation. J Clean Prod 367:133110

Cunningham JE, Kuiack C (1992) Production of citric and oxalic acids and solubilization of calcium phosphate by *Penicillium bilaii*. Appl Environ Microbiol 58(5):1451–1458

D'Alo F, Odriozola I, Baldrian P, Zucconi L, Ripa C, Cannone N, Malfasi F, Brancaleoni L, Onofri S (2021) Microbial activity in alpine soils under climate change. Sci Total Environ 783:147012

De Vries FT, Griffiths RI, Knight CG, Nicolitch O, Williams A (2020) Harnessing rhizosphere microbiomes for drought-resilient crop production. Science 368(6488):270–274

Demain AL (1999) Pharmaceutically active secondary metabolites of microorganisms. Appl Microbiol Biotechnol 52:455–463

Deng J, Yin Y, Zhu W, Zhou Y (2018) Variations in soil bacterial community diversity and structures among different revegetation types in the Baishilazi Nature Reserve. Front Microbiol 9:2874

Diels L, Van Roy S, Taghavi S, Van Houdt R (2009) From industrial sites to environmental applications with Cupriavidus metallidurans. Antonie Van Leeuwenhoek 96:247–258

Dischinger J, Josten M, Szekat C, Sahl HG, Bierbaum G (2009) Production of the novel two-peptide lantibioticlichenicidin by Bacillus licheniformis DSM 13. PLoS One 4(8):e6788

Doran JW, Parkin TB (1994) Defining and assessing soil quality. Defining Soil Qual Sustain Environ 35:1–21

Doran JW, Parkin TB (1996) Quantitative indicators of soil quality: a minimum data set. Methods Assess Soil Qual 49:25–37. Soil Science Society of America, Madison

Doran JW, Zeiss MR (2000) Soil health and sustainability: managing the biotic component of soil quality. Appl Soil Ecol 15(1):3–11

Egamberdieva D, Wirth SJ, Alqarawi AA, Abd-Allah EF, Hashem A (2017) Phytohormones and beneficial microbes: essential components for plants to balance stress and fitness. Front Microbiol 8:2104

Egamberdiyeva D, Höflich G (2003) The effect of associative bacteria from different climates on plant growth of pea in different soils and temperatures. Arch Agron Soil Sci 49(2):203–212

Etesami H, Jeong BR, Glick BR (2021) Contribution of arbuscular mycorrhizal fungi, phosphate–solubilizing bacteria, and silicon to P uptake by plant. Front Plant Sci 12:699618

Fall AF, Nakabonge G, Ssekandi J, Founoune-Mboup H, Apori SO, Ndiaye A, Badji A, Ngom K (2022) Roles of arbuscular mycorrhizal fungi on soil fertility: contribution in the improvement of physical, chemical, and biological properties of the soil. Front Fungal Biol 3:3

Fitter AH (2005) Darkness visible: reflections on underground ecology. J Ecol 93(2):231–243

Frac M, Hannula SE, Bełka M, Jedryczka M (2018) Fungal biodiversity and their role in soil health. Front Microbiol 9:707

Furtak K, Gajda AM (2018) Activity and variety of soil microorganisms depending on the diversity of the soil tillage system. Sustain Agroecosyst 45:45–61. https://doi.org/10.5772/intechopen.72966

Geisen S, Mitchell EA, Adl S, Bonkowski M, Dunthorn M, Ekelund F, Fernandez LD, Jousset A, Krashevska V, Singer D, Spiegel FW (2018) Soil protists: a fertile frontier in soil biology research. FEMS Microbiol Soc 42(3):293–323

Glick BR (2018) Soil microbes and sustainable agriculture. Pedosphere 28(2):167–169

Goldstein AH (1995) Recent progress in understanding the molecular genetics and biochemistry of calcium phosphate solubilization by gram negative bacteria. Biol Agric Hortic 12(2):185–193

Gougoulias C, Clark JM, Shaw LJ (2014) The role of soil microbes in the global carbon cycle: tracking the below-ground microbial processing of plant-derived carbon for manipulating carbon dynamics in agricultural systems. J Sci Food Agric 94(12):2362–2371

Gregorich EG, Carter MR (1997) Soil Quality for Crop Production and Ecosystem Health (eds). In: Developments in Soil Science 25. Elsevier, Amsterdam, pp 115–141

Gupta MM (2020) Arbuscular mycorrhizal fungi: the potential soil health indicators. In: Giri B, Varma A (eds) Soil health. Springer, Cham, pp 183–195

Hodge A (2017) Accessibility of inorganic and organic nutrients for mycorrhizas. In: Mycorrhizal mediation of soil. Elsevier, pp 129–148

Hurek T, Reinhold-Hurek B, Van Montagu M, Kellenberger E (1994) Root colonization and systemic spreading of *Azoarcus sp.* strain BH72 in grasses. J Bacteriol 176(7):1913–1923

Ibraheem I, Abdel-Raouf N (2007) Allelopathic activity of some local cyanobacterial extra-metabolites against some pathogenic bacteria. Egypt J Phycol 8(1):119–129

Ingham ER (2009) Soil biology primer, Chapter 4: Soil fungus. Soil and Water Conservation; Soil and Water Conservation Society, Ankeny, IA, pp 22–23

Jacoby R, Peukert M, Succurro A, Koprivova A, Kopriva S (2017) The role of soil microorganisms in plant mineral nutrition—current knowledge and future directions. Front Plant Sci 8:1617

Jamir E, Kangabam RD, Borah K, Tamuly A, DekaBoruah HP, Silla Y (2019) Role of soil microbiome and enzyme activities in plant growth nutrition and ecological restoration of soil health. In:

Kumar A, Sharma S (eds) Microbes and enzymes in soil health and bioremediation. Springer, Singapore, pp 99–132
Khalid KA (2012) Biological fertilization and its effect on medicinal and aromatic plants. Nusantara Biosci 4(3):124–133
Khatoon Z, Huang S, Rafique M, Fakhar A, Kamran MA, Santoyo G (2020) Unlocking the potential of plant growth-promoting rhizobacteria on soil health and the sustainability of agricultural systems. J Environ Manag 273:111118
Kinkel LL, Bakker MG, Schlatter DC (2011) A coevolutionary framework for managing disease-suppressive soils. Annu Rev Phytopathol 49:47–67
Kohl J, Kolnaar R, Ravensberg WJ (2019) Mode of action of microbial biological control agents against plant diseases: relevance beyond efficacy. Front Plant Sci 10:845
Kumawat TK, Kumawat V, Sharma S, Kandwani N, Biyani M (2021) Applications of EPS in environmental bioremediations. In: Nadda AK, Sajna KV, Sharma S (eds) Microbial exopolysaccharides as novel and significant biomaterials, 1st edn. Springer, Cham, Switzerland, pp 285–302
Kuzyakov Y, Mason-Jones K (2018) Viruses in soil: nano-scale undead drivers of microbial life, biogeochemical turnover and ecosystem functions. Soil Biol Biochem 127:305–317
Lavelle P, Spain A (2002) Soil ecology. Springer Science & Business Media
Lin W, Okon Y, Hardy RW (1983) Enhanced mineral uptake by Zea mays and Sorghum bicolor roots inoculated with Azospirillum brasilense. Appl Environ Microbiol 45(6):1775–1779
Liu L, Chu KK, Houser GH, Diephuis BJ, Li Y, Wilsterman EJ, Shastry S, Dierksen G, Birket SE, Mazur M, Byan-Parker S (2013) Method for quantitative study of airway functional micro-anatomy using micro-optical coherence tomography. PLoS One 8(1):e54473
Lombard N, Prestat E, van Elsas JD, Simonet P (2011) Soil-specific limitations for access and analysis of soil microbial communities by metagenomics. Federation of European Microbiological Societies and Microbiology. Ecology 78(1):31–49
Malboobi MA, Owlia P, Behbahani M, Sarokhani E, Moradi S, Yakhchali B, Deljou A, Morabbi Heravi K (2009) Solubilization of organic and inorganic phosphates by three highly efficient soil bacterial isolates. World J Microbiol Biotechnol 25:1471–1477
Mandal A, Dhaliwal SS, Mani PK, Toor AS (2021) Conservation agricultural practices under organic farming. In: Advances in organic farming. Woodhead Publishing, pp 17–37
Mathivanan K, Chandirika JU, Mathimani T, Rajaram R, Annadurai G, Yin H (2021) Production and functionality of exopolysaccharides in bacteria exposed to a toxic metal environment. Ecotoxicol Environ Saf 208:111567
Molinatto G, Puopolo G, Sonego P, Moretto M, Engelen K, Viti C, Ongena M, Pertot I (2016) Complete genome sequence of *Bacillus amyloliquefaciens* subsp. plantarum S499, a rhizobacterium that triggers plant defences and inhibits fungal phytopathogens. J Biotechnol 238:56–59
Nair A, Juwarkar AA, Singh SK (2007) Production and characterization of siderophores and its application in arsenic removal from contaminated soil. Water Air Soil Pollut 180:199–212
Naseem H, Ahsan M, Shahid MA, Khan N (2018) Exopolysaccharides producing rhizobacteria and their role in plant growth and drought tolerance. J Basic Microbiol 58(12):1009–1022
Ongena M, Jacques P (2008) Bacillus lipopeptides: versatile weapons for plant disease biocontrol. Trends Microbiol 16(3):115–125
Pahalvi HN, Lone LR, Rashid S, Nisar B (2021a) Chemical fertilizers and their impact on soil health. Microbiota Biofertil 2:1–20
Pahalvi HN, Rafiya L, Rashid S, Nisar B, Kamili AN (2021b) Chemical fertilizers and their impact on soil health. In: Dar GH, Bhat RA, Mehmood MA, Hakeem KR (eds) Microbiota and biofertilizers, vol 2. Springer, Cham, pp 1–20
Pandey HP, Maaren IE, Shah KK, Maraseni TN (2020) Response of topographic and biodiversity variables on biomass and carbon density in community forests of himalayan foot-hills. J For Livelihood 19(1):51–65
Parikh SJ, James BR (2012) Soil: the foundation of agriculture. Nat Educ Knowl 3(10):2

Parisi V, Menta C, Gardi C, Jacomini C, Mozzanica E (2005) Microarthropod communities as a tool to assess soil quality and biodiversity: a new approach in Italy. Agric Ecosyst Environ 105(1-2):323–333

Park S, Kim AL, Hong YK, Shin JH, Joo SH (2021) A highly efficient auxin-producing bacterial strain and its effect on plant growth. J Genet Eng Biotechnol 19:1–9

Patel T, Saraf M (2017) Biosynthesis of phytohormones from novel rhizobacterial isolates and their *in vitro* plant growth-promoting efficacy. J Plant Interact 12(1):480–487

Poveda J, Gonzalez-Andres F (2021) Bacillus as a source of phytohormones for use in agriculture. Appl Microbiol Biotechnol 105(23):8629–8645

Pozo MJ, Lopez-Raez JA, Azcon-Aguilar C, Garcia-Garrido JM (2015) Phytohormones as integrators of environmental signals in the regulation of mycorrhizal symbioses. New Phytol 205(4):1431–1436

Raaijmakers JM, Mazzola M (2012) Diversity and natural functions of antibiotics produced by beneficial and plant pathogenic bacteria. Annu Rev Phytopathol 50:403–424

Radzki W, Gutierrez Mañero FJ, Algar E, Lucas García JA, García-Villaraco A, Ramos Solano B (2013) Bacterial siderophores efficiently provide iron to iron-starved tomato plants in hydroponics culture. Antonie Van Leeuwenhoek 104:321–330

Raouf AN, Al-Homaidan AA, Ibraheem IBM (2012) Agricultural importance of algae. Afr J Biotechnol 11(54):11648–11658

Rasul G, Mirza MS, Latif F, Malik KA (1998) Identification of plant growth hormones produced by bacterial isolates from rice, wheat and kallar grass. In: Nitrogen fixation with non-legumes: Proceedings of the 7th international symposium on nitrogen fixation with non-legumes, held 16–21 October 1996 in Faisalabad, Pakistan. Springer Netherlands, Dordrecht, pp 25–37

Richardson AE (1994) Soil microorganisms and phosphorus availability. In: Soil biota: management in sustainable farming systems. CSIRO, Australia, pp 50–62

Riggs PJ, Chelius MK, Iniguez AL, Kaeppler SM, Triplett EW (2001) Enhanced maize productivity by inoculation with diazotrophic bacteria. Funct Plant Biol 28(9):829–836

Riley J (2001) Multidisciplinary indicators of impact and change: key issues for identification and summary. Agric Ecosyst Environ 87(2):245–259

Roberson EB, Firestone MK (1992) Relationship between desiccation and exopolysaccharide production in a soil *Pseudomonas* sp. Appl Environ Microbiol 58(4):1284–1291

Saha M, Maurya BR, Meena VS, Bahadur I, Kumar A (2016a) Identification and characterization of potassium solubilizing bacteria (KSB) from Indo-Gangetic Plains of India. Biocatal Agric Biotechnol 7:202–209

Saha M, Sarkar S, Sarkar B, Sharma BK, Bhattacharjee S, Tribedi P (2016b) Microbial siderophores and their potential applications: a review. Environ Sci Pollut Res 23:3984–3999

Samaddar S, Karp DS, Schmidt R, Devarajan N, McGarvey JA, Pires AF, Scow K (2021) Role of soil in the regulation of human and plant pathogens: soils contributions to people. Philos Trans R Soc B 376(1834):20200179

Sapountzis P, De Verges J, Rousk K, Cilliers M, Vorster BJ, Poulsen M (2016) Potential for nitrogen fixation in the fungus-growing termite symbiosis. Front Microbiol 7:1993

Sasaki M, Takagi A, Ota S, Kawano S, Sasaki D, Asayama M (2020) Coproduction of lipids and extracellular polysaccharides from the novel green alga *Parachlorella* sp. BX1.5 depending on cultivation conditions. Biotechnol Rep 25:e00392

Shah KK, Tripathi S, Tiwari I, Shrestha J, Modi B, Paudel N, Das BD (2021) Role of soil microbes in sustainable crop production and soil health: a review. Agric Sci Technol 13(2):109–118

Shameer S (2016) Biosorption of lead, copper and cadmium using the extracellular polysaccharides of *Bacillus sp.,* from solar salterns. 3 Biotech 6(2):194

Singer MJ, Ewing S (2000) Soil quality. In: Handbook of soil science, Sumner (Ed), CRC Press, Boca Raton, pp 271–298

Singh R, Kumar M, Mittal A, Mehta PK (2017) Microbial metabolites in nutrition, healthcare and agriculture. 3 Biotech 7:1–14

Singh P, Chauhan PK, Upadhyay SK, Singh RK, Dwivedi P, Wang J, Jain D, Jiang M (2022) Mechanistic insights and potential use of siderophores producing microbes in rhizosphere for mitigation of stress in plants grown in degraded land. Front Microbiol 13:898979

Sinsabaugh RL (2010) Phenol oxidase, peroxidase and organic matter dynamics of soil. Soil Biol Biochem 42(3):391–404

Sobucki L, Ramos RF, Meireles LA, Antoniolli ZI, Jacques RJS (2021) Contribution of enzymes to soil quality and the evolution of research in Brazil. Revista Brasileira de Ciencia do Solo 45:e0210109

Song X, Bo Y, Feng Y, Tan Y, Zhou C, Yan X, Ruan R, Xu Q, Cheng P (2022) Potential applications for multifunctional microalgae in soil improvement. Front Environ Sci 10:1035332

Sukanya K, Kumutha K, Subhashini R, Kannan P (2022) Isolation and screening of soil bacteria for EPS production. Pharma Innov J 11(8S):1620–1626

Teintze M, Hossain MB, Barnes CL, Leong J, Van der Helm D (1981) Structure of ferric pseudobactin: a siderophore from a plant growth promoting Pseudomonas. Biochemistry 20(22):6446–6457

Thirkell TJ, Charters MD, Elliott AJ, Sait SM, Field KJ (2017) Are mycorrhizal fungi our sustainable saviours? Considerations for achieving food security. J Ecol 105(4):921–929

Tsavkelova EA, Cherdyntseva TA, Klimova SY, Shestakov AI, Botina SG, Netrusov AI (2007) Orchid-associated bacteria produce indole-3-acetic acid, promote seed germination, and increase their microbial yield in response to exogenous auxin. Arch Microbiol 188:655–664

Ul Hassan T, Bano A (2015) The stimulatory effects of L-tryptophan and plant growth promoting rhizobacteria (PGPR) on soil health and physiology of wheat. J Soil Sci Pant Nutr 15(1):190–201

Ullah S, Ai C, Huang S, Zhang J, Jia L, Ma J, Zhou W, He P (2019) The responses of extracellular enzyme activities and microbial community composition under nitrogen addition in an upland soil. PLoS One 14(9):e0223026

Upadhyay SK, Singh DP (2015) Effect of salt-tolerant plant growth-promoting rhizobacteria on wheat plants and soil health in a saline environment. Plant Biol 17(1):288–293

Usharani KV, Roopashree KM, Naik D (2019) Role of soil physical, chemical and biological properties for soil health improvement and sustainable agriculture. J Pharmacog Phytochem 8(5):1256–1267

Uwituze Y, Nyiraneza J, Fraser TD, Dessureault-Rompre J, Ziadi N, Lafond J (2022) Carbon, nitrogen, phosphorus and extracellular soil enzyme responses to different land use. Front Soil Sci 2:814554

Vessey JK (2003) Plant growth promoting rhizobacteria as biofertilizers. Plant Soil 255:571–586

Wallenstein MD, Burns RG (2011) Ecology of extracellular enzyme activities and organic matter degradation in soil: a complex community-driven process. Methods Soil Enzymol 9:35–55

Wang X, Li Q, Sui J, Zhang J, Liu Z, Du J, Xu R, Zhou Y, Liu X (2019) Isolation and characterization of antagonistic bacteria *Paenibacillus jamilae* HS-26 and their effects on plant growth. BioMed Res Int 3638926. https://doi.org/10.1155/2019/3638926

Willey JM, Sherwood L, Woolverton CJ (2011) Prescott's microbiology, 7th edn. McGraw-Hill, New York

Wu L, Shang H, Wang Q, Gu H, Liu G, Yang S (2016) Isolation and characterization of antagonistic endophytes from *Dendrobium candidum* Wall ex Lindl., and the biofertilizing potential of a novel *Pseudomonas saponiphila* strain. Appl Soil Ecol 105:101–108

Zahran HH (1999) Rhizobium-legume symbiosis and nitrogen fixation under severe conditions and in an arid climate. Microbiol Mol Biol Rev 63(4):968–989

Zhang MK, Liu ZY, Wang H (2010) Use of single extraction methods to predict bioavailability of heavy metals in polluted soils to rice. Commun Soil Sci Plant Anal 41(7):820–831

Zhao Y, Cartabia A, Lalaymia I, Declerck S (2022) Arbuscular mycorrhizal fungi and production of secondary metabolites in medicinal plants. Mycorrhiza 32(3–4):221–256

Zou YN, Srivastava AK, Wu QS (2016) Glomalin: a potential soil conditioner for perennial fruits. Int J Agric Biol 18:293–297

Chapter 6
Microbial Enzymes for Soil Health

Gitika Bhardwaj, Anjali Chauhan, Abhishek Walia, and Perminder Singh Brar

6.1 Introduction

A certain balance of biological, chemical, and physical factors maintains the health of the soil. Hence, indications of each of these elements are necessary for evaluating soil health. Healthy soils are essential for terrestrial ecosystems to remain intact or recover from disturbances such as severe weather, changes in the climate, and infestations of insects, environmental damage, and exploitation by human beings, which include agriculture (Ellert et al. 1997). Since poorly managed soil and contaminated soil may negatively influence air, groundwater, and surface water that people consume, soil degradation is an issue for the health of humans, animals, and plants (Singer and Ewing 2000). Being a vital component of the terrestrial environment, soil must be maintained and preserved since it is necessary for the existence of every kind of terrestrial organisms. A complete understanding of the functions of the soil enzymes is therefore essential to ensuring the continued soil's health. Although enzymes of soil constitute subject of in-depth research, little is known about how they contribute to the health of soil. As a result, it becomes essential to comprehend how these enzymes function and how well they preserve soil health for the advancement of soil later on. The capability of the soil to contribute to the well-being of the organisms determines the health of the soil. There are other

G. Bhardwaj · A. Chauhan
Department of Soil Science and Water Management, Dr Yashwant Singh Parmar University of Horticulture and Forestry, Nauni, Solan, Himachal Pradesh, India

A. Walia
Department of Microbiology, College of Basic Sciences, CSK Himachal Pradesh Agricultural University, Palampur, Himachal Pradesh, India

P. S. Brar (✉)
MS Swaminathan School of Agriculture, Shoolini University, Solan, Himachal Pradesh, India

R. K. Bhatia, A. Walia (eds.), *Advancements in Microbial Biotechnology for Soil Health*, Microorganisms for Sustainability 50,
https://doi.org/10.1007/978-981-99-9482-3_6

bacteria, plants, and animals involved in the filtration and management of water, the breakdown of chemical compounds, and the decay of organic materials (da silva et al. 2021).

6.2 Soil Enzymes

To preserve soil biology, chemical and physical characteristics, crop growth, and health of the soil, soil enzymes, which typically live in soil, play an important role. These enzymes perform vital biochemical processes that are essential to the complete organic material degradation in soil's environment (Sinsabaugh and Linkins 1989). Soil enzymes catalyze various essential reactions required for soil microorganisms' living activities as well as the improvement of the soil's structure, organic matter degradation, organic material production, and the cycling of nutrients (Dick et al. 1994; Dick 1997). According to McLaren (1975), a collection of enzymes found in all soils control the metabolic processes that depend on biochemical, microbiological, physical, and chemical characteristics of soil. Every kind of soil contains a different amount of soil organic matter content, the composition, organism's activity, and biological intensity. The biochemical reactions are mostly catalyzed by enzymes and a wide range of compounds that are used as source of energy for microorganisms (Kiss et al. 1978). Soil enzymes include arylsulfatases, amylase, cellulase, β-glucosidase, dehydrogenase, chitinase, protease, phosphatase, and urease generated by plants, biological material, animals, microbes, and soils (Miwa et al. 1937; James et al. 1991). The biochemical activities that occur in plants, animals, and microorganisms as well as organic matter breakdown, stability of structure of soil, nutrient cycling, etc. depend on soil enzymes. In an ecosystem, several microorganisms and enzymes have a role in the cycling of C, N, P, and S. Nutrient cycles occur in soil because of biological and physicochemical processes. Soil microorganisms and activities of enzymes control the nutrient cycling in natural environment largely (Neemisha and Sharma 2022).

Soil health is the soil's ability to perform the whole set of ecological activities in a manner that is suited to its surroundings. The ability of a particular type of soil to persist is defined as its ability to work as a key component of life under unmanaged or managed ecosystem boundaries, while maintaining productivity of plants and animals, to sustainably enhance the quality of water and air environments, as well as to contribute to the well-being of humans and their habitation. The flow chart of soil health factors and their impacts is given in Fig. 6.1.

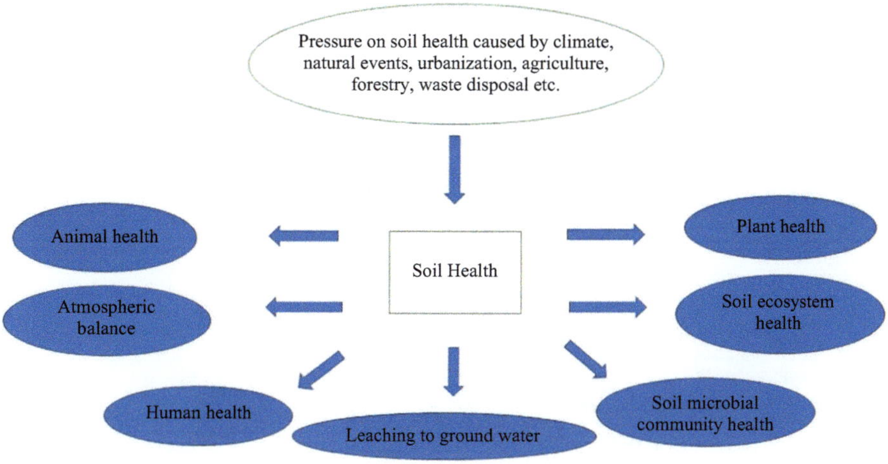

Fig. 6.1 The flow chart indicating factors of soil health as well as their influence (Das and Varma 2011)

6.3 Indicators of Soil Health

6.3.1 Microorganisms as Soil Health Indicators

Microorganisms have the capacity to provide a comprehensive measure of health of soil. Modifications in the number of microbes and activity might therefore be a good indicator of changes in the condition of the soil since microorganisms adapt fast to environmental conditions and respond quickly to changes (Kennedy and Papendick 1995; Pankhurst et al. 1995). As a result of their high surface-to-volume ratio and intimate relationship to their surroundings, microorganisms also react to environmental stress more quickly. Some alterations in the number of microbes or activities might take place before noticeable alterations in soil chemical and physical features, providing either an early indication of improvement in soil or a preliminary sign of soil degradation (Pankhurst et al. 1995). Microbial activity determines how some substances affect the condition of the soil. Soil microorganisms serve as significant soil indicators in this way.

6.3.2 Soil Enzymes as Indices of Good Soil Health

Enzymes are directly involved in the biological breakdown of soil organic and mineral components. As a result, these catalysts provide a useful analysis of rates of reaction for essential processes in soil. The physical characteristics, activity of microbes or soil biomass, and organic matter are frequently closely correlated with soil enzyme activities. Furthermore, soil enzyme activities can be used to quantify

Table 6.1 List of soil enzymes as indicator of soil health

Enzyme	Reaction catalyzed	Indicator of microbial activity
α-amylase	Starch hydrolysis	C cycling
β-amylase		
Dehydrogenase	Oxidation of organic compound	C cycling, microbial oxidative activity
Endo-1,4-β-glucanase	Cellulose endohydrolysis	C cycling
Exo-1,4-β-glucanase	Cellulose cleavage at the ends	
Alkaline phosphatase	Hydrolysis of ester and anhydrides of phosphoric acid	P cycling
Acid phosphatase		
Protease	N mineralization	N cycling
Chitinase	Degradation and hydrolysis of chitin	C and N cycling
β-glucosidase	Cellobiose hydrolysis	C cycling
Cellulase	Cellulose hydrolysis	C cycling
Phenol oxidase	Lignin hydrolysis	C cycling
Urease	Urea hydrolysis	N cycling
Amidase	N mineralization	N cycling
Phosphatase	Release of PO_4^-	P cycling
Arylsulfatase	Release of SO_4^-	S cycling

(Source: Rao et al. 2014; Das and Varma 2011)

the activity of microbes, productivity of soil, and the restraining impact caused by pollutants (Tate 1995). Several soil enzyme activities have simple, well-established tests available (Dick et al. 1994; Tabatabai 1994a, b). A list of several soil enzymes that acts as indicator of microbial activity with type of reaction catalyzed is given in Table 6.1.

6.4 Potential Roles of Soil Enzymes in Maintaining Soil health

Soil enzymes' possible significance in maintaining soil health is stated as follows:

6.4.1 Amylase

According to Ross (1976), the starch-hydrolyzing enzyme amylase is composed of alpha- and beta-amylase (King 1967; Thoma et al. 1971). Animals, plants, and microbes all produce alpha-amylase, but mostly plants produce beta-amylase (Pazur 1965). The enzyme beta-amylase that transform starch to maltose plays a crucial part in the action, which is extensively found in plants and soils (Thoma et al. 1971). The functions and actions of the α-amylase and β-amylase enzymes, however, appear to have been altered by a variety of variables, such as cultural techniques, flora types, surrounding conditions, and kinds of soil (Pancholy and Rice 1973; Ross 1975). For instance, plants might produce enzymes directly from excreted substances, or through indirect means supply substrate for microorganisms' metabolic processes. To maximize the advantages that may be received from such enzymes, it is important for having a better knowledge of the importance of these enzymes that exist in soil which is further required for the development of appropriate management approaches.

6.4.2 Arylsulfatases

Soil profiles contain varying amounts of sulfate, some of which are bonded into organic substances and therefore through indirect means available to plants (Gupta et al. 1993; Ganeshamurthy et al. 1995). Arylsulfatases usually are abundant in nature as well as soil (Dodgson et al. 1982). These enzymes are responsible for hydrolyzing sulfate esters in soil and are released into the environment in reaction against sulfur limitation by bacteria (Kertesz and Mirleau 2004; McGill and Colle 1981). Its presence in various ecosystems in soil is frequently associated with biomass of microbes and rate of S immobilization (Klose and Tabatabai 1999; Vong et al. 2003). This enzyme aids in the hydrolysis of aromatic sulfate esters (R-O-SO_3) into phenols (R-OH) and sulfate, or sulfate sulfur (SO_4^{2-} or SO_4-S). According to Burns (1982), soil is impacted by a number of environmental factors, including heavy metal contamination (Tyler 1981), pH variations in the soil solution (Acosta-Martınez and Tabatabai 2000), the amount and kind of organic matter present (Sarathchandra and Perrott 1981), the absorption of extracellular arylsulfatases into soil particles, and soil particle surface absorption. Knowing more about how arylsulfatases contribute to sulfur mobilization in agricultural soils is essential for sulfur in plant nutrition. According to Kertesz and Mirleau (2004), arylsulfatases is the major enzyme in soil organosulfur cycle, although not much is known about the individual microbial genera or species that serve an integral part in this cycle.

6.4.3 β-Glucosidase

Glucosidase is a common and widespread enzyme in soils (Eivazi and Tabatabai 1988). Depending on kind of bond it hydrolyzes, it is termed accordingly. Since it is engaged in accelerating the hydrolysis and biodegradation, this enzyme plays an important soil functions (Ajwa and Tabatabai 1994; Martinez and Tabatabai 1997). Glucose is the end result and an extremely valuable supply of carbon energy for soil microorganisms (Esen 1993). Because of its capacity to represent historical biological processes, retain organic matter in the soil, and reveal management impacts on soils, b–glucosidase is a useful soil quality indicator (Bandick and Dick 1999; Ndiaye et al. 2000). When this enzyme is active, this property might be used to serve as a dependable biochemical indicator to follow ecological alterations caused by acidification of the soil. A greater understanding of the b-glucosidase enzyme activity and the environmental factors that control the enzyme might considerably help soil health studies.

6.4.4 Cellulases

Cellulose is the most common organic component in the biosphere, it accounts for more than half of the biomass produced by CO_2 photosynthetic fixation (Eriksson et al. 1990). In the majority of agricultural soils, the carbon source found in cellulose influences the development and existence of microorganisms (Deng and Tabatabai 1994). Cellulases enzymes must first break down cellulose into the sugar glucose, cellobiose, and high-molecular-weight oligosaccharides before carbon can be freed for use as an energy source by microbes (White 1982). Cellulases are a kind of enzyme that aids in the degradation of cellulose, a polysaccharide composed of β-1,4-linked glucose units (Deng and Tabatabai 1994). Cellulase activity in agricultural soils is influenced by the organic matter quality and soil mineral components (Sinsabaugh and Linkins 1989), trace elements from fungicides, and soil pH, water, and oxygen contents (abiotic conditions) (Deng and Tabatabai 1994). As comparison to red soil, black soil stimulated cellulases more strongly (Srinivasulu and Rangaswamy 2006). All of these results point to the possibility of using cellulase activity to provide an early indication of some of the chemical and physical characteristics of soil, hence are beneficial in agricultural soil management practices.

6.4.5 Chitinase

Chitinases or chitinolytic enzymes are the principal enzymes involved in the breakdown and degradation of chitin (polyb-1-4-(2-ncetamido-2-deoxy)-D-glucoside). They are also thought to be an important structural component of certain walls of fungi that use hyper parasitism defenses toward pathogens and insects attack (Chet and Henis 1975; Chet 1987). By utilizing additional processes like antibiosis or competition mechanisms, these biological agents also decrease disease-causing substances (Park 1960). This important agricultural enzyme is produced or released by a range of organisms, including the plants and microorganisms (Deshpande 1986). Its presence in the ecosystem, in various forms, is important in preventing and treating soil-borne illnesses from *Sclerotium rolfsii* and *Rhizoctonia solani* in beans and cotton, respectively (Ordentlich et al. 1988; Shapira et al. 1989). One approach that is recommended is the use of lytic enzymes such as chitinase, which results in the breakdown of pathogenic fungal cell walls (Ordentlich et al. 1988; Chet et al. 1990; Singh et al. 1999). In addition to its role in biological pest management, this enzyme has a wide variety of uses because of its ecofriendly nature, which helps to sustain maintain healthy soil and, as consequences, enhance plant growth and development and yields.

6.4.6 *Dehydrogenase*

Dehydrogenase enzyme activity oftenly used to measure soil biological activity (Burns 1978). This enzyme is recognized to oxidize soil organic substances by transferring electrons and protons from substrate to acceptors. These processes are part of the respiration pathways of soil microorganisms and are closely associated with the soil type and soil air-water conditions (Kandeler 1996; Glinski and Stepniewski 1985). It is essential to conduct studies on the dehydrogenase enzyme's activity in the soil because they may provide idea regarding the soil's ability to sustain biochemical activities that are essential for preserving soil fertility and soil health. Temperature and soil water content exert a passive impact on dehydrogenase activity by changing the soil's redox state (Brzezinska et al. 1998). The oxygen content in the soil quickly depletes after floods, causing the activity of aerobic microbes to switch to anaerobic microorganisms. These redox changes are connected to soil microorganism respiratory activity. It might be used as markers for the microbial redox systems in soils and as a potential evidence of oxidative activity of microorganisms (Tabatabai 1982; Trevors 1984). A shortage of oxygen might lead the facultative anaerobes to start activities of dehydrogenase enzymes and enable the utilization of Fe (III) forms as terminal electron acceptors in their metabolic processes (Bromfield 1954; Galstian and Awungian 1974), affecting the ecosystem's ability to supply plants with iron (Benckiser et al. 1984). The enzyme dehydrogenase is extensively utilized to quantify the impact of chemical pesticides, trace element, and farming practice on soil; in addition, it is a direct indicator of soil microbial activity (Garcia and Hernandez 1997; Reddy and Faza 1989; Wilke 1991). Moreover, it can show the kind and degree of soil pollution. The fly ash-affected soils had lower levels of dehydrogenase enzyme than soils contaminated with papers and pulp mill waste products (McCarthy et al. 1994). It has been demonstrated that dehydrogenases are more active at low pesticide doses and less active at higher pesticide dosages (Baruah and Mishra 1986).

6.4.7 *Proteases*

According to Ladd and Jackson (1982), proteases in the soil are vital in the process of N mineralization, which controls the quantity of nitrogen availability and development of plant. Proteases is commonly associated to inorganic and organic colloidal materials (Burns 1982; Nannipieri et al. 1996). The quantity associated with this extracellular activity of enzymes might have a significant role in the ecological functioning of microorganisms in the ecosystem, in addition to indicating the soil biological capacity for the conversion of the substrate, which is unaffected by the amount of microbial activity in the soil (Burns 1982).

6.4.8 Urease

The urease enzyme is responsible for the breakdown of urea fertilizers that are applied to the soil into NH_3 and CO_2, resulting in a rise in soil pH (Andrews et al. 1989; Byrnes and Amberger 1989). As a result, there is a fast loss of N to the environment because of NH_3 volatilization (Simpson et al. 1984; Simpson and Freney 1988). According to Polacco (1977), the primary sources of soil urease enzymes are microorganisms and plants that produce extra- and intracellular enzymes (Mobley and Hausinger 1989). Several variables influence urease activity in soils, including farming history, concentration of organic matter in the soil, depth of soil, soil additives, heavy metals, and environmental parameters such as temperature (Tabatabai 1977; Yang et al. 2006). The urease enzyme is extremely sensitive to detrimental amounts of heavy metals (Yang et al. 2006). In general, the activity of urease increases with temperature, and high temperatures increase the urease enzyme activity coefficient. Therefore, it is recommended that urea be sprayed during periods of low temperature during the day.

6.5 Sources of Soil Enzymes

Soil enzymes are key components of the soil system that are required to ensure functioning of ecosystems and recycling of nutrients. They are capable of effectively catalyzing the organic material breakdown in the soil system assists in the preservation of soil microorganisms' life activities and the stabilization of soil structural integrity. The bulk of the enzymes present in soil are microbial in origin, while they can also be derived from plants or animals. Depending on the position, these enzymes can be extracellular or intracellular, and they can be free enzymes, intracellular or cell-associated depending on their origin (Rahul et al. 2022). Sources of soil enzymes are given in Fig. 6.2.

Microorganisms in soil serve as critical biological markers for measuring soil health and are required for carbon (C)-climate feedback (Qu et al. 2023). Soil pH level, soil organic matter, quantity of nutrients, and particulate fraction all influence the diversity and composition of the microbial population (Cheng et al. 2021; Zheng et al. 2021). The soil enzymes also contribute in the breakdown of organic materials (Bilen and Turan 2022). A list of several significant soil enzymes and their roles in the cycling of nutrients and soil organic matter breakdown is given in Table 6.2.

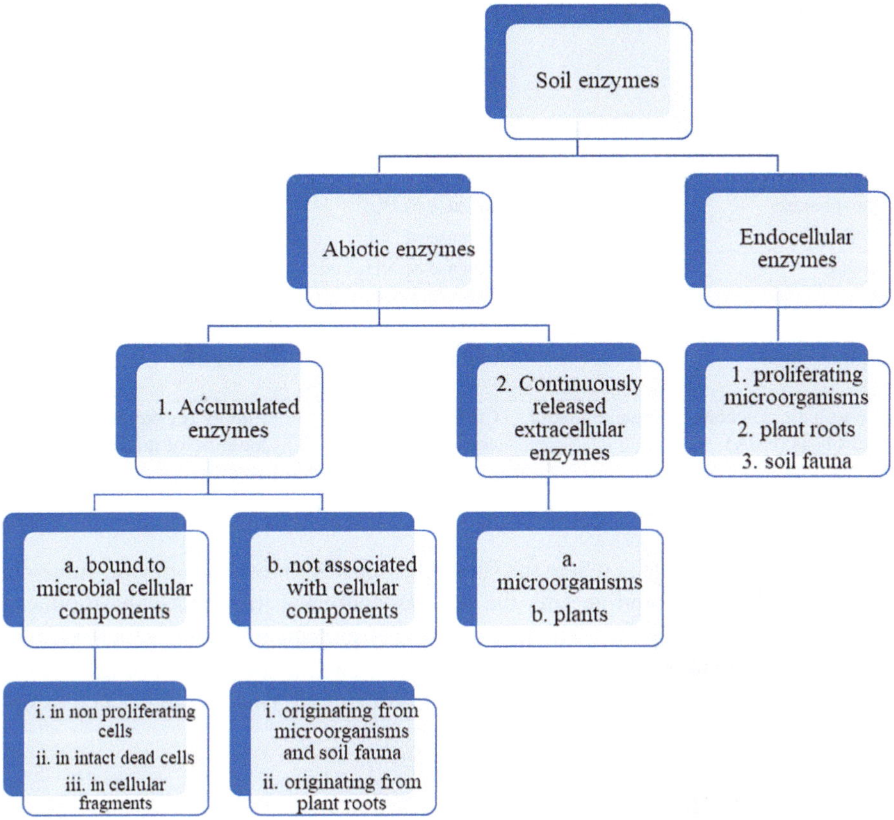

Fig. 6.2 Sources of soil enzymes (Dick and Tabatabai 1992; Skujins 1978)

The most active components of the soil system, soil enzymes, are essential for controlling biogeochemical cycling. A few enzymes can also alter the physical appearance of soil nutrients (Xie et al. 2021). According to earlier research, soil enzyme activity are influenced by the quantity of organic carbon, nitrogen, and phosphorus in the soil, as well as other soil parameters such as pH, mechanical arrangement, and porosity (Qiang et al. 2020). Soil enzymes are an excellent indication of the condition of the soil. Soil quality parameter changes must be measured in order to evaluate the impact of soil and crop management practices (Jat et al. 2021).

6.5.1 Mode of Action of Enzymes

In a chemical reaction, the mechanisms by which enzymes work include substrate binding, catalysis, substrate presentation, and allosteric regulation. But the binding of the substrate is the most frequent way that enzymes work. A specific active site

Table 6.2 Some significant soil enzymes and their roles in nutrient cycling and soil organic matter breakdown

Enzyme	Substrate	Enzyme reaction	Significance
1. Nutrient cycling			
Amidase	Carbon and N compounds	N-mineralization	Plant -available NH_4^+
Phosphatase	Phosphorus	Release of PO_4^{3+}	Plant-available P
Sulfatase	Sulphur	Release of SO_4^{2-}	Plant-available S
Urease	Nitrogen	Release of NH_3 and CO_2 compounds	Plant-available NH_4^+
2. Decomposition of organic matter			
β-glucosidase	Carbon compounds	Cellulose hydrolysis	Energy for microorganisms
Fluorescein diacetate hydrolysis (FDA)	Organic matter	Carbon and various nutrients	Energy for organisms as measure of microbial biomass

Source: Yang et al. (2012)

on an enzyme molecule is where the substrate attaches to form an enzyme-substrate combination. At the binding site, the process continues to produce the products, which are momentarily linked to the enzyme. The enzyme molecule is then released in an active state to start a new round of catalysis, and the product is then liberated. Various models to explain the mechanism of action of enzymes are given as follows:

6.5.1.1 The Lock-and-Key Hypothesis

The lock-and-key theory, also referred to as the template model, was proposed by Emil Fischer in 1898. According to this theory, the formation of an enzyme-substrate complex is stimulated by an interaction of the substrate with the enzyme at the active site, which operates in a similar way as a key fits into a lock. In fact, the molecular makeup of the enzyme and substrate must be compatible with each other for the enzyme and substrate to form a bond. The resultant enzyme-substrate complex is very fragile and breaks down nearly rapidly, liberating the reaction's byproducts and restoring the free enzyme. This approach enables energy to be generated and released, which causes the energy level of the substrate molecule to increase and produces the active or transition state. In this active state, some of the substrate molecule's bonds become cleavable. One of the model's limitations is that it cannot explain the stability of the enzyme's transition state or the concept of the stiffness of the active site.

As shown in Fig. 6.3, the presence of complementary structures between the enzyme and the substrate explains the general framework of the enzymatic specialty, where the substrate is entered with fittings that match the complementary position on the enzyme as the key gets into the lock.

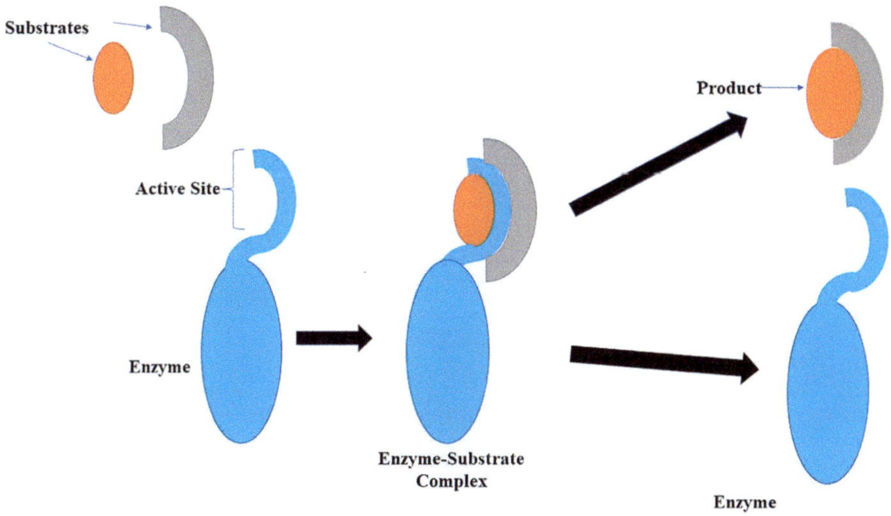

Fig. 6.3 Lock and key hypothesis

6.5.1.2 Induced Fit Hypothesis

The induced fit hypothesis modifies the lock-and-key hypothesis proposed by Koshland in 1958. According to this theory, the enzyme molecule loses its former structure and form. Instead, the contact with the substrate produces configurational or geometrical changes in the enzyme's active site. The structure and active centers of the substrate are thus perfectly matched by the enzyme molecule. Other amino acid residues continue to remain hidden within the molecule. It's possible that a distinct set of events triggered the conformational transition. Some enzymes might alter conformation before approaching the substrate. The alternative procedures might include binding the substrate first, followed by a modification in conformation in the active site. Third, additional isomerization that results in ultimately confirmation might trigger both processes to occur consecutively. The comparison of the induced fit and lock-and-key models for the binding of a substrate to an enzyme's active site is shown in Fig. 6.4 given below.

6.5.2 Mode of Action of Some Important Enzymes

6.5.2.1 Urease Enzyme (In Vivo Urease Activation)

Urease hydrolyzes urea to generate ammonia and carbamate, which then spontaneously hydrolyzes to make carbonic acid and an extra molecule of ammonia, as illustrated below:

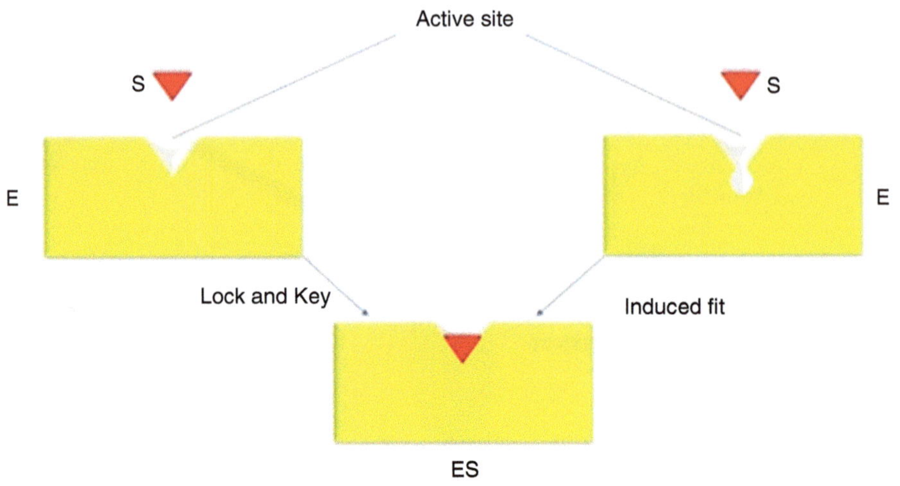

Fig. 6.4 Comparison of the induced fit and lock-and-key models for the binding of a substrate to an enzyme's active site

$$NH_2 - CO - NH_2 + H_2O \rightarrow NH_3 + NH_2 - CO - OH$$

$$NH_2 - CO - OH + H_2O \rightarrow NH_3 + H_2CO_3$$

In solution, an equilibrium exists between the two molecules of protonated ammonia and the released carbonic acid. An increase in pH is the result of these reactions as a whole.

$$H_2CO_3 \rightleftharpoons H^+ + HCO_3$$

$$2NH_3 + H_2O \rightleftharpoons 2NH_4^+ + 2OH$$

The activation of a urease enzyme requires the involvement of several additional proteins that deposit nickel into the urease's catalytic region. The existence of multiple genes in urease gene clusters that support the construction of a functional metallocentre in the protein was shown by sequence data. Many species' genetic investigations indicated the possibility of removing urease function by interrupting genes apart from those generating urease subunits (Collins and Gutman 1992; Kinghorn and Fluri 1984). Urea is catalytically broken down by the hydrolytic enzyme urease into volatile ammonia and CO_2 (Karplus et al. 1997). The final products are then produced spontaneously from carbamate and ammonia that are released by the enzyme. Various factors led to the interest in urease enzyme, which was first examined by Sumner (1926). The enzyme's action has also been proved to have agricultural implications through the volatilization of urea, an extensively used fertilizer, which is produced by the enzyme (Bremner 1995). The first nickel-dependent enzyme was discovered to be urease (Dixon et al. 1975) (Fenton 2004).

6.5.2.2 Phosphatase

Phosphatases activate the breakdown of a phosphomonoester, which results in the removal of the phosphate group from the substrate. Water is divided throughout the process, with the −OH group attaching the phosphate ion and the H+ protonating the hydroxyl group of the additional product. The process eventually culminates in the breakdown of a phosphomonoester, which is replaced by a phosphate ion and a molecule with a free hydroxyl group (Li et al. 2013). Reaction is given as follow:

$$P-B+H-OH \rightleftharpoons P-OH+H-B$$

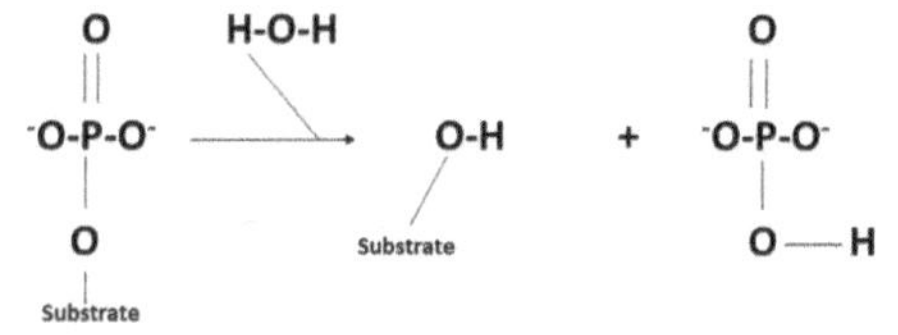

6.5.2.3 Dehydrogenase

A dehydrogenase is a type of oxidoreductase, or enzyme, which reduces an electron acceptor to oxidize a substrate, commonly NAD+/NADP+ or a flavin coenzyme such as FAD or FMN (Alberty 1994). They have the ability to catalyze both forward and reverse processes, much like all catalysts, and this can sometimes have physiological effects. In animals, alcohol dehydrogenase, for example, stimulates the conversion of ethanol to acetaldehyde.

In order to oxidize a substrate, dehydrogenases function by transferring hydrogen to an electron acceptor, often NAD^+ or FAD. It would be considered an oxidation of the substrate since it either loses hydrogen atoms or gains one (from water) (Clark 2002). The word "dehydrogenase" relates to both of these characteristics since it aids in the removal of hydrogen (de-) and is an enzyme (−ase).

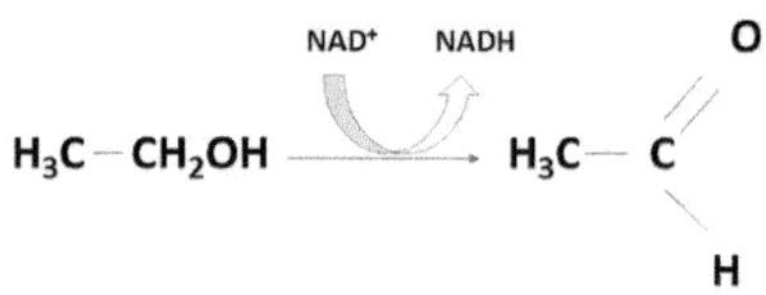

Alcohol Dehydrogenase

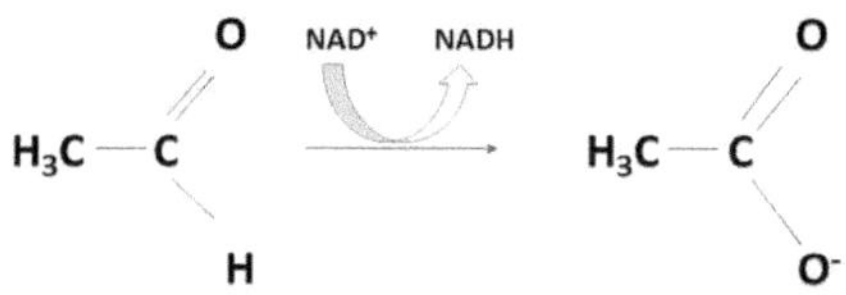

Alcohol Dehydrogenase

The exchange of a hydride and release of a proton (usually with water as a second reactant) and the exchange of two hydrogens are the two types of dehydrogenase reactions that take place most commonly. The term "dehydrogenase" refers to both of these features since it assists in the elimination of hydrogen (de-) and also an enzyme (−ase). The most prevalent forms of dehydrogenase reactions are the exchange of a hydride and proton release (usually with water as another reactant) and the exchange of a pair of hydrogens.

6.5.2.4 α-Amylase

α-Amylases are endo-acting amylases that hydrolyze the internal -(1–4) glycosidic bonds of starch polymers. A variety of models have been used to characterize the amylase action pattern, including unpredictable action and the numerous attack action. Random behavior has been described to as one attack or a multichain attack activity (Pandey et al. 2000). Before the enzyme-substrate complex splits in the first example, the polymer molecule goes through multiple full hydrolysis processes. The latter, on the other hand, hydrolyzes just one bond per effective their encounter. The enzyme executes a series of hydrolytic attacks following the initial (random) attack before destroying its connection with the substrate in the multiple attack action, which is an intermediate phase between the single-chain and multichain actions. The multiple attack action, as described by Svensson and Sogaard (1993), is a widely accepted concept to explain why amylases behave differently from one another. The majority of endoamylases possess minimal to extremely low levels of various attack activity, which is increased by temperature depending on the amylase.

6.5.2.5 Chitinase

Chitinases may break down polymer chitin into N-acetylglucosamine, low-molecular-weight COS (chitooligosaccharides), and GlcNAc (chitooligosaccharides) (Dean et al. 2012). Depending on how they work, chitinases can have an endo- or an exo-directed function. Exochitinases demonstrate increasing catalytic activity beginning at the nonreducing end of chitin, as opposed to endochitinases, which act indiscriminately on the chitin chain at an interior site. Examples of these enzymes include chitobiodidases and 1,4-glucosaminidases. Chitinases have been demonstrated to play a range of functions in morphogenesis, pathogenicity, parasitism, feeding, growth regulation, immunology, and defense, according to a study by Kuranda and Robbins (1991). Chitinase-induced degradation products of chitin, such as COS and GlcNAc, are gaining popularity because of its extensive applications in plant protection and growth promotion (Stoykov et al. 2015).

6.5.2.6 Cellulase

Cellulase attacks cellulose, causing significant physical changes to the substance. Among these modifications include fragmentation, swelling, a marked decline in tensile strength, transverse cracking, and a drop in the degree of polymerization. According to Tangnu (1982), the swelling and fragmentation of cellulose fibers are signs of purified C_x enzyme activity as opposed to C_1 enzyme activity.

The first process for the enzymatic breakdown of cellulose that includes a component C_1 was studied by Reese (1956). They suggested that the conversion of native cellulase was a two-step process: C_1 activates the cellulose chains, and the subsequent hydrolysis reaction is finished by the enzyme known as C_x. Early research mostly concentrated on C_x component examinations. Since 1964, a comprehensive search for the C_1 component has been made, and important data that point to the existence of a C_1-like component have been uncovered. Numerous researches have backed the hypothesis that the component C_1 has nonhydrolytic functions. As opposed to that, it has been suggested that, in light of the results of recent fractionation investigations, the (C_1–C_x) idea be reexamined and that the cellulase action mechanism be rebuilt. Arguments put up by a number of publications to prove that C_1 is a cellobiohydrolase seem to be convincing. However, there are still significant differences in how much the various C_1-type components hydrolyze highly organized cellulose. The mechanism of action for each cellulase component is listed below:

I. **Endo-β-1,4-glucanase**

This comprises of several components that were produced randomly in varying degrees. Although it only sporadically affects cellodextrin, phosphoric acid, swollen cellulose, and crystalline cellulose, this component has no impact on cellobiose. Cellotriose and cellobiose are the two main products (Zhang and Zhang 2013).

II. **Exo-β-1,4-glucanase**

When cellodextrin chains of 4–7 units long are generated by activity of endo-glucanase, this enzyme works on them, but it has difficulties breaking down insoluble cellulose (Tsuru et al. 1972). Despite being unable to significantly damage crystalline cellulose, it can degrade cellulose substrates by gradually eliminating cellobiose residues from the chain ends.

III. **β-Glucosidase**

Cellobiose and short-chain cello-oligosaccharides are hydrolyzed into glucose in this process, while cellulose is left unaffected. Cellobiose and cellotriose are hydrolyzed quickly, but as the polymerization level rises, attack also increases.

6.6 Conclusion

Understanding the potential functions of various enzymes in soil is crucial for preserving health of soil and managing ecosystem fertility. These enzymes, which are often present in soil, might have a big impact on environmental management, biology of soil, plant growth and development, and absorption of nutrients in ecosystems and sustaining soil health. The effectiveness and the actions of enzymes may be slightly or significantly impacted by unidentified cultural management practices. The identification of novel enzymes from microorganisms in soil may be the best technique that might have favorably affect the activities of soil enzymes for better growth and development and giving suitable conditions to support further living things. The preservation, maintenance, and enhancement of soil is important since it supports all the connected biological populations. Soil health is the capacity of soil to perform in a dynamic balance along with healthy environment to support animal, plant, and human life. Only biological, chemical, and physical markers may be used to evaluate quality of soil since they all have major contribution in preserving soil health. Examining the soil enzymes allows for a direct assessment of soil quality. Soil enzymes are the basic elements of the soil system that are necessary for sustaining ecosystem function and nutrient recycling. They can efficiently encourage the breakdown of organic compounds in the soil framework, protecting the life processes of soil microbes and stabilizing the structural integrity of soil. The bulk of the enzymes present in soil are derived by microbes, while they can also be found in plants and animals. Many soil enzymes are quite effective in maintaining and improving soil quality.

References

Acosta-Martınez V, Tabatabai MA (2000) Enzyme activities in a limed agricultural soil. Biol Fertil Soils 31:85–91

Ajwa HA, Tabatabai MA (1994) Decomposition of different organic materials in soils. Biol Fertil Soils 18:175–182

Alberty RA (1994) Recommendations for nomenclature and tables in biochemical thermodynamics. Pure Appl Chem 66(8):1641–1666. https://doi.org/10.1351/pac199466081641

Andrews RK, Blakeley RL, Zerner B (1989) Urease: a Ni (II) metalloenzyme. In: Lancaster JR (ed) The bioinorganic chemistry of nickel. VCH, New York, pp 141–166

Bandick AK, Dick RP (1999) Field management effects on soil enzyme activities. Soil Biol Biochem 31:1471–1479

Baruah M, Mishra RR (1986) Effect of herbicides butachlor, 2,4-d and oxyfluorfen on enzyme activities and CO_2 evolution in submerged paddy field soil. Plant Soil 96:287–291

Benckiser G, Santiago S, Neue HU, Watanabe I, Ottow JCG (1984) Effect of fertilization and exudation, dehydrogenase activity, iron-reducing populations and Fe^{2+} formation in the rhizosphere of rice (*Oryza sativa* L.) in relation to iron toxicity. Plant Soil 79:305–316

Bilen S, Turan V (2022) Enzymatic analyses in soils. In: Practical handbook on agricultural microbiology. Humana, New York, NY, pp 377–385

Bremner JM (1995) Recent research on problems in the use of urea as a nitrogen fertilizer. Fertil Res 42:321–329

Bromfield SM (1954) Reduction of ferric compounds by soil bacteria. J Gen Microbiol 11:1–6

Brzezinska M, Stepniewska Z, Stepniewski W (1998) Soil oxygen status and dehydrogenase activity. Soil Biol Biochem 30:1783–1790

Burns RG (1978) Enzyme activity in soil: some theoretical and practical considerations. In: Bums RG (ed) Soil enzymes. Academic, London, pp 295–340

Burns RG (1982) Enzyme activity in soil: location and possible role in microbial ecology. Soil Biol Biochem 14:423–427

Byrnes BH, Amberger A (1989) Fate of broadcast urea in a flooded soil when treated with N-(n-butyl)thiophospheric triamide, a urease inhibitor. Fertil Res Pract 18:221–231

Cheng JM, Han ZJ, Cong J, Yu JJ, Zhou JZ, Zhao MX (2021) Edaphic variables are better indicators of soil microbial functional structure than plant-related ones in subtropical broad-leaved forests. Sci Total Environ 773(10):145630

Chet I (1987) Trichoderma-application, mode of action, and potential as biocontrol agent of soil borne pathogenic fungi. In: Chet I (ed) Innovative approaches to plant disease control. Wiley, New York, pp 137–349

Chet I, Henis Y (1975) Sclerotial morphogenesis in fungi. Annu Rev Phytopathol 13:169–192

Chet I, Ordentlich A, Shapira R, Oppenheim A (1990) Mechanism of biocontrol of soil borne plant pathogen by rhizobacteria. Plant Soil 129:85–92

Clark, J. 2002. Definitions of oxidation and reduction (redox) Chemguide. Retrieved February 14, 2016.

Collins CM, Gutman DM (1992) Insertional inactivation of an Escherichia coli urease gene by IS3411. J Bacteriol 174:883–888

da Silva GJ, Rivadavea WR, de Lima JD, Monteiro PHR, da Silva FB (2021) Microbial enzymes and soil health. In: Mohamed HI, El-Beltagi HEDS, Abd-Elsalam KA (eds) Plant growth-promoting microbes for sustainable biotic and abiotic stress management. Springer, Cham. https://doi.org/10.1007/978-3-030-66587-66

Das SK, Varma A (2011) Role of enzymes in maintaining soil health. Soil Enzymol:25–42

Dean R, Van Kan JA, Pretorius ZA, Hammond-Kosack KE, Di Pietro A, Spanu PD, Rudd JJ, Dickman M, Kahmann R, Ellis J (2012) The top 10 fungal pathogens in molecular plant pathology. Mol Plant Pathol 13:414–430

Deng SP, Tabatabai MA (1994) Cellulase activity of soils. Soil Biol Biochem 26:1347–1354

Deshpande MV (1986) Enzymatic degradation of chitin and its biological applications. J Sci Ind Res 45:273–281

Dick RP (1997) Soil enzyme activities as integrative indicators of soil health. In: Pankhurst CE, Doube BM, Gupta VVSR (eds) Biological indicators of soil health. CABI, Wellingford, pp 121–156

Dick WA, Tabatabai MA (1992) Significance and potential uses of soil enzymes. In: Meting FB (ed) Soil microbial ecology. Marcel Dekker, New York, Basel, Hong Kong, pp 95–127

Dick RP, Sandor JA, Eash NS (1994) Soil enzyme activities after 1500 years of terrace agriculture in the Colca valley. Agric Ecosyst Environ 50:123–131

Dixon NE, Gazzola C, Blakeley RL, Zerner B (1975) Jack Bean urease (EC 3.5.1.5). A metalloenzyme. A simple biological role for nickel? J Am Chem Soc 97(14):4131–4133

Dodgson KS, White G, Fitzgerald JW (1982) Sulphatase enzyme of microbial origin. Afr J Biotechnol 1:156–159

Eivazi F, Tabatabai MA (1988) Glucosidases and galactosidases in soils. Soil Biol Biochem 20:601–606

Ellert BH, Clapperton MJ, Anderson DW (1997) An ecosystem perspective of soil quality. In: Gregorich EG, Carter MR (eds) Soil quality for crop production and ecosystem health. Elsevier, Amsterdam, pp 115–141

Eriksson KEL, Blancbette RA, Ander P (1990) Biodegradation of cellulose. In: Eriksson KEL, Blanchette RA, Ander P (eds) Microbial and enzymatic degradation of wood and wood components. Springer, New York, pp 89–180
Esen A (1993) Beta-glucosidases: overview. In: Esen A (ed) β-glucosidases and molecular biology. American Chemical Society, Washington, DC, pp 9–17
Fenton DE (2004) Urease—a muse for the coordination chemist. Proc Indian Natl Sci Acad Part A 70(2):311–328
Galstian AS, Awungian ZS (1974) Significance of the enzymes in oxidation of Fe and Mn oxides in soil (in Russian). In: Trans 10th International congress soil science III. Nauka Publishing House, Moscow, pp 130–135
Ganeshamurthy AM, Singh G, Singh NT (1995) Sulphur status and response of rice to sulphur on some soils of Andaman and Nicobar Islands. J Indian Soc Soil Sci 43:637–641
Garcia C, Hernandez T (1997) Biological and biochemical indicators in derelict soils subject to erosion. Soil Biol Biochem 29:171–177
Glinski J, Stepniewski W (1985) Soil aeration and its role for plants. CRC, Boca Raton, FL
Gupta VVSR, Farrell RE, Germida JJ (1993) Activity of arylsuphatases in Saskatchewan soils. Can J Soil Sci 73:341–347
James ES, Russel LW, Mitrick A (1991) Phosphate stress response in hydroponically grown maize. Plant Soil 132:85–90
Jat HS, Datta A, Choudhary M, Sharma PC, Dixit B, Jat ML (2021) Soil enzymes activity: effect of climate smart agriculture on rhizosphere and bulk soil under cereal based systems of north-west India. Eur J Soil Biol 103:103292
Kandeler E (1996) Nitrate. In: Schinner FO, Linger R, Kandeler E, Margesin R (eds) Methods in soil biology. Springer, Berlin, pp 408–410
Karplus PA, Pearson MA, Hausinger RP (1997) 70 years of crystalline urease: what have we learned. Acc Chem Res 30(8):330–337
Kennedy AC, Papendick RI (1995) Microbial characteristics of soil quality. J Soil Water Conserv 50(3):243–248
Kertesz MA, Mirleau P (2004) The role of soil microbes in plant sulphur nutrition. J Exp Bot 55:1939–1945
King NJ (1967) Glucoamylase of Coniophora cerebella. Biochem J 105:577–583
Kinghorn JR, Fluri R (1984) Genetic studies of purine breakdown in the fission yeast *Schizo saccharomyces* pombe. Curr Genet 8:99–105
Kiss S, Dragan-Bularda M, Radulescu D (1978) Soil polysaccharides: activity and agricultural importance. In: Burns RG (ed) Soil enzymes. Academic, London, pp 117–147
Klose S, Tabatabai MA (1999) Arylsulphatase activity of microbial biomass in soils. Soil Sci Soc Am J 63:569–574
Kuranda MJ, Robbins PW (1991) Chitinase is required for cell separation during growth of *Saccharomyces cerevisiae*. J Biol Chem 266:19758–19767
Ladd JN, Jackson RB (1982) Biochemistry of ammonification. In: Stevenson FJ (ed) Nitrogen in agricultural soils. American Society of Agronomy, Madison, WI, pp 173–228
Li X, Wilmanns M, Thornton J, Kohn M (2013) Elucidating human phosphatase-substrate networks. Sci Signal 6(275):rs10. https://doi.org/10.1126/scisignal.2003203
Martinez CE, Tabatabai MA (1997) Decomposition of biotechnology by-products in soils. J Environ Qual 26:625–632
McCarthy GW, Siddaramappa R, Reight RJ, Coddling EE, Gao G (1994) Evaluation of coal combustion by products as soil liming materials: their influence on soil pH and enzyme activities. Biol Fertil Soils 17:167–172
McGill WB, Colle CV (1981) Comparative aspects of cycling of organic C, N, S and P through soil organic matter. Geoderma 26:267–286
McLaren AD (1975) Soil as a system of humus and clay immobilised enzymes. Chem Scr 8:97–99

Miwa T, Ceng CT, Fujisaki M, Toishi A (1937) Zur Frage der Spezifitat der Glykosidasen. I. Verhalted vonb-d-glucosidases verschiedener Herkunft gegenuberdenb-d-Glucosiden mit verschiedenen Aglykonen. Acta Phytochim (Tokyo) 10:15–170
Mobley HLT, Hausinger RP (1989) Microbial urease: significance, regulation and molecular characterization. Microbiol Rev 53:85–108
Nannipieri P, Sequi P, Fusi P (1996) Humus and enzyme activity. In: Piccolo A (ed) Humic substances in terrestrial ecosystems. Elsevier, New York, pp 293–328
Ndiaye EL, Sandeno JM, McGrath D, Dick RP (2000) Integrative biological indicators for detecting change in soil quality. Am J Altern Agric 15:26–36
Neemisha, Sharma S (2022) Soil enzymes and their role in nutrient cycling. In: Giri B, Kapoor R, Wu QS, Varma A (eds) Structure and functions of pedosphere. Springer, Singapore. https://doi.org/10.1007/978-981-16-8770-98
Ordentlich A, Elad Y, Chet I (1988) The role of chitinase of Serratia marcescens in biocontrol of *Sclerotium rolfsii*. Phytopathology 78:84–88
Pancholy SK, Rice EL (1973) Soil enzymes in relation to old field succession; amylase, cellulose, invertase, dehydrogenase and urease. Soil Sci Soc Am J 37:47–50
Pandey A, Nigam P, Soccol VT, Singh D, Mohan R (2000) Advances in microbial amylases. Biotechnol Appl Biochem 31(2):135–152
Pankhurst CE, Hawke BG, McDonald HJ, Kirkby CA, Buckerfield JC, Michelsen P, O'Brien KA, Gupta VVSR, Doube BM (1995) Evaluation of soil biological properties as potential bioindicators of soil health. Aust J Exp Agric 35:1015–1028
Park D (1960) Antagonism- the background of soil fungi. In: Parkinson D, Waid JS (eds) The ecology of soil fungi. Liverpool University Press, Liverpool, pp 148–159
Pazur JH (1965) Enzymes in the synthesis and hydrolysis of starch. In: Whistler R, Paschall EF (eds) Starch: chemistry and technology, Fundamental aspects, vol 1. New York, Academic, pp 133–175
Polacco C (1977) Is nickel a universal component of plant ureases? Plant Sci Lett 10:249–255
Qiang W, Yang B, Liu Y, Qi K, Yang T, Pang X (2020) Effects of reclamation age on soil microbial communities and enzymatic activities in the sloping citrus orchards of southwestern China. Appl Soil Ecol 152:103566., ISSN 0929-1393. https://doi.org/10.1016/j.apsoil.2020.103566
Qu R, Liu G, Yue M, Wang G, Peng C, Wang K, Gao X (2023) Soil temperature, microbial biomass and enzyme activity are the critical factors affecting soil respiration in different soil layers in Ziwuling Mountains, China. Front Microbiol 14:339
Rahul R, Sharma P, Singh A, Singh J, Kumar M (2022) Soil enzymes and their role in soil health improvement. In: Nayak SK, Baliyarsingh B, Mannazzu I, Singh A, Mishra BB (eds) Advances in agricultural and industrial microbiology. Springer, Singapore, pp 39–61
Rao MA, Scelza R, Gianfreda L (2014) Soil enzymes. Enzymes in agricultural sciences. OMICS Group eBooks, Foster City, pp 10–43
Reddy GB, Faza A (1989) Dehydrogenase activity in sludge amended soil. Soil Biol Biochem 21:327
Reese ET (1956) Enzymatic hydrolysis of cellulose. Appl Microbiol 4(1):39–45
Ross DJ (1975) Studies on a climosequence of soils in tussock grasslands-5. Invertase and amylase activities of topsoils and their relationships with other properties. N Z J Soc Sci 18:511–518
Ross DJ (1976) Invertase and amylase activities in ryegrass and white clover plants and their relationships with activities in soils under pasture. Soil Biol Biochem 8:351–356
Sarathchandra SU, Perrott KW (1981) Determination of phosphatase and arylsulphatase activity in soils. Soil Biol Biochem 13:543–545
Shapira R, Ordentlich A, Chet I, Oppenheim AB (1989) Control of plant diseases by chitinase expressed from cloned DNA in Escherichia coli. Phytopathology 79:1246–1249
Simpson JR, Freney JR (1988) Interacting processes in gaseous nitrogen loss from urea applied to flooded rice fields. In: Pushparajah E, Husin A, Bachik AT (eds) Proceedings of international symposium on urea technology and utilization. Malaysian Society of Soil Science, Kuala Lumpur, pp 281–290

Simpson JR, Freney JR, Wetselaar R, Muirhead WA, Leuning R, Denmead OT (1984) Transformations and losses of urea nitrogen after application to flooded rice. Aust J Agric Res 35:189–200

Singer MJ, Ewing S (2000) Soil quality. In: Sumner ME (ed) Handbook of soil science. CRC, Boca Raton, FL, pp 271–298

Singh PP, Shin YC, Park CS, Chung YR (1999) Biological control of Fusarium wilt of cucumber by chitinolytic bacteria. Phytopathology 89:92–99

Sinsabaugh RL, Linkins AE (1989) Natural disturbance and the activity of *Trichoderma viride* cellulase complex. Soil Biol Biochem 21:835–839

Skujins J (1978) History of abiotic soil enzyme research. In: Burns RG (ed) Soil enzymes. Academic, London, pp 1–49

Srinivasulu M, Rangaswamy V (2006) Activities of invertase and cellulase as influenced by the application of tridemorph and captan to groundnut (*Arachis hypogaea*) soil. Afr J Biotechnol 5:175–180

Stoykov YM, Pavlov AI, Krastanov AI (2015) Chitinase biotechnology: production, purification, and application. Eng Life Sci 15:30–38

Sumner JB (1926) The isolation and crystallization of the enzyme urease. J Biol Chem 69:435–441

Svensson B, Sogaard M (1993) Mutational analysis of glycosylase function. J Biotechnol 29(1):1–37

Tabatabai MA (1977) Effect of trace elements on urease activity in soils. Soil Biol Biochem 9:9–13

Tabatabai MA (1982) Soil enzyme. In: Page AL (ed) Methods of soil analysis, part 2. American Society of Agronomy, Madison, WI, pp 903–948

Tabatabai MA (1994a) Soil enzymes. In: Weaver RW, Angle JS, Bottomley PS (eds) Methods of soil analysis, part 2, Microbiological and biochemical properties. SSSA Book Series No. 5. Soil Science Society of America, Madison, WI, pp 775–833

Tabatabai MA (1994b) Soil enzymes. In: Mickelson SH (ed) Methods of soil analysis, part 2. Microbiological and biochemical properties. Soil Science Society of America, Madison, WI, pp 775–833

Tangnu SK (1982) Process development for ethanol production based on enzymatic hydrolysis of cellulosic biomass. Process Biochem 17(3):36–45

Tate RL (1995) Soil microbiology. John Wiley, New York

Thoma JA, Spradlin JE, Dygert S (1971) Plant and animal amylases. In: Boyer PD (ed) The enzymes, 5th edn. Academic, New York, pp 115–189

Trevors JT (1984) Dehydrogenase activity in soil: a comparison between the INT and TTC assay. Soil Biol Biochem 16:673–674

Tsuru D, Hiraoka N, Fukumoto J (1972) Studies on mold dextranases: IV. Substrate specificity of *Aspergillus carneus* dextranase. J Biochem 71(4):653–660

Tyler G (1981) Heavy metals in soil biology and biochemistry. In: Paul EA, Ladd JN (eds) Soil biochemistry, vol 5. Marcel Dekker, New York, pp 371–414

Vong PC, Dedourge O, Lasserre-Joulin F, Guckert A (2003) Immobilized-S, microbial biomass-S and soil arylsulphatase activity in the rhizosphere soil of rape and barley as affected by labile substrate C and N additions. Soil Biol Biochem 35:1651–1661

White AR (1982) Visualization of cellulases and cellulose degradation. In: Brown RM (ed) Cellulose and other natural polymer systems: biogenesis, structure, and degradation. Plenum, New York, pp 489–509

Wilke BM (1991) Effect of single and successive additions of cadmium, nickel and zinc on carbon dioxide evolution and dehydrogenase activity in a sandy Luvisol. Biol Fertil Soils 11:34–37

Xie X, Pu L, Zhu M, Meadows M, Sun T, Wu XB, Yan X (2021) Differential effects of various reclamation treatments on soil characteristics: an experimental study of newly reclaimed tidal mudflats on the east China coast. Sci Total Environ 68:144996., ISSN 0048-9697,. https://doi.org/10.1016/j.scitotenv.2021.144996

Yang Z, Liu S, Zheng D, Feng S (2006) Effects of cadmium, zinc and lead on soil enzyme activities. J Environ Sci 18:1135–1141

Yang L, Zhang Y, Li F (2012) Soil enzyme activities and soil fertility dynamics. In: Advances in citrus nutrition. Springer, pp 143–156. https://doi.org/10.1007/978-94-007-4171-3_11

Zhang XZ, Zhang YHP (2013) Cellulases: characteristics, sources, production, and applications. In: Bioprocessing technologies in biorefinery for sustainable production of fuels, chemicals, and polymers, pp 131–146. https://doi.org/10.1002/9781118642047.ch8

Zheng W, Zhao ZY, Lv FL, Yin YN, Wang ZH, Zhao ZY, Li Z, Zhai B (2021) Fungal alpha diversity influence stochasticity of bacterial and fungal community assemblies in soil aggregates in an apple orchard. Appl Soil Ecol 162(11):103878

Chapter 7
Microbial Biofertilizers for Soil Health

Shagun Sharma, Sidharth Sharma, Sivasakthivelan Panneerselvam, Arivukkarasu Kamaraj, Gomathi Selvaraj, and Pankaj Kumar

7.1 Introduction

Soil is regarded as part of the Earth and as a medium on which plants grow. It is composed of three layers: topsoil, subsoil, and parent material (Itelima et al. 2018). The top soil consists of air, water, minerals, living organism (roots, microorganisms, insects, and worms), inorganic matter, and high concentration of organic matter (humus) that are indispensable for the process of plant growth (Bohn et al. 2015; Jangir et al. 2019). In the composition of typical soils, minerals constitute approximately 45% of the overall volume. Concomitantly, water and air are present in comparable proportions, each accounting for approximately 25%. Concurrently, the content of organic matter falls within the range of 2–5% (Rasulov and Uzbekistan 2019). Within the matrix of soil, an immense microbial populace thrives, numbering in the hundreds of millions to billions of microorganisms within every gram of soil (Devi 2022). It contains varieties of microorganisms, including bacteria, fungi, actinomycetes, soil algae, and soil protozoa (Johns 2017). Among the various types of microorganisms present in soil, bacteria stand out as the most abundant with 10^8–10^9 bacterial cells per gram (Putrie et al. 2020). Actinomycetes are the second most abundant with 10^7–10^8 microbes per gram of soil, followed by fungi and algae with

Shagun Sharma, Sidharth Sharma and Pankaj Kumar contributed equally to this work and should be considered as the first author.

S. Sharma · S. Sharma · P. Kumar (✉)
Dr. Yashwant Singh Parmer University of Horticulture and Forestry, Nauni, Solan, Himachal Pradesh, India

S. Panneerselvam · A. Kamaraj · G. Selvaraj
Annamalai University, Chidambaram, Tamil Nadu, India

R. K. Bhatia, A. Walia (eds.), *Advancements in Microbial Biotechnology for Soil Health*, Microorganisms for Sustainability 50,
https://doi.org/10.1007/978-981-99-9482-3_7

10^5–10^6 and 10^4–10^5 per gram, respectively (Javed et al. 2021; El Barnossi and Housseini 2023; Chen et al. 2003).

Soil microbes constitute a pivotal element of ecosystem services (Miransari 2013). These microorganisms play a fundamental role in the soil ecosystem by transforming organic matter into nutrients that are essential for the growth and development of plants (Griffiths et al. 2021). Their contribution lies in enhancing the accessibility of vital nutrients within the soil, thereby augmenting both soil quality and plant productivity, consequently fostering robust plant growth (Hill et al. 2000). The term "biofertilizers" pertains to biological formulations containing active or dormant microorganisms or their byproducts present within the root zone or internal plant tissues (Naveed et al. 2015). Comprising bacterial, fungal, and algal strains, biofertilizers facilitate plant growth by colonizing the rhizosphere and internal plant regions. Moreover, they engender improvements in soil physicochemical attributes and influence the configuration and operations of microbial communities. This is achieved by modulating factors like microbial carbon levels, diversity, and community-wide physiological patterns (Allouzi et al. 2022; Thomas and Singh 2019). Microbial strains employ diverse mechanisms to enhance nutrient uptake, soil fertility, and crop yields. These mechanisms encompass nitrogen fixation (Gupta et al. 2012), phosphorus solubilization and mobilization, and potassium solubilization (Youssef and Eissa 2014). Further, they engage in the secretion of phytohormones (such as auxins, gibberellins, and cytokinins) (Singh et al. 2021), synthesis of compounds that inhibit phytopathogens (Mącik et al. 2020), tolerance development against abiotic stressors (such as salinity, heat, drought, flooding, and metal toxicity) (Abideen et al. 2022), and resistance to biotic stress factors (including viruses, bacteria, fungi, nematodes, insects, arachnids, and weeds) (Bhardwaj et al. 2014; Gupta et al. 2021). Furthermore, these microbes partake in detoxifying subterranean pollutants (Mącik et al. 2020). Around the plant roots, rhizobacteria are known as plant growth-promoting rhizobacteria (PGPR), while fungi are termed plant growth-promoting fungi (PGPF). Both these categories exhibit beneficial impacts on plant growth, productivity, and physiological attributes (Bhattacharyya and Jha 2012).

Fertilizers hold paramount significance within the realm of agriculture by significantly enhancing crop yields, and the fertilizer used can either be inorganic (chemical fertilizers) or organic (biofertilizers) (Kour et al. 2020a, b). It depends entirely on the type of fertilizers used to grow healthy crops that will satisfy the demands of the world's expanding population (Snyder et al. 2009; Kour et al. 2020a, b). Chemical fertilizers are employed to rapidly fulfill the nutritional requirements of plants and yield prompt outcomes (Timsina 2018). These fertilizers provide high concentrations of nutrition, and thus increased reliance on chemical fertilizers that are harming the environment's ecology and impairing human health (Kour et al. 2020a, b). Their excessive or unbalanced use can lead to negative consequences like soil erosion, soil pollution, water contamination, pesticide poisoning, and depletion in the groundwater table (Lal 2009; Atieno et al. 2020). In view of the above-stated facts, biofertilizers should be used, as these microbes are organic agricultural inputs that improved soil health and sustainability from their byproducts that make them eco-friendly, cost-effective, and nontoxic, making them the best substitute for synthetic fertilizers (De Corato 2020; Mącik et al. 2020).

7.2 Types of Biofertilizers

A list of different biofertilizers with microorganisms involved is represented in Table 7.1.

7.2.1 Nitrogen-Fixing Biofertilizers (NFB)

Nitrogen gas occupies 78% of the total atmospheric gas volume, and it's also the source of nitrogen in the soil (Leghari et al. 2016). For all forms of life, nitrogen is one of the most important elements (Ohyama 2010). It serves as the primary component in the synthesis of proteins, nucleic acids, chlorophyll, coenzymes, phytohormones, and secondary metabolites (Ohyama 2010; Hawkesford et al. 2023). While plants lack the capacity to directly convert atmospheric dinitrogen into ammonia for their growth and development, they possess no mechanism to facilitate

Table 7.1 List of different biofertilizers with microorganisms involved

Type of biofertilizer	Microorganisms involved	References
Nitrogen fixing biofertilizers	*Azotobacter*, *Rhizobium*, *Anabaena*, *Nostoc*, *Azospirillum*	Asoegwu et al. (2020), Saha et al. (2020), Sahu et al. (2012)
Phosphorus-solubilizing biofertilizers	*Bacillus*, *Pseudomonas*, *Penicillium*, *Aspergillus*, *Trichoderma*	Rawat et al. (2021), Liang et al. (2020)
Phosphate-mobilizing biofertilizers	*Glomus*, *Scutellospora*, *Laccaria*, *Pisolithus*, *Amanita*, *Pezizella*	Khan et al. (2009a, b), De La Providencia et al. (2007), Hanif et al. (2023)
Potassium biofertilizers	*Bacillus*, *Aminobacter*, *Arthrobacter*	Kour et al. (2020a, b), Nosheen et al. (2021)
Iron siderophore synthesizers	*Pseudomonas*, *Bacillus*	Bhatt and Denny (2004)
Zinc-solubilizing biofertilizers	*Pseudomonas*, *Burkholderia*, *Bacillus*, *Gluconacetobacter*, *Serratia*	Naz et al. (2016), Haroon et al. (2022)
Sulfur-oxidizing biofertilizers	*Acidithiobacillus*, *Aquaspirillum*, *Bacillus*, *Pseudomonas*, *Xanthobacter*, *Thiobacillus*	Chaudhary et al. (2022), Kumar et al. (2020)
Silicate-solubilizing biofertilizers	*Bacillus*, *Pseudomonas*	Etesami and Jeong (2022)
Plant growth-promoting biofertilizers	*Azospirillum*, *Azotobacter*, *Pseudomonas*, *Serratia*, *Streptomyces*, *Rhizobium*	Kamran et al. (2010), Ruiu (2020)

this process (Gupta et al. 2015). Numerous free-living, symbiotic, and asymbiotic nitrogen-fixing microbes exist in soil that aid in biological nitrogen fixation (BNF) (Santi et al. 2013; Mukherjee and Sen 2021). BNF is the term used to describe the enzymatic process wherein living organisms convert atmospheric dinitrogen gas into ammonia. This transformation is facilitated by the nitrogenase enzyme (Barney and Plunkett 2022). Microbes like aerobic and anaerobic bacteria (Poudel et al. 2018), as well as cyanobacteria (Kamshybayeva et al. 2022), produce nitrogenase enzyme, which are termed as diazotrophs (Fig. 7.1) (Harding et al. 2022).

Currently, 65% of the N_2 used in agriculture is produced through the BNF process, and it will continue to be crucial for crop productivity in the future, especially in sustainable systems (Bello et al. 2021; Hayat et al. 2010). Utilizing energy in the form of ATPs, nitrogen is converted to ammonia (Lee et al. 2020). The following is the reaction that the enzyme nitrogenase catalyzes:

$$N_2 + 8H^+ + 16ATP + 8e^? \, ? \, 2NH_3 + H_2 + 16ADP + 16Pi$$

7.2.1.1 Azotobacter

Azotobacter are gram-negative, aerobic, free-living, oval or spherical bacteria that live in soil and produce thick-walled cysts (Jnawali et al. 2015). These bacteria use nitrogen gas from the atmosphere to create the proteins that make up their cells

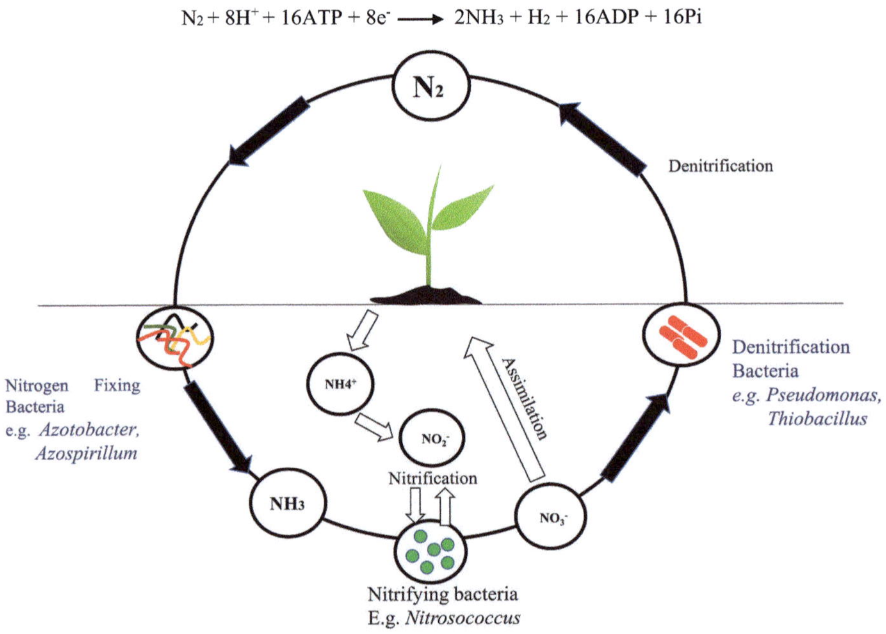

Fig. 7.1 Atmospheric nitrogen fixation cycle

(Chaudhary et al. 2020). After the death of *Azotobacter* cells, this cell protein is then mineralized in the soil, allowing crop plants to access nitrogen (Basit et al. 2021; Jnawali et al. 2015). The genus *Azotobacter* comprises around six species, including *A. chroococcum*, which most commonly inhabits soils worldwide (Wani et al. 2016; Kumar and Kumar 2019). The bacterium generates large amounts of slime, which aids in soil aggregation (Saha et al. 2020). Through the biosynthesis of biologically active compounds, stimulation of rhizospheric microbes, and production of phytopathogenic inhibitors, *Azotobacter* has positive effects on crop growth and yield (Aasfar et al. 2021; Sumbul et al. 2020). *Azotobacter* also makes thiamin, riboflavin, indole acetic acid, and gibberellins in addition to fixing nitrogen (Seenivasagan and Babalola 2021).

Azotobacter, a type of bacteria classified as gram-negative and aerobic, exist as free-living entities in soil. These microorganisms display an oval or spherical shape and have the capability to form durable cysts with robust walls (Jnawali et al. 2015). Fascinatingly, *Azotobacter* employs atmospheric nitrogen gas to synthesize the essential proteins constituting their cellular structure (Chaudhary et al. 2020). Upon the demise of *Azotobacter* cells, the proteins within them undergo mineralization within the soil. This process subsequently facilitates the accessibility of nitrogen by crop plants (Basit et al. 2021; Jnawali et al. 2015). Within the *Azotobacter* genus, approximately six distinct species are identified, with *A. chroococcum* being particularly prevalent across diverse soil environments globally (Wani et al. 2016; Kumar and Kumar 2019). These bacteria possess the remarkable ability to generate substantial quantities of slimy substances, contributing to the aggregation of soil particles (Saha et al. 2020). The influence of *Azotobacter* extends beyond mere soil interactions. The bacteria engage in the biosynthesis of biologically active compounds, stimulate the growth of rhizospheric microbes, and produce agents that inhibit phytopathogens. Consequently, *Azotobacter* exerts a positive impact on crop development and yield (Aasfar et al. 2021; Sumbul et al. 2020). Notably, *Azotobacter's* capabilities encompass more than nitrogen fixation alone. In addition to this crucial function, the bacteria also synthesize important compounds such as thiamin, riboflavin, indole acetic acid, and gibberellins (Seenivasagan and Babalola 2021). This multifaceted contribution underscores the significance of *Azotobacter* as a pivotal player in soil ecosystems and agriculture.

7.2.1.2 Rhizobium

The *Rhizobium* bacterium exhibits a gram-negative characteristic and takes on a rod-shaped morphology. It possesses motility, allowing it to move, and thrives in an aerobic environment. This soil-dwelling bacterium does not generate spores. Remarkably, it demonstrates the capacity to establish itself within the rhizosphere, the soil region proximal to the roots of leguminous plants. Through a symbiotic relationship with these plants, *Rhizobium* plays a vital role in the fixation of atmospheric nitrogen. The size of *Rhizobium* bacteria is 0.5–1.0 × 1.2–3.0 μm. Every known strain of *Rhizobium* causes plants to develop root nodules known as hypertrophisms. The *Rhizobium* bacteria attach to the epidermal root hairs of the host and typically infiltrate by causing the hair to deform and then forming a specialized invasion structure called the "infection thread." The bacteria divide into infection threads, which slowly invade the root cortex. *Rhizobium* stands out as the most proficient biofertilizer in terms of its exceptional capacity for nitrogen fixation.

7.2.1.3 Azospirillum

Azospirillum stands out as a frequently investigated type of plant growth-promoting bacteria (PGPB) (Bashan and De-Bashan 2010). Notably, *Azospirillum brasilense* garners significant recognition as the most extensively studied species (Cassán et al. 2020). These bacteria inhabit the rhizosphere vicinity around the roots or the intercellular gaps within the root cortex of graminaceous plants, establishing symbiotic relationships through association (Panwar and Laxmi 2005). A fundamental avenue through which *Azospirillum* species exert their influence involves the synthesis and metabolism of various phytohormones and growth-regulating molecules (Cassán and Diaz-Zorita 2016). The positive impacts of *Azospirillum* species on the growth of diverse plant species have spurred researchers' interest in cultivating specific strains tailored for agricultural use. Beyond their role in nitrogen fixation, *Azospirillum* bacteria additionally secrete growth-promoting substances like indole-3-acetic acid (IAA), exhibit disease resistance, and demonstrate resilience in the face of drought conditions (Bashan and De-Bashan 2010; Hayat et al. 2010).

7.2.2 Phosphorus Biofertilizers

Next to nitrogen, phosphorus is the second-most essential macronutrient that plants require for growth and development (Chan et al. 2021). While phosphorus is present in ample quantities within the soil, a noteworthy proportion of this element exists in an insoluble state. As a result of this insolubility, the greater part of phosphorus remains inaccessible, impeding its potential to foster and facilitate plant growth ((Daniel et al. 2022). In the context of plant nutrient uptake, it is noteworthy that phosphorus is absorbed by plants exclusively in two soluble configurations, namely, monobasic and dibasic forms (Dordipour et al. 2018). Phosphorus is required at the molecular level and also in many physiological and biochemical processes including photosynthesis (Pandey et al. 2015). The natural phosphorus cycle relies greatly on microorganisms (Liang et al. 2020). Mineralization, solubilization, and immobilization are the predominant ways of phosphorus within the soil involves a process facilitated by phosphate-solubilizing microorganisms (PSMs) and phosphate mobilization microorganisms (PMMs) (Rawat et al. 2021; Yadav 2022). Two bacterial species that are frequently employed as potassium biofertilizers are *Aminobacter* and *Arthrobacter*. These microorganisms can solubilize potassium from nonavailable forms in the soil, enhancing its availability to plants. When potassium-containing minerals are broken down by *Aminobacter* and *Arthrobacter's* enzymes, potassium ions are released that can be absorbed by plant roots.

7.2.2.1 Glomus

Glomus belong to the category of arbuscular mycorrhizal fungi, a group of fungi that exhibit a symbiotic relationship with the roots of various plants (Kirk et al. 2008). These fungi have the unique ability to generate spores at the tips of their hyphae, whether within the host root or in the surrounding soil (Błaszkowski et al. 2008). Notably, the hyphae of *Glomus* species have been scientifically demonstrated to enhance the absorption of phosphate from the soil by the host plant (Fasusi et al. 2021). In the realm of practical applications, certain species of *Glomus* are intentionally cultivated and marketed as mycorrhizal inoculants (Bhattacharjee and Dey 2014). These products are designed to facilitate the establishment of mycorrhizal partnerships between the fungi and plant roots, thereby fostering improved nutrient uptake and overall plant health.

7.2.2.2 Trichoderma

Trichoderma are endophytic and symbiotic fungi belonging to the family *Hypocreaceae* which can be readily cultivated using synthetic growth mediums (Harman et al. 2004). *Trichoderma* is reported to safeguard plants against soil-borne pathogens and also improve their growth by boosting nutrient absorption (Sood et al. 2020; Shafique et al. 2016). Species like *T. asperellum*, *T. harzianum*, *T. viride*, and *T. hamatum* are used as biocontrol agents for fungal root diseases (Harman 2006).

7.2.3 Potassium Biofertilizers

Apart from nitrogen (N) and phosphorus (P), potassium (K) stands as the third essential macronutrient imperative for plant growth (Zahedi 2016). The significance of potassium extends to numerous biochemical and physiological mechanisms within plants. These encompass pivotal functions such as photosynthesis, the activation of ATP synthase enzymes, the synthesis of proteins, the breakdown of sugars, and the regulation of stomatal openings (Meena et al. 2014; Hasanuzzaman et al. 2018; Das and Pradhan 2016). Additionally, potassium is essential for the movement of nutrients and water through the xylem in plants (Jaiswal et al. 2016). Even though there is an abundance of K in the soil, only 1–2% of it is available to plants; the remainder is present as mineral K, which plants cannot absorb (Etesami et al. 2017). Several bacterial strains have been identified for their ability to transform insoluble potassium (K) into soluble forms through various mechanisms. These mechanisms encompass the production of acids, chelation, acidolysis, complexolysis, and exchange reactions (Meena et al. 2014, 2015; Etesami et al. 2017). In addition, certain fungi, such as species belonging to *Aspergillus* and *Penicillium* genera, demonstrate the capability to solubilize and release K from both organic and inorganic sources (Soumare et al. 2023). Within the rhizosphere, a multitude of microorganisms like *Bacillus* sp., *Aminobacter* sp., and *Arthrobacter* sp. function as potassium biofertilizers (Nosheen et al. 2021; Kour et al. 2020a, b).

7.2.3.1 Bacillus

Bacillus represents a type of soil microorganism characterized by its gram-positive nature and ability to form spores (Azizoglu 2019). It is a group of bacteria frequently used for biocontrol in agriculture (Sivasakthi et al. 2014). In addition to help in protecting plants from soil-borne pathogens, these microorganisms enhance the uptake of nutrients by plants, thus enhancing their growth (Pineda et al. 2010; Nanjundappa et al. 2019). This microorganism exhibits remarkable resilience to a diverse array of challenging environmental circumstances while also generating durable spores with enduring capabilities (Seifan et al. 2016). It holds a significant position among biofilm-forming rhizobacteria known for their pivotal role in facilitating plants' nutrient absorption and establishing root colonization (Mahapatra et al. 2022).

7.2.4 Zinc-Solubilizing Biofertilizers (ZSB)

Zinc deficiency stands out as the predominant scarcity among micronutrients (Naveed et al. 2020), exerting significant influence on diverse metabolic pathways reliant on zinc (Zn). Notably, zinc plays a pivotal role in facilitating the synthesis of vital compounds such as chlorophyll, enzymes, and proteins (Singh et al. 2018; Mousavi et al. 2013). Addressing this concern, zinc-solubilizing bacteria offer a promising avenue, effectively converting insoluble zinc into its soluble form (Kushwaha et al. 2020). These bacteria hold substantial potential in promoting plant growth, achieved through the solubilization of zinc and the secretion of phytohormones like kinetin, indole-3-acetic acid (IAA), and gibberellic acid (Rana et al. 2020; Devi et al. 2020). Their biological activities encompass the production of organic acids, inducing a decrease in pH levels, which subsequently interact with zinc complexes in the soil, leading to the sequestration of zinc cations (Kumar 2018). Mechanisms employed by these microbes encompass the generation of chelated ligands, siderophores, and the redox system within cell membranes, all contributing to the solubilization of zinc (Rawat et al. 2018). Prominent members of the zinc-solubilizing microbial community encompass *Pseudomonas* sp.,

Gluconacetobacter sp., *Serratia* sp., *Thiobacillus* sp., *Burkholderia* sp., and *Azospirillum* sp. (Sammauria et al. 2020; Fasusi et al. 2021; Rana et al. 2020).

7.2.5 Sulfur-Oxidizing Biofertilizers (SOB)

Sulfur is a crucial nutrient for plants, along with N, P, and K, as it is a component of many secondary metabolites, including amino acids, proteins, metal cofactors, vitamins, and coenzymes (Narayan et al. 2022; Becana et al. 2018; Nazar et al. 2011). The insufficiency of sulfur has a detrimental impact on both the growth of plants and their photosynthesis processes. Additionally, it hinders the production of seeds in both leguminous and nonleguminous species (Becana et al. 2018). Sulfur-oxidizing bacteria (SOB) or sulfur biofertilizers are microbes that are prepared to oxidize sulfur (Chaudhary et al. 2022). SOB plays a vital role in the detoxification of reduced sulfide in sediments (Li et al. 2021). In the process of decomposing organic substances, various microorganisms act simultaneously (Golchin et al. 2018). First, polymeric material is hydrolyzed into monomeric compounds that are soluble (Behera et al. 2014). Fermentative bacteria then transform these byproducts into short-chain fatty acids, carbon dioxide, and hydrogen ions (Itelima et al. 2018). Major sulfur-oxidizing microbes include *Aquaspirillum* sp., *Acidithiobacillus* sp., *Aquifer* sp., *Beggiatoa* sp., *Bacillus* sp., *Pseudomonas* sp., *Thiobacillus* sp., and *Rhodobacter* sp., etc. (Mohapatra et al. 2008; Behera et al. 2014; Reineke and Schlömann 2023).

7.2.6 Silicate-Solubilizing Biofertilizers (SSB)

Silicates undergo a transformative process into soluble silica through the activity of silicate-solubilizing bacteria (Cruz et al. 2022). These particular bacterial strains are equipped with the enzyme silicase, responsible for converting silicates into a soluble form of silica. This enzymatic action has the notable consequence of rendering silica accessible for absorption by plants (Raturi et al. 2022; Etesami et al. 2021). The impact of these silicate-solubilizing bacteria goes beyond silica conversion alone. These microorganisms exhibit a capacity to effectively solubilize phosphates, potassium, and other previously insoluble silicate compounds. This multifaceted action contributes significantly to enhancing soil fertility and consequently amplifying plant productivity (Maleva et al. 2017). The ecological distribution of silicate-solubilizing bacteria spans various environments including sediments, water bodies, soils, and silicate mineral substrates (Raturi et al. 2021). Their presence plays a pivotal role in the process of solubilizing the otherwise recalcitrant forms of silicates, thereby not only bolstering plant defense mechanisms but also heightening soil fertility levels (Naureen et al. 2015). Among the prominent microbial players in silicate oxidation are bacterial strains such as *Bacillus* sp., *Pseudomonas* sp., etc. (Vasanthi et al. 2018; Ashour and Afify 2023).

7.2.7 Plant Growth-Promoting Biofertilizers (PGPB)

Any bacterial organism, spanning from free-living types to those engaging in specific symbiotic associations with plants, as well as bacterial endophytes and cyanobacteria (commonly referred to as blue-green algae), that possesses the capability to induce and enhance plant growth is classified as a plant growth-promoting bacteria (PGPB) (Daniel et al. 2022; Glick 2020). It has been observed that PGPB exert an influence on various aspects of plant development, including growth, yield, and nutrient assimilation, employing a diverse array of strategies (Ramakrishna et al. 2019). These bacteria are alternatively termed plant growth-promoting *Rhizobacteria* (PGPR) (Vessey 2003). PGPR functions by employing two principal mechanisms to support plant growth (Oleńska et al. 2020): (1) direct mechanisms: these involve the direct provision of resources and nutrients, such as nitrogen (N), phosphorus (P), iron (Fe), and other vital minerals, to plants (Jha and Saraf 2015). Additionally, PGPR can regulate plant hormone levels, encompassing cytokinins, gibberellins, indole-3-acetic acid, and ethylene, thereby further influencing plant growth (Fahad et al. 2015). (2) Indirect mechanisms, PGPR also operate through indirect pathways, such as acting as biocontrol agents to mitigate the detrimental impacts of various pathogens on plant growth and yield (Ahemad and Kibret 2014; Sivasakthi et al. 2014). Prominent examples of commonly employed PGPB encompass species like *Azospirillum* sp., *Azotobacter* sp., *Pseudomonas* sp., *Serratia* sp., *Streptomyces* sp., *Rhizobium* sp., *Bacillus* sp., and *Delftia* sp., which have been documented to significantly enhance plant growth (Tahami et al. 2017; Ruiu 2020; Bhattacharyya and Jha 2012).

7.2.7.1 Streptomyces

Streptomyces is a member of the bacterial family *Streptomycetaceae*, characterized by its aerobic nature, ability to form spores, and its classification as a gram-positive organism (Sousa and Olivares 2016). Streptomyces is a sporulating filamentous organism that can survive in unfavorable environmental conditions, which enables them to compete more effectively with other rhizosoil-dwelling microorganisms (Vurukonda et al. 2018, 2021). *Streptomycetes* produce lytic enzymes during their metabolic processes (Pacios-Michelena et al. 2021). The enzymes have the capability to enzymatically degrade recalcitrant organic polymers such as chitin and cellulose, rendering them into alternative saccharide moieties. These resultant saccharides serve as substrates that plant roots can readily associate with and uptake (Shivlata and Satyanarayana 2015).

7.2.7.2 Pseudomonas

Pseudomonas, a gram-negative bacterium categorized within the *Pseudomonadaceae* family (Gani et al. 2019), exhibits distinct characteristics. Members of this bacterial genus are generally anaerobic, adopt a rod-shaped morphology, and display aerobic behavior while refraining from spore formation (Fakhar et al. 2022). In the realm of plant growth-promoting *Rhizobacteria* (PGPR), *Pseudomonas* stands out due to its multifaceted attributes. These encompass root colonization, nitrogen fixation, and the synthesis of exopolysaccharides, siderophores, hydrogen cyanide (HCN), and phytohormones (Jha and Saraf 2015; Singh et al. 2022; Meena et al. 2020). Additionally, *Pseudomonas* facilitates the solubilization of essential nutrients like phosphorus, potassium, and zinc. It further demonstrates antioxidant capabilities and the capacity to adapt to stressful conditions (Singh et al. 2022). Within the soil rhizosphere, two notably robust strains of PGPR, namely, *Pseudomonas fluorescens* and *Pseudomonas aeruginosa*, have established a significant presence (Ganeshan and Kumar 2005).

7.2.7.3 Serratia

Serratia bacteria exhibit a gram-negative nature, characterized by their anaerobic metabolism and rod-shaped morphology belonging to the family *Yersiniaceae* (Friedrich et al. 2021). They are 1–5 μm long and do not produce spores (Khanna et al. 2013). *Serratia* is nitrogen-fixing PGPRs having the capacity to augment the mineral nitrogen concentration within the soil solution is facilitated by their nitrogen-fixing attributes, while concurrently inhibiting the leaching of nitrogen within the soil (Zaidi et al. 2017).

7.3 Classification

The classification of biofertilizers is represented in Fig. 7.2.

7.3.1 Nitrogen-Fixing Biofertilizers

7.3.1.1 Free-Living Nitrogen Fixation

The term "nitrogen-fixing free-living microorganisms" (commonly referred as "plant growth-promoting bacteria" or "PGPB") (Cortes et al. 2020) refers to naturally existing bacteria with the capability to perform nitrogen fixation. These microorganisms are found in the soil, rhizosphere, phyllosphere, and other plant-friendly

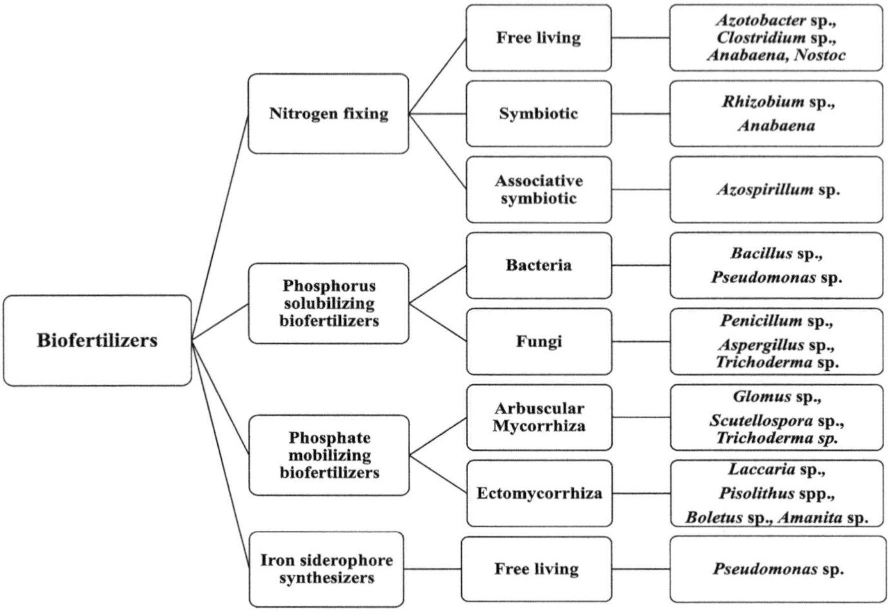

Fig. 7.2 Classification of biofertilizers

environments (Shridhar 2012). Without having any direct contact with plants, these free-living heterotrophic bacteria fix significant amounts of atmospheric nitrogen in the soil (Smercina et al. 2019). These organisms do not have their own source of energy (Cruz et al. 2022); hence, their energy acquisition occurs through the process of oxidizing organic compounds, which are either liberated by other organisms or originate from decomposition activities (Basu et al. 2021; Wu et al. 2020). Some free-living organisms have the ability to use inorganic compounds as a source of energy due to their chemolithotrophic nature (Kazemi et al. 2021). Examples are *Azotobacter*, *Clostridium*, *Beijerinckia*, *Rhodospirillum*, *Desulfovibrio*, *Anabaena*, *Nostoc*, *Oscillatoria*, etc. (Smercina et al. 2019; Rajkumar and Kurinjimalar 2021; Nosheen et al. 2021; Bon et al. 2021).

7.3.1.2 Symbiotic Nitrogen Fixation

In a symbiotic nitrogen fixation process termed mutualistic, plants offer a suitable habitat for bacteria and fixed carbon compounds. In reciprocation, these bacteria supply nitrogen to the plants (Mus et al. 2016). These bacteria allow plants to grow in soils with low levels of available nitrogen (Soumare et al. 2020). Numerous diazotrophic bacterial species have developed beyond surface colonization (Liu et al. 2020) to spread and multiply inside plant tissues forming nodules without causing harm to the plant (Shamseldin 2022; Rehmaan et al. 2022). Legumes are the primary crops that benefited by this process in agricultural systems (Kirova and Kocheva 2021). The microorganism that fixes nitrogen receives plant-derived sugars via photosynthesis, utilizing them to energize the nitrogen fixation process (Doty 2011). In exchange for these sugars, the microbes convert atmospheric nitrogen into a usable form for plants, aiding their growth (Wagner 2011). Examples are *Rhizobium*, *Frankia*, *Klebsiella*, *Anabaena*, etc. (Sheoran et al. 2021; Nosheen et al. 2021).

7.3.1.3 Associative Symbiotic Nitrogen Fixation

Associative symbiosis characterizes a relationship where one organism gains advantages from another without inflicting harm (Tejani et al. 2018). In this symbiotic relationship, bacteria establish proximity (loose mutualism) with plant roots to facilitate nitrogen fixation (Schloter and Hartmann 1998; Beattie 2007). The relationship is referred to as associative symbiosis, and the nitrogen fixation is termed as associative symbiotic nitrogen fixation (Hemati et al. 2021; Pankievicz et al. 2019). In this type of association, bacterial cells attach to plant roots without creating root nodules or other specialized structures (Knights et al. 2021). These microorganisms can be found on roots or in the space between the soil and roots in the rhizosphere region. Nitrogen is fixed by the bacteria and supplied to the roots, while carbohydrates are supplied by the roots to the bacteria for nourishment (Dellagi et al. 2020; Alawiye and Babalola 2019). Examples are *Azospirillum*, *Azotobacter*, *Beijerinckia*, *Enterobacter*, *Acetobacter*, etc. (Dellagi et al. 2020; Flora et al. 2021; Nosheen et al. 2021).

7.3.2 Phosphorus-Solubilizing Biofertilizers

Microbes primarily enhance soil phosphorus solubilization through the generation of both organic and inorganic acids. This process reduces soil pH, facilitating the dissolution of otherwise insoluble phosphates within the soil. As a result, these phosphates become accessible for uptake by plants (Khan et al. 2009a, b; Alori et al. 2017; Malhotra et al. 2018). Decomposition contributes to the creation of various organic acids, including lactic acid, citric acid, malic acid, gluconic acid, isovaleric acid, oxalic acid, tartaric acid, and fumaric acid (Liu et al. 2012; Rajput et al. 2013; Li et al. 2020; Kalayu 2019). These organic compounds have low molecular weights and use their hydroxyl and carboxyl groups to chelate the phosphorus-bound cations (Al, Ca, Fe) (Rawat et al. 2021; Kumar et al. 2018). This resulted into decrease in the rhizosphere's pH and releasing the bound phosphorus to plants in a form they can use (Sharma et al. 2011). The solubilization of phosphorus also involves other mechanisms, such as the generation of CO_2, the generation of H_2S, the production of siderophores, and the direct oxidation pathway (Rawat et al. 2021; Yadav 2022). Numerous saprophytic bacteria perform the process of phosphorus solubilization, viz., *Bacillus*, *Pseudomonas*, *Enterobacter*, *Klebsiella*, *Flavobacterium*, *Acinetobacter*, etc. (Mukhtar et al. 2020; Srinivasan et al. 2020); fungi, viz., *Penicillium*, *Aspergillus*, *Trichoderma*, *Acremonium*, etc. (Ichriani et al. 2018; Nosheen et al. 2021); and Actinomycetes, viz., *Streptomyces*, *Microbacterium*, etc. (Nandimath et al. 2017).

7.3.3 Phosphorus-Mobilizing Biofertilizers

Phosphate-mobilizing microorganisms facilitate the movement of previously immobile forms of phosphorous (Krishnaprabu 2020). These microorganisms, particularly mycorrhizal fungi, extract insoluble phosphate from soil layers and transport it to the root cortex (Asoegwu et al. 2020). Mycorrhizal fungi, also referred to as phosphate absorbers, form a heterogeneous taxonomic group that resides within plant root systems, establishing symbiotic relationships with plants (Melnichuk et al. 2021; Richardson et al. 2009). Another group, phosphorus-solubilizing microbes (PSMs), including fungi, infiltrate roots, enhancing root surface area (Bagyaraj et al. 2015). This action stimulates metabolic processes, promoting nutrient absorption by the roots (Etesami 2020). Additionally, phosphorus-solubilizing bacteria (PSB) occasionally exhibit phosphate mobilization capabilities (Nosheen et al. 2021). Mycorrhizal fungi engage in symbiosis with over 90% of vascular plant species, encompassing vital crops like maize, wheat, rice, and potato (Fig. 7.3) (Bianciotto et al. 2018; Hijri 2016; Bennett et al. 2012).

There exist two primary categories of mycorrhizae: ectomycorrhizal fungi (EM) and endomycorrhizal fungi (AM), both facilitating root colonization (Fig. 7.4; created in the BioRender) (Mohammadi et al. 2011).

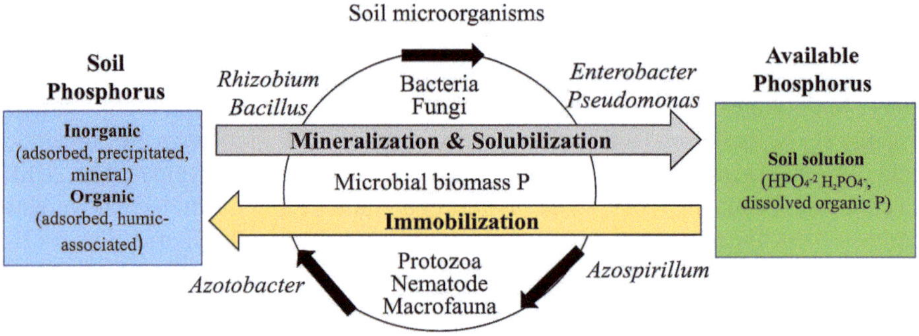

Fig. 7.3 Phosphorus mobilization cycle (source: Richardson and Simpson 2011, with modification)

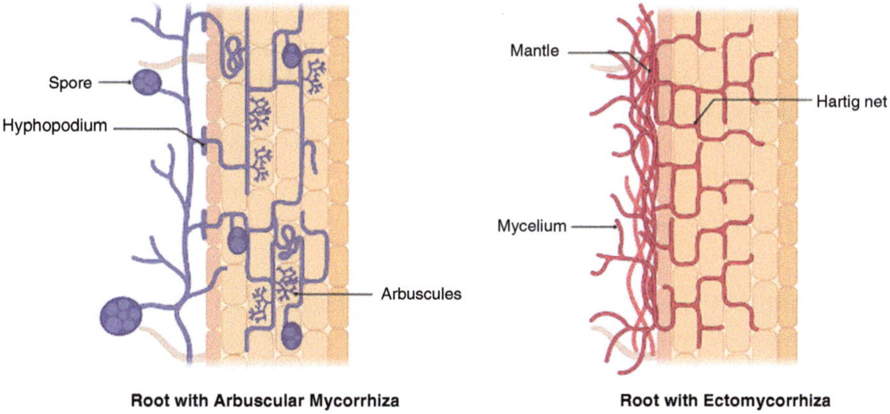

Fig. 7.4 Root colonization structure of endomycorrhiza and ectomycorrhiza

7.3.3.1 Ectomycorrhiza

Ectomycorrhizal fungi establish a multifaceted symbiotic association with plants, characterized by their avoidance of penetration into the host's cell walls (Plett and Martin 2011). It creates a structure around the root tips of fine roots called mantle (Rooney et al. 2009). Mantle is a dense layer of highly branched hyphae (Agerer 2006). These hyphae penetrate the external cortex of the plant's roots without infiltrating the epidermal cells (Harrison 2012; Brundrett and Tedersoo 2020). They create a structure known as the Hartig net that is entirely intercellular (Vaario and Matsushita 2021). Most ectomycorrhizal fungi belong to the *Basidiomycota* class (Pérez-Moreno et al. 2021). Arbutoid mycorrhiza is a type of ectomycorrhiza (De Valgaz et al. 2022).

7.3.3.2 Endomycorrhiza

Endomycorrhizal fungi, the most abundant variety, establish a presence across a diverse range of plant species, encompassing grasses, shrubs, certain trees, and various other plant types (Simon et al. 1993; Crush 1973). These fungi engage in a mutually beneficial partnership with plants, wherein the fungus infiltrates the cellular walls. In contrast, ectomycorrhizal associations are confined to the external layer of the root, without delving into the internal cellular structure (Luginbuehl and Oldroyd 2017). Endomycorrhiza fungi inhabits the roots and retrieve the nutrients that plants need (Harrison 2005). Such associations are caused by *Glomeromycota* members (Dhiman et al. 2022). Endomycorrhiza are of three different types: arbuscular, ericoid, and orchid (Sizonenko et al. 2020). Arbuscular mycorrhizal fungi (AMF) are widespread in almost all naturally occurring terrestrial communities and form obligate symbiotic associations with more than 80% of the vascular plants (Albornoz et al. 2021; Uwamungu et al. 2022). The term "arbuscular mycorrhiza

(AM)" was previously referred to as "vesicular-arbuscular mycorrhiza" (VAM) (Nair et al. 2022). Arbuscular, ericoid and orchid mycorrhizas are examples of endomycorrhizas (Valadares et al. 2021), while arbutoid mycorrhizas are classified as ectoendomycorrhizas (Islam et al. 2020).

7.3.4 Iron Siderophore Synthesis

Although iron (Fe) is abundant in the Earth's surface, it's not easily available to plants in soil (Schmidt 2003). Siderophores, small iron-binding proteins, help gather ferric iron (Fe^{3+}) on the planet's surface (Saha et al. 2016). Microbes producing siderophores have diverse uses like bioremediation, biocontrol, biosensors, and chelation (Ahmed and Holmström 2014; Saha et al. 2016), aiding soil mineral weathering (Ahmed and Holmström 2014). In low-iron soils, microbial siderophores transport iron to plants, promoting faster growth (Rajkumar et al. 2010). These molecules dissolve iron from minerals or organics when iron is scarce (Pérez-Montaño et al. 2014). Typically, siderophores bind 1:1 with Fe^{3+}, creating complexes absorbed by bacterial cells; there, Fe^{3+} transforms to Fe^{2+} and is released within the cell (Raymond et al. 1984; Sheldon et al. 2016). Studies highlight bacteria's role in plant siderophore production (Weyens et al. 2009), like *Pseudomonas aeruginosa* generating pyochelin (PCH) and pyoverdin (PVD) as iron collectors (Swayambhu et al. 2021).

7.4 Role of Biofertilizer in Soil Health

An optimally balanced soil creates an environment that supports and enhances the growth as well as the overall development of plants, which is essential for sustainable agriculture (Mahanty et al. 2017). In order to promote sustainable agriculture, biofertilizers are essential (Reddy et al. 2020). They support maintaining natural soil fertility while enhancing the soil's texture, structure, and water retention capacity (Ammar et al. 2022; Singh et al. 2020; Wang et al. 2015). A diverse range of microbial biofertilizers has proven effective in promoting the growth and maturation of plants (Basu et al. 2021). The use of biofertilizer promotes root development, high vegetative growth, high decomposition rate, endurance in the face of both living organism-related challenges (biotic stress) and nonliving environmental pressures (abiotic stress), and increased mineral and water uptake (Asoegwu et al. 2020; Sharma et al. 2021). Biofertilizers also make it easier to stop pathogenic microorganisms from growing close to crop plant's roots (Mahanty et al. 2017). They enhance the crop's resistance to pathogen-caused diseases by interactions with a plant's root systems (Rosier et al. 2018). In general, biofertilizers use organic material, which is considered a viable approach to maintaining soil health and fostering long-term sustainability while also reducing environmental pollution (Tyagi et al.

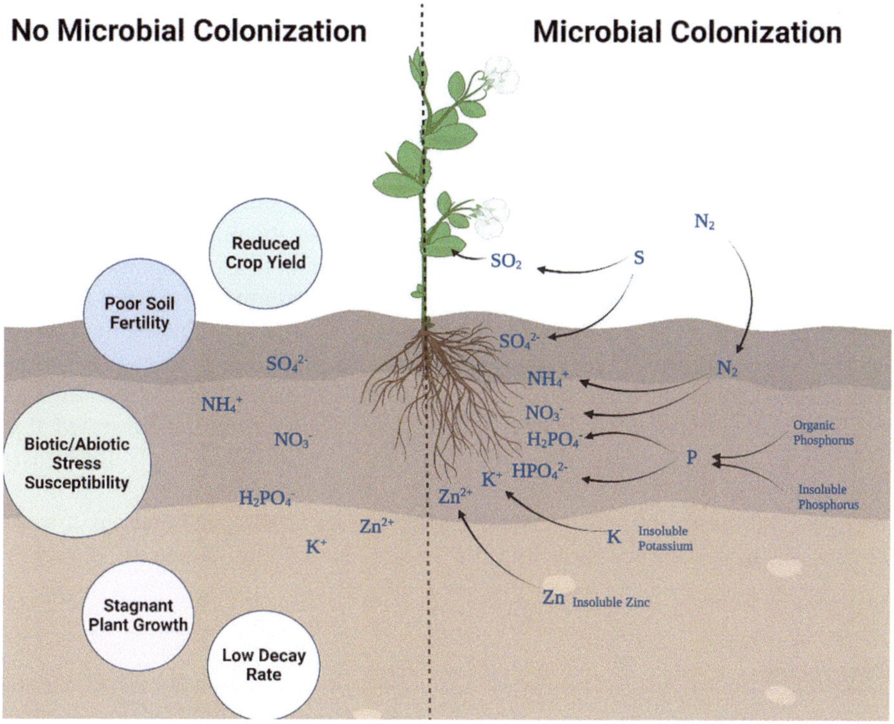

Fig. 7.5 The difference between microbial colonization and nonmicrobial colonization of plant (image was created in the BioRender)

2019; Atieno et al. 2020). The difference between microbial colonization and non-microbial colonization of plants is described in Fig. 7.5.

7.5 Conclusion

Biofertilizers represent natural fertilizers comprised of living microbial inoculants encompassing bacteria, algae, and fungi, either singularly or in synergistic combinations. These biological agents play a pivotal role in augmenting the accessibility of nutrients to plants. Recognized as an indispensable instrument for fostering sustainable agricultural practices, biofertilizers host a diversity of microorganisms. These organisms contribute to the fixation of atmospheric nitrogen, the solubilization and mobilization of essential elements such as phosphorus, potassium, and zinc, and the secretion of phytohormones. The impact of biofertilizers extends to root development, instigating profound enhancements in plant growth and, consequently, yielding superior crop outputs. A pressing concern in contemporary agriculture involves the overzealous utilization of chemical fertilizers. This excessive reliance has, in turn, brought about significant alterations in soil composition, shifting its pH balance toward acidity or alkalinity. This shift has led to a decline in the population of naturally occurring soil microorganisms, which has, in turn, impeded agricultural productivity. By incorporating biofertilizers, the scenario takes a positive turn. These bioagents not only foster enriched soil microbial activity but also elevate organic matter content. Consequently, this dual effect not only fosters healthier soil conditions but also engenders enhancements in the physical structure of the soil. This, in turn, amplifies water retention capacity, mitigates risks of erosion, and collectively contributes to overall soil health improvement. Different microbes use a different mechanisms to increase nutrient uptake, soil fertility, and crop yields. These microbes interact with plant root systems by colonizing the surface of roots. Biofertilizers play a pivotal role in mitigating the environmental ramifications associated with the use of chemical fertilizers, particularly concerning soil and water ecosystems. Comparative studies have consistently demonstrated that biofertilizers offer a superior long-term cost-effectiveness. Additionally, their environmental compatibility, efficiency, and productivity outshine those of chemical fertilizers. Moreover, biofertilizers cater to the needs of marginal and small-scale farmers, ensuring their accessibility to sustainable agricultural practices.

References

Aasfar A, Bargaz A, Yaakoubi K, Hilali A, Bennis I, Zeroual Y, Kadmiri I (2021) Nitrogen fixing *Azotobacter* species as potential soil biological enhancers for crop nutrition and yield stability. Front Microbiol 12:628379

Abideen Z, Waqif H, Munir N, El-Keblawy A, Hasnain M, Radicetti E, Haider G (2022) Algal-mediated nanoparticles, phycochar, and biofertilizers for mitigating abiotic stresses in plants: a review. Agronomy 12(8):1788

Agerer R (2006) Fungal relationships and structural identity of their ectomycorrhizae. Mycol Progr 5:67–107

Ahemad M, Kibret M (2014) Mechanisms and applications of plant growth promoting rhizobacteria: current perspective. J King Saud Univ-Sci 26(1):1–20

Ahmed E, Holmström SJ (2014) Siderophores in environmental research: roles and applications. Microb Biotechnol 7(3):196–208

Alawiye TT, Babalola OO (2019) Bacterial diversity and community structure in typical plant rhizosphere. Diversity 11(10):179

Albornoz FE, Orchard S, Standish RJ, Dickie IA, Bending GD, Hilton S, Ryan MH (2021) Evidence for niche differentiation in the environmental responses of co-occurring Mucoromycotinian fine root endophytes and Glomeromycotinian arbuscular mycorrhizal fungi. Microb Ecol 81:864–873

Allouzi MMA, Allouzi SMA, Keng ZX, Supramaniam CV, Singh A, Chong S (2022) Liquid biofertilizers as a sustainable solution for agriculture. Heliyon 8(12):e12609

Alori ET, Glick BR, Babalola OO (2017) Microbial phosphorus solubilization and its potential for use in sustainable agriculture. Front Microbiol 8:971

Ammar EE, Aioub AA, Elesawy AE, Karkour AM, Mouhamed MS, Amer AA, El-Shershaby NA (2022) Algae as bio-fertilizers: between current situation and future prospective. Saudi J Biol Sci 29:3083–3096

Ashour A, Afify AH (2023) Mini review: silicate bacteria and plant diseases. J Agric Chem Biotechnol 14(2):13–17

Asoegwu CR, Awuchi CG, Nelson K, Orji CG, Nwosu OU, Egbufor UC, Awuchi CG (2020) A review on the role of biofertilizers in reducing soil pollution and increasing soil nutrients. Himalayan J Agric 1:34–38

Atieno M, Herrmann L, Nguyen HT, Phan HT, Nguyen NK, Srean P, Lesueur D (2020) Assessment of biofertilizer use for sustainable agriculture in the Great Mekong Region. J Environ Manag 275:111300

Azizoglu U (2019) *Bacillus thuringiensis* as a biofertilizer and biostimulator: a mini-review of the little-known plant growth-promoting properties of Bt. Curr Microbiol 76(11):1379–1385

Bagyaraj DJ, Sharma MP, Maiti D (2015) Phosphorus nutrition of crops through arbuscular mycorrhizal fungi. Curr Sci 108:1288–1293

Barney BM, Plunkett MH (2022) Rnf1 is the primary electron source to nitrogenase in a high-ammonium-accumulating strain of *Azotobacter vinelandii*. Appl Microbiol Biotechnol 106:5051–5061

Bashan Y, De-Bashan LE (2010) How the plant growth-promoting bacterium *Azospirillum* promotes plant growth-a critical assessment. Adv Agron 108:77–136

Basit A, Shah ST, Ullah I, Muntha ST, Mohamed HI (2021) Microbe-assisted phytoremediation of environmental pollutants and energy recycling in sustainable agriculture. Arch Microbiol 203:5859–5885

Basu A, Prasad P, Das SN, Kalam S, Sayyed RZ, Reddy MS, El Enshasy H (2021) Plant growth promoting rhizobacteria (PGPR) as green bioinoculants: recent developments, constraints, and prospects. Sustainability 13(3):1140

Beattie GA (2007) Plant-associated bacteria: survey, molecular phylogeny, genomics and recent advances. In: Gnanamanickam SS (ed) Plant-associated bacteria. Springer, Dordrecht, pp 1–56

Becana M, Wienkoop S, Matamoros MA (2018) Sulfur transport and metabolism in legume root nodules. Front Plant Sci 9:1434

Behera BC, Mishra RR, Dutta SK, Thatoi HN (2014) Sulphur oxidising bacteria in mangrove ecosystem: a review. Afr J Biotechnol 13(29):2897–2907

Bello SK, Shobayo AB, Ibrahim MM, Alasinrin SY, Aliyu IA, Yusuf AA (2021) Biological nitrogen fixation contributes to soil productivity in tropical agroecologies. Nigerian J Soil Sci 31(1):1–14

Bennett AJ, Bending GD, Chandler D, Hilton S, Mills P (2012) Meeting the demand for crop production: the challenge of yield decline in crops grown in short rotations. Biol Rev 87(1):52–71

Bhardwaj D, Ansari MW, Sahoo RK, Tuteja N (2014) Biofertilizers function as key player in sustainable agriculture by improving soil fertility, plant tolerance and crop productivity. Microb Cell Factor 13:1–10

Bhatt G, Denny TP (2004) *Ralstonia solanacearum* iron scavenging by the siderophore staphyloferrin B is controlled by PhcA, the global virulence regulator. J Bacteriol 186(23):7896–7904

Bhattacharjee R, Dey U (2014) Biofertilizer, a way towards organic agriculture: a review. Afr J Microbiol Res 8(24):2332–2343

Bhattacharyya PN, Jha DK (2012) Plant growth-promoting rhizobacteria (PGPR): emergence in agriculture. World J Microbiol Biotechnol 28:1327–1350

Bianciotto V, Victorino I, Scariot V, Berruti A (2018) Arbuscular mycorrhizal fungi as natural biofertilizers: current role and potential for the horticulture industry. Acta Hortic 1191:207–216

Błaszkowski J, Czerniawska B, Wubet T, Schäfer T, Buscot F, Renker C (2008) *Glomus irregulare*, a new arbuscular mycorrhizal fungus in the Glomeromycota. Mycotaxon 106:247–267

Bohn HL, O'Connor GA, Strawn DG (2015) Soil chemistry. John Wiley & Sons

Bon IC, Salvatierra LM, Lario LD, Morató J, Pérez LM (2021) Prospects in cadmium-contaminated water management using free-living cyanobacteria (*Oscillatoria* sp.). Water 13(4):542

Brundrett MC, Tedersoo L (2020) Resolving the mycorrhizal status of important northern hemisphere trees. Plant Soil 454:3–34

Cassán F, Diaz-Zorita M (2016) *Azospirillum* sp. in current agriculture: from the laboratory to the field. Soil Biol Biochem 103:117–130

Cassán F, Coniglio A, López G, Molina R, Nievas S, de Carlan CLN, Mora V (2020) Everything you must know about *Azospirillum* and its impact on agriculture and beyond. Biol Fertil Soils 56:461–479

Chan C, Liao YY, Chiou TJ (2021) The impact of phosphorus on plant immunity. Plant Cell Physiol 62(4):582–589

Chaudhary IJ, Neeraj A, Siddiqui MA, Singh V (2020) Nutrient management technologies and the role of organic matrix-based slow-release biofertilizers for agricultural sustainability: a review. Agric Rev 41(1):1–13

Chaudhary S, Dhanker R, Kumar R, Goyal S (2022) Importance of legumes and role of sulphur oxidizing bacteria for their production: a review. Legume Res Int J 45(3):275–284

Chen G, Zhu H, Zhang Y (2003) Soil microbial activities and carbon and nitrogen fixation. Res Microbiol 154(6):393–398

Cortes AD, Opulencia RB, Aggangan NS (2020) Characterization of plant growth promoting diazotrophic bacteria isolated from cacao (*Theobroma cacao* l.) rhizosphere treated with bamboo biochar and arbuscular mycorrhizal fungi. Philipp J Sci 149(4):1063–1070

Crush JR (1973) Significance of endomycorrhizas in tussock grassland in Otago, New Zealand. N Z J Bot 11(4):645–660

Cruz JA, Tubana BS, Fultz LM, Dalen MS, Ham JH (2022) Identification and profiling of silicate-solubilizing bacteria for plant growth-promoting traits and rhizosphere competence. Rhizosphere 23:100566

Daniel AI, Fadaka AO, Gokul A, Bakare OO, Aina O, Fisher S, Klein A (2022) Biofertilizer: the future of food security and food safety. Microorganisms 10(6):1220

Das I, Pradhan M (2016) Potassium-solubilizing microorganisms and their role in enhancing soil fertility and health. In: Meena V, Maurya B, Verma J, Meena R (eds) Potassium solubilizing microorganisms for sustainable agriculture. Springer, New Delhi, pp 281–291

De Corato U (2020) Agricultural waste recycling in horticultural intensive farming systems by on-farm composting and compost-based tea application improves soil quality and plant health: a review under the perspective of a circular economy. Sci Total Environ 738:139840

De La Providencia IE, Fernandez F, Declerck S (2007) Hyphal healing mechanism in the arbuscular mycorrhizal fungi *Scutellospora reticulata* and *Glomus clarum* differs in response to severe physical stress. FEMS Microbiol Lett 268(1):120–125

de Valgaz APF, Barcos-Arias M, Naranjo-Morán J, Tapia DP, Moreira-Gómez R (2022) Ericaceous plants: a review for the bioprospecting of ericoid Mycorrhizae from Ecuador. Diversity 14(8):648

Dellagi A, Quillere I, Hirel B (2020) Beneficial soil-borne bacteria and fungi: a promising way to improve plant nitrogen acquisition. J Exp Bot 71(15):4469–4479

Devi D (2022) Chapter-5 Soil microorganisms and their function to maintain soil fertility. Chief Editor Dr Neeraj Kumar, pp. 61–75.

Devi R, Kaur T, Kour D, Rana KL, Yadav A, Yadav AN (2020) Beneficial fungal communities from different habitats and their roles in plant growth promotion and soil health. Microb Biosyst 5(1):21–47

Dhiman M, Sharma L, Kaushik P, Singh A, Sharma MM (2022) Mycorrhiza: an ecofriendly bio-tool for better survival of plants in nature. Sustainability 14(16):10220

Dordipour E, Bastamikojour Z, Baranimotlagh M, Gharanjiki A, Olamaee M (2018) Effect of low-molecular-weight organic acids on growth and phosphorus uptake via maize (*Zea mays* L. cv. SC704) in some soils of Golestan province. J Agric Eng Soil Sci Agric Mech (Sci J Agric) 40(2):137–152

Doty SL (2011) Nitrogen-fixing endophytic bacteria for improved plant growth. In: Maheshwari D (ed) Bacteria in agrobiology: plant growth responses. Springer, Berlin, Heidelberg, pp 183–199

El Barnossi A, Housseini AI (2023) Characterization of the microbiological effects of pomegranate, banana, and mandarin peels on water under laboratory conditions. Heliyon 9(2):e13402

Etesami H (2020) Enhanced phosphorus fertilizer use efficiency with microorganisms. In: Meena R (ed) Nutrient dynamics for sustainable crop production. Springer, Singapore, pp 215–245

Etesami H, Jeong BR (2022) Biodissolution of silica by rhizospheric silicate-solubilizing bacteria. In: Etesami H, Al Saeedi AH, El-Ramady H, Fujita M, Pessarakli M, Anwar Hossain M (eds) Silicon and nano-silicon in environmental stress management and crop quality improvement. Academic Press, pp 265–276

Etesami H, Emami S, Alikhani HA (2017) Potassium solubilizing bacteria (KSB): mechanisms, promotion of plant growth, and future prospects. a review. J Soil Sci Plant Nutr 17(4):897–911

Etesami H, Jeong BR, Glick BR (2021) Contribution of arbuscular mycorrhizal fungi, phosphate–solubilizing bacteria, and silicon to P uptake by plant. Front Plant Sci 12:699618

Fahad S, Hussain S, Bano A, Saud S, Hassan S, Shan D, Huang J (2015) Potential role of phytohormones and plant growth-promoting rhizobacteria in abiotic stresses: consequences for changing environment. Environ Sci Pollut Res 22:4907–4921

Fakhar A, Gul B, Gurmani AR, Khan SM, Ali S, Sultan T, Chaudhary HJ, Rafique M, Rizwan M (2022) Heavy metal remediation and resistance mechanism of *Aeromonas, Bacillus*, and *Pseudomonas*: a review. Crit Rev Environ Sci Technol 52(11):1868–1914

Fasusi OA, Cruz C, Babalola OO (2021) Agricultural sustainability: microbial biofertilizers in rhizosphere management. Agriculture 11(2):163

Flora Y, Rabha P, Shinde A, Jha P, Jobby R (2021) Non-symbiotic bacteria for soil nitrogen fortification. Sustain Agric Rev 52:417–435

Friedrich I, Bodenberger B, Neubauer H, Hertel R, Daniel R (2021) Down in the pond: isolation and characterization of a new *Serratia marcescens* strain (LVF3) from the surface water near frog's lettuce (*Groenlandia densa*). PLoS One 16(11):e0259673

Ganeshan G, Kumar MA (2005) Pseudomonas fluorescens, a potential bacterial antagonist to control plant diseases. J Plant Interact 1:123–134

Gani M, Rao S, Miller M, Scoular S (2019) *Pseudomonas mendocina* bacteremia: a case study and review of literature. Am J Case Rep 20:453

Glick BR (2020) Introduction to plant growth-promoting bacteria. Springer International Publishing, Cham, Switzerland, pp 1–28

Golchin A, Baldock JA, Oades JM (2018) A model linking organic matter decomposition, chemistry, and aggregate dynamics. In: Soil processes and the carbon cycle. CRC Press, pp 245–266

Griffiths HM, Ashton LA, Parr CL, Eggleton P (2021) The impact of invertebrate decomposers on plants and soil. New Phytol 231(6):2142–2149

Gupta G, Panwar J, Akhtar MS, Jha PN (2012) Endophytic nitrogen-fixing bacteria as biofertilizer. In: Lichtfouse E (ed) Sustainable agriculture reviews. Springer, Dordrecht, pp 183–221

Gupta G, Parihar SS, Ahirwar NK, Snehi SK, Singh V (2015) Plant growth promoting rhizobacteria (PGPR): current and future prospects for development of sustainable agriculture. J Microb Biochem Technol 7(2):96–102

Gupta A, Bano A, Rai S, Dubey P, Khan F, Pathak N, Sharma S (2021) Plant Growth Promoting Rhizobacteria (PGPR): a sustainable agriculture to rescue the vegetation from the effect of biotic stress: a review. Lett Appl Nanobiosci 10:2459–2465

Hanif M, Ashraf Z, Bashir S, Riaz F, Amanat R, Yousaf N, Sarwar S (2023) Ectomycorrhizal fungi as biofertilizers in forestry. In: Arbuscular Mycorrhizal fungi in agriculture—new insights. IntechOpen. https://doi.org/10.5772/intechopen.110090

Harding KJ, Turk-Kubo KA, Mak EWK, Weber PK, Mayali X, Zehr JP (2022) Cell-specific measurements show nitrogen fixation by particle-attached putative non-cyanobacterial diazotrophs in the North Pacific Subtropical Gyre. Nat Commun 13(1):6979

Harman GE (2006) Overview of mechanisms and uses of *Trichoderma* spp. Phytopathology 96(2):190–194

Harman GE, Howell CR, Viterbo A, Chet I, Lorito M (2004) *Trichoderma* species-opportunistic avirulent plant symbionts. Nat Rev Microbiol 2(1):43–56

Haroon M, Khan ST, Malik A (2022) Zinc-Solubilizing Bacteria: an option to increase zinc uptake by plants. In: Khan ST, Malik A (eds) Microbial biofertilizers and micronutrient availability. Springer, Cham, pp 207–238

Harrison MJ (2005) Signaling in the arbuscular mycorrhizal symbiosis. Ann Rev Microbiol 59:19–42

Harrison MJ (2012) Cellular programs for arbuscular mycorrhizal symbiosis. Curr Opin Plant Biol 15(6):691–698

Hasanuzzaman M, Bhuyan MB, Nahar K, Hossain MS, Mahmud JA, Hossen MS, Fujita M (2018) Potassium: a vital regulator of plant responses and tolerance to abiotic stresses. Agronomy 8(3):31

Hawkesford MJ, Cakmak I, Coskun D, De Kok LJ, Lambers H, Schjoerring JK, White PJ (2023) Functions of macronutrients. In: Rengel Z, Cakmak I, White PJ (eds) Marschner's mineral nutrition of plants. Academic Press, San Diego, pp 201–281

Hayat R, Ali S, Amara U, Khalid R, Ahmed I (2010) Soil beneficial bacteria and their role in plant growth promotion: a review. Ann Microbiol 60:579–598

Hemati A, Delangiz N, Lajayer BA, Ghorbanpour M (2021) A comprehensive review of commercial production of microbial fertilizers. In: Agri-based bioeconomy, 1st edn. CRC Press, pp 173–178

Hijri M (2016) Analysis of a large dataset of mycorrhiza inoculation field trials on potato shows highly significant increases in yield. Mycorrhiza 26(3):209–214

Hill GT, Mitkowski NA, Aldrich-Wolfe L, Emele LR, Jurkonie DD, Ficke A, Nelson EB (2000) Methods for assessing the composition and diversity of soil microbial communities. Appl Soil Ecol 15(1):25–36

Ichriani GI, Syehfani N, Y. and Handayanto, E. (2018) Formulation of biochar-compost and phosphate solubilizing fungi from oil palm empty fruit bunch to improve growth of maize in an ultisol of central kalimantan. J Ecol Eng 19:45–55

Islam W, Noman A, Naveed H, Huang Z, Chen HY (2020) Role of environmental factors in shaping the soil microbiome. Environ Sci Pollut Res 27:41225–41247

Itelima JU, Bang WJ, Onyimba IA, Sila MD, Egbere OJ (2018) Bio-fertilizers as key player in enhancing soil fertility and crop productivity: a review. Direct Res J Agric Food Sci 6(3):73–83

Jaiswal DK, Verma JP, Prakash S, Meena VS, Meena RS (2016) Potassium as an important plant nutrient in sustainable agriculture: a state of the art. In: Meena VS, Maurya BR, Verma PJ, Meena RS (eds) Potassium solubilizing microorganisms for sustainable agriculture. Springer, New Delhi, pp 21–29

Jangir CK, Kumar S, Meena RS (2019) Significance of soil organic matter to soil quality and evaluation of sustainability. In: Meena RS (ed) Sustainable agriculture. Scientific Publisher, Jodhpur, pp 357–381

Javed Z, Tripathi GD, Mishra M, Dashora K (2021) Actinomycetes—the microbial machinery for the organic-cycling, plant growth, and sustainable soil health. Biocatal Agric Biotechnol 31:101893

Jha CK, Saraf M (2015) Plant growth promoting rhizobacteria (PGPR): a review. J Agric Res Dev 5(2):108–119

Jnawali AD, Ojha RB, Marahatta S (2015) Role of azotobacter in soil fertility and sustainability—a review. Adv Plants Agric Res 2(6):1–5

Johns C (2017) Living soils: the role of microorganisms in soil health. Fut Direct Int:1–7

Kalayu G (2019) Phosphate solubilizing microorganisms: promising approach as biofertilizers. Int J Agron 2019:1–7. https://doi.org/10.1155/2019/4917256

Kamran S, Shazia A, Shahida H (2010) Growth responses of *Triticum aestivum* to plant growth promoting Rhizobacteria used as a biofertilizer. Res J Microbiol 5(10):1022–1030

Kamshybayeva GK, Kossalbayev BD, Sadvakasova AK, Zayadan BK, Bozieva AM, Dunikov D, Allakhverdiev SI (2022) Strategies and economic feasibilities in cyanobacterial hydrogen production. Int J Hydrog Energy 47(69):29661–29684

Kazemi E, Soofiyani SR, Ahangari H, Eyvazi S, Hejazi MS, Tarhriz V (2021) Chemolithotroph bacteria: from biology to application in medical sciences. Crescent J Med Biol Sci 8(2):81–89
Khan AA, Jilani G, Akhtar MS, Naqvi SMS, Rasheed M (2009a) Phosphorus solubilizing bacteria: occurrence, mechanisms and their role in crop production. J Agric Biol Sci 1(1):48–58
Khan MS, Zaidi A, Wani PA (2009b) Role of phosphate solubilizing microorganisms in sustainable agriculture—a review. In: Lichtfouse E, Navarrete M, Debaeke P, Véronique S, Alberola C (eds) Sustainable agriculture. Springer, Dordrecht, pp 551–570
Khanna A, Khanna M, Aggarwal A (2013) Serratia marcescens—a rare opportunistic nosocomial pathogen and measures to limit its spread in hospitalized patients. J Clin Diagn Res 7(2):243–246
Kirk PM, Cannon PF, Minter DW, Stalpers JA (2008) Dictionary of the fungi, 10th edn. CABI, Wallingford, UK, p 287
Kirova E, Kocheva K (2021) Physiological effects of salinity on nitrogen fixation in legumes—a review. J Plant Nutr 44(17):2653–2662
Knights HE, Jorrin B, Haskett TL, Poole PS (2021) Deciphering bacterial mechanisms of root colonization. Environ Microb Rep 13(4):428–444
Kour D, Rana KL, Kaur T, Yadav N, Halder SK, Yadav AN, Saxena AK (2020a) Potassium solubilizing and mobilizing microbes: biodiversity, mechanisms of solubilization, and biotechnological implication for alleviations of abiotic stress. In: Rastegari AS, Yadav AN, Yadav N (eds) New and future developments in microbial biotechnology and bioengineering. Elsevier, pp 177–202
Kour D, Rana KL, Yadav AN, Yadav N, Kumar M, Kumar V, Saxena AK (2020b) Microbial biofertilizers: bioresources and eco-friendly technologies for agricultural and environmental sustainability. Biocatal Agric Biotechnol 23:101487
Krishnaprabu S (2020) Liquid microbial consortium: a potential tool for sustainable soil health. J Pharmacogn Phytochem 9(2):2191–2199
Kumar VV (2018) Biofertilizers and biopesticides in sustainable agriculture. In: Meena V (ed) Role of rhizospheric microbes in soil. Springer, Singapore, pp 377–398
Kumar M, Kumar K (2019) Role of bio-fertilizers in vegetables production: a review. J Pharmacogn Phytochem 8(1):328–334
Kumar A, Kumar A, Patel H (2018) Role of microbes in phosphorus availability and acquisition by plants. Int J Curr Microbiol App Sci 7(5):1344–1347
Kumar M, Zeyad MT, Choudhary P, Paul S, Chakdar H, Rajawat MVS (2020) Thiobacillus. In: Amaresan N, Senthil Kumar M, Annapurna K, Kumar K, Sankaranarayanan A (eds) Beneficial microbes in agro-ecology. Academic Press, pp 545–557
Kushwaha P, Kashyap PL, Pandiyan K, Bhardwaj AK (2020) Zinc-solubilizing microbes for sustainable crop production: current understanding, opportunities, and challenges. In: Solanki M, Kashyap P, Kumari B (eds) Phytobiomes: current insights and future vistas. Springer, Singapore, pp 281–298
Lal R (2009) Soil degradation as a reason for inadequate human nutrition. Food Security 1:45–57
Lee YS, Yuan M, Cai R, Lim K, Minteer SD (2020) Nitrogenase bioelectrocatalysis: ATP-independent ammonia production using a redox polymer/MoFe protein system. ACS Catal 10(12):6854–6861
Leghari SJ, Wahocho NA, Laghari GM, HafeezLaghari A, MustafaBhabhan G, HussainTalpur K, Bhutto TA, Wahocho SA, Lashari AA (2016) Role of nitrogen for plant growth and development: a review. Adv Environ Biol 10(9):209–219
Li WL, Wang JF, Lv Y, Dong HJ, Wang LL, He T, Li QS (2020) Improving cadmium mobilization by phosphate-solubilizing bacteria via regulating organic acids metabolism with potassium. Chemosphere 244:125475
Li M, Fang A, Yu X, Zhang K, He Z, Wang C, Yan Q (2021) Microbially-driven sulfur cycling microbial communities in different mangrove sediments. Chemosphere 273:128597

Liang JL, Liu J, Jia P, Yang TT, Zeng QW, Zhang SC, Liao B, Shu WS, Li JT (2020) Novel phosphate-solubilizing bacteria enhance soil phosphorus cycling following ecological restoration of land degraded by mining. ISME J 14(6):1600–1613

Liu W, He Y, Zhang K, Fan J, Cao H (2012) Isolation, identification and characterization of a strain of phosphate-solubilizing bacteria from red soil. Acta Microbiol Sin 52(3):326–333

Liu W, Bai X, Li Y, Min J, Kong Y, Hu X (2020) CheY1 and CheY2 of *Azorhizobium caulinodans* ORS571 regulate chemotaxis and competitive colonization with the host plant. Appl Environ Microbiol 86(15):e00599–e00520

Luginbuehl LH, Oldroyd GE (2017) Understanding the arbuscule at the heart of endomycorrhizal symbioses in plants. Curr Biol 27(17):R952–R963

Mącik M, Gryta A, Frąc M (2020) Biofertilizers in agriculture: an overview on concepts, strategies and effects on soil microorganisms. Adv Agron 162:31–87

Mahanty T, Bhattacharjee S, Goswami M, Bhattacharyya P, Das B, Ghosh A, Tribedi P (2017) Biofertilizers: a potential approach for sustainable agriculture development. Environ Sci Pollut Res 24:3315–3335

Mahapatra S, Yadav R, Ramakrishna W (2022) Bacillus subtilis impact on plant growth, soil health and environment: Dr. Jekyll and Mr. Hyde. J Appl Microbiol 132(5):3543–3562

Maleva M, Borisova G, Koshcheeva O, Sinenko O (2017) Biofertilizer based on silicate solubilizing bacteria improves photosynthetic function of *Brassica juncea*. AGROFOR Int J 2:13–19

Malhotra H, Vandana SS, Pandey R (2018) Phosphorus nutrition: plant growth in response to deficiency and excess. In: Hasanuzzaman M, Fujita M, Oku H, Nahar K, Hawrylak-Nowak B (eds) Plant nutrients and abiotic stress tolerance. Springer, Singapore, pp 171–190

Meena VS, Maurya BR, Verma JP (2014) Does a rhizospheric microorganism enhance K^+ availability in agricultural soils? Microbiol Res 169:337–347

Meena VS, Maurya BR, Verma JP, Aeron A, Kumar A, Kim K, Bajpai VK (2015) Potassium solubilizing rhizobacteria (KSR): isolation, identification, and K-release dynamics from waste mica. Ecol Eng 81:340–347

Meena M, Swapnil P, Divyanshu K, Kumar S, Tripathi YN, Zehra A, Marwal A, Upadhyay RS (2020) PGPR-mediated induction of systemic resistance and physiochemical alterations in plants against the pathogens: current perspectives. J Basic Microbiol 60(10):828–861

Melnichuk TN, Abdurashytov SF, Egovtseva AY, Abdurashytova ER, Turin EN, Gongalo AA, Zubochenko AA, Pashtetskiy VS (2021) Taxonomic structure of the southern chernozem under application of biological preparations and different farming systems. In: IOP conference series: earth and environmental science, vol 937, No. 3. IOP Publishing, p 032011

Miransari M (2013) Soil microbes and the availability of soil nutrients. Acta Physiol Plantarum 35(11):3075–3084

Mohammadi K, Khalesro S, Sohrabi Y, Heidari G (2011) A review: beneficial effects of the mycorrhizal fungi for plant growth. J Appl Environ Biol Sci 1(9):310–319

Mohapatra BR, Gould WD, Dinardo O, Koren DW (2008) An overview of the biochemical and molecular aspects of microbial oxidation of inorganic sulfur compounds. Clean-Soil Air Water 36:823–829

Mousavi SR, Galavi M, Rezaei M (2013) Zinc (Zn) importance for crop production—a review. Int J Agron Plant Prod 4(1):64–68

Mukherjee R, Sen S (2021) Role of biological nitrogen fixation (BNF) in sustainable agriculture: a review. Int J Adv Life Sci Res 4(3):01–07

Mukhtar S, Zareen M, Khaliq Z, Mehnaz S, Malik KA (2020) Phylogenetic analysis of halophyte-associated rhizobacteria and effect of halotolerant and halophilic phosphate-solubilizing biofertilizers on maize growth under salinity stress conditions. J Appl Microbiol 128(2):556–573

Mus F, Crook MB, Garcia K, Garcia Costas A, Geddes BA, Kouri ED, Paramasivan P, Ryu MH, Oldroyd GED, Poole PS, Udvardi MK, Voigt CA, Ané JM, Peters JW (2016) Symbiotic nitrogen fixation and the challenges to its extension to nonlegumes. Appl Environ Microbiol 82(13):3698–3710

Nair A, Rao AS, Bhanu L, More VS, Anantharaju KS, More SS (2022) *Arbuscular mycorrhizae*, a treasured symbiont to agriculture. Elsevier, pp 45–62

Nandimath AP, Karad DD, Gupta SG, Kharat AS (2017) Consortium inoculum of five thermo-tolerant phosphate solubilizing Actinomycetes for multipurpose biofertilizer preparation. Iran J Microbiol 9(5):295

Nanjundappa A, Bagyaraj DJ, Saxena AK, Kumar M, Chakdar H (2019) Interaction between arbuscular mycorrhizal fungi and *Bacillus* spp. in soil enhancing growth of crop plants. Fungal Biol Biotechnol 6:1–10

Narayan OP, Kumar P, Yadav B, Dua M, Johri AK (2022) Sulfur nutrition and its role in plant growth and development. Plant Signal Behav:2030082

Naureen Z, Aqeel M, Hassan MN, Gilani SA, Bouqellah N, Mabood F, Hussain J, Hafeez FY (2015) Isolation and screening of silicate bacteria from various habitats for biological control of phytopathogenic fungi. Am J Plant Sci 6(18):2850–2859

Naveed M, Mehboob I, Shaker MA, Hussain MB, Farooq M (2015) Biofertilizers in Pakistan: initiatives and limitations. Int J Agric Biol 17(3):411–420

Naveed M, Khalid H, Ayub MA, Rehman MZU, Rizwan M, Rasul A, Haq MAU (2020) Biofortification of cereals with zinc and iron: recent advances and future perspectives. Resources Use Efficiency Agric:615–646

Naz I, Ahmad H, Khokhar SN, Khan K, Shah AH (2016) Impact of zinc solubilizing bacteria on zinc contents of wheat. Am Eurasian J Agric Environ Sci 16:449–454

Nazar R, Iqbal N, Masood A, Syeed S, Khan NA (2011) Understanding the significance of sulfur in improving salinity tolerance in plants. Environ Exp Bot 70:80–87

Nosheen S, Ajmal I, Song Y (2021) Microbes as biofertilizers, a potential approach for sustainable crop production. Sustainability 13(4):1868

Ohyama T (2010) Nitrogen as a major essential element of plants. Nitrogen Assim Plants 37:1–17

Oleńska E, Małek W, Wójcik M, Swiecicka I, Thijs S, Vangronsveld J (2020) Beneficial features of plant growth-promoting rhizobacteria for improving plant growth and health in challenging conditions: a methodical review. Sci Total Environ 743:140682

Pacios-Michelena S, Aguilar Gonzalez CN, Alvarez-Perez OB, Rodriguez-Herrera R, Chávez-González M, Valdes RA, Valdes JAA, Salas MG, Ilyina A (2021) Application of Streptomyces antimicrobial compounds for the control of phytopathogens. Front Sustain Food Syst 5:696518

Pandey R, Zinta G, AbdElgawad H, Ahmad A, Jain V, Janssens IA (2015) Physiological and molecular alterations in plants exposed to high [CO_2] under phosphorus stress. Biotechnol Adv 33:303–316

Pankievicz V, Irving TB, Maia LG, Ané JM (2019) Are we there yet? The long walk towards the development of efficient symbiotic associations between nitrogen-fixing bacteria and non-leguminous crops. BMC Biol 17(1):1–17

Panwar J, Laxmi V (2005) Biological nitrogen fixation in pulses and cereals. Dev Physiol Biochem Mol Biol Plants 1:125

Pérez-Montaño F, Alías-Villegas C, Bellogín RA, del Cerro P, Espuny MR, Jiménez-Guerrero I, López-Baena FJ, Ollero FJ, Cubo T (2014) Plant growth promotion in cereal and leguminous agricultural important plants: from microorganism capacities to crop production. Microbiol Res 169:325–336

Pérez-Moreno J, Guerin-Laguette A, Rinaldi AC, Yu F, Verbeken A, Hernández-Santiago F, Martínez-Reyes M (2021) Edible mycorrhizal fungi of the world: what is their role in forest sustainability, food security, biocultural conservation and climate change? Plants People Planet 3(5):471–490

Pineda A, Zheng SJ, van Loon JJ, Pieterse CM, Dicke M (2010) Helping plants to deal with insects: the role of beneficial soil-borne microbes. Trends Plant Sci 15(9):507–514

Plett JM, Martin F (2011) Blurred boundaries: lifestyle lessons from ectomycorrhizal fungal genomes. Trends Genet 27(1):14–22

Poudel S, Colman DR, Fixen KR, Ledbetter RN, Zheng Y, Pence N, Seefeldt LC, Peters JW, Harwood CS, Boyd ES (2018) Electron transfer to nitrogenase in different genomic and metabolic backgrounds. J Bacteriol 200(10):e00757–e00717
Putrie RFW, Aryantha INP, Iriawati I, Antonius S (2020) Diversity of endophytic and rhizosphere bacteria from pineapple (Ananas comosus) plant in semi-arid ecosystem. Biodivers J Biol Divers 21(7):3084–3093
Rajkumar R, Kurinjimalar C (2021) Microbes and plant mineral nutrition. In: Soni R, Suyal DC, Bhargava P, Goel R (eds) Microbiological activity for soil and plant health management. Springer, Singapore, pp 111–132
Rajkumar M, Ae N, Prasad MNV, Freitas H (2010) Potential of siderophore-producing bacteria for improving heavy metal phytoextraction. Trends Biotechnol 28(3):142–149
Rajput MS, Naresh Kumar G, Rajkumar S (2013) Repression of oxalic acid-mediated mineral phosphate solubilization in rhizospheric isolates of Klebsiella pneumoniae by succinate. Arch Microbiol 195:81–88
Ramakrishna W, Yadav R, Li K (2019) Plant growth promoting bacteria in agriculture: two sides of a coin. Appl Soil Ecol 138:10–18
Rana KL, Kour D, Kaur T, Devi R, Yadav AN, Yadav N, Dhaliwal HS, Saxena AK (2020) Endophytic microbes: biodiversity, plant growth-promoting mechanisms and potential applications for agricultural sustainability. Antonie Van Leeuwenhoek 113:1075–1107
Rasulov X, Uzbekistan TSAU (2019) Principles and characteristics of sustainable soils. In: Устойчивое развитие сельского хозяйства Международная онлайн интернет конференция, 14–15 апреля 2019 г. Материалы конференции.–М. Изд. ВНИИА, 2019, 262 c, vol 14, p 235
Raturi G, Sharma Y, Rana V, Thakral V, Myaka B, Salvi P, Singh M, Dhar H, Deshmukh R (2021) Exploration of silicate solubilizing bacteria for sustainable agriculture and silicon biogeochemical cycle. Plant Physiol Biochem 166:827–838
Raturi G, Sharma Y, Mandlik R, Kumawat S, Rana N, Dhar H, Tripathi DK, Sonah H, Sharma TR, Deshmukh R (2022) Genomic landscape highlights molecular mechanisms involved in silicate solubilization, stress tolerance, and potential growth-promoting activity of bacterium Enterobacter sp. LR6 Cells 11(22):3622
Rawat J, Sanwal P, Saxena J (2018) Towards the mechanisms of nutrient solubilization and fixation in soil system. In: Meena V (ed) Role of rhizospheric microbes in soil. Springer, Singapore, pp 229–257
Rawat P, Das S, Shankhdhar D, Shankhdhar SC (2021) Phosphate-solubilizing microorganisms: mechanism and their role in phosphate solubilization and uptake. J Soil Sci Plant Nutr 21:49–68
Raymond KN, Muller G, Matzanke BF (1984) Complexation of iron by siderophores a review of their solution and structural chemistry and biological function. Top Curr Chem 123:49–102
Reddy GC, Goyal RK, Puranik S, Waghmar V, Vikram KV, Sruthy KS (2020) Biofertilizers toward sustainable agricultural development. In: Varma A, Tripathi S, Prasad R (eds) Plant microbe symbiosis. Springer, Cham, pp 115–128
Rehmaan IU, Jan B, Khan NF, Islam T, Rehman S, Sheergojri IA, Rashid I, War AF, Nanda SA, Wani AH (2022) Nitrogen biofertilizers: role in sustainable agriculture. In: Yousuf PY, Shabir PA, Hakeem KR (eds) Advances in plant nitrogen metabolism. CRC Press, Boca Raton, pp 170–184
Reineke W, Schlömann M (2023) Cycles of sulfur, iron and manganese. Springer Spektrum, Berlin, Heidelberg, pp 307–329
Richardson AE, Simpson RJ (2011) Soil microorganisms mediating phosphorus availability update on microbial phosphorus. Plant Physiol 156:989–996
Richardson AE, Barea JM, McNeill AM, Prigent-Combaret C (2009) Acquisition of phosphorus and nitrogen in the rhizosphere and plant growth promotion by microorganisms. Plant Soil 321:305–339
Rooney DC, Killham K, Bending GD, Baggs E, Weih M, Hodge A (2009) Mycorrhizas and biomass crops: opportunities for future sustainable development. Trends Plant Sci 14(10):542–549

Rosier A, Medeiros FH, Bais HP (2018) Defining plant growth promoting rhizobacteria molecular and biochemical networks in beneficial plant-microbe interactions. Plant Soil 428:35–55

Ruiu L (2020) Plant-growth-promoting bacteria (PGPB) against insects and other agricultural pests. Agronomy 10(6):861

Saha M, Sarkar S, Sarkar B, Sharma BK, Bhattacharjee S, Tribedi P (2016) Microbial siderophores and their potential applications: a review. Environ Sci Pollut Res 23:3984–3999

Saha I, Datta S, Biswas D (2020) Exploring the role of bacterial extracellular polymeric substances for sustainable development in agriculture. Curr Microbiol 77:3224–3239

Sahu D, Priyadarshani I, Rath B (2012) Cyanobacteria—as potential biofertilizer. CIBTech J Microbiol 1:20–26

Sammauria R, Kumawat S, Kumawat P, Singh J, Jatwa TK (2020) Microbial inoculants: potential tool for sustainability of agricultural production systems. Arch Microbiol 202(4):677–693

Santi C, Bogusz D, Franche C (2013) Biological nitrogen fixation in non-legume plants. Ann Bot 111(5):743–767

Schloter M, Hartmann A (1998) Endophytic and surface colonization of wheat roots (*Triticum aestivum*) by different *Azospirillum brasilense* strains studied with strain-specific monoclonal antibodies. Symbiosis 25:159–179

Schmidt W (2003) Iron solutions: acquisition strategies and signaling pathways in plants. Trends Plant Sci 8(4):188–193

Seenivasagan R, Babalola OO (2021) Utilization of microbial consortia as biofertilizers and biopesticides for the production of feasible agricultural product. Biology 10(11):1111

Seifan M, Samani AK, Berenjian A (2016) Bioconcrete: next generation of self-healing concrete. Appl Microbiol Biotechnol 100:2591–2602

Shafique HA, Sultana V, Ehteshamul-Haque S, Athar M (2016) Management of soil-borne diseases of organic vegetables. J Plant Protect Res 56(3):221–230

Shamseldin A (2022) Future outlook of transferring biological nitrogen fixation (BNF) to cereals and challenges to retard achieving this dream. Curr Microbiol 79(6):171

Sharma S, Kumar V, Tripathi RB (2011) Isolation of phosphate solubilizing microorganism (PSMs) from soil. J Microbiol Biotechnol Res 1(2):90–95

Sharma M, Delta AK, Kaushik P (2021) *Glomus mosseae* and *Pseudomonas fluorescens* application sustains yield and promote tolerance to water stress in *Helianthus annuus* L. Stress 1(4):305–316

Sheldon JR, Laakso HA, Heinrichs DE (2016) Iron acquisition strategies of bacterial pathogens. Microbiol Spectr 4:43–85

Sheoran S, Kumar S, Kumar P, Meena RS, Rakshit S (2021) Nitrogen fixation in maize: breeding opportunities. Theor Appl Genet 134:1263–1280

Shivlata L, Satyanarayana T (2015) Thermophilic and alkaliphilic Actinobacteria: biology and potential applications. Front Microbiol 6:1014

Shridhar BS (2012) Nitrogen fixing microorganisms. Int J Microbiol Res 3(1):46–52

Simon L, Bousquet J, Lévesque RC, Lalonde M (1993) Origin and diversification of endomycorrhizal fungi and coincidence with vascular land plants. Nature 363(6424):67–69

Singh A, Singh NÁ, Afzal S, Singh T, Hussain I (2018) Zinc oxide nanoparticles: a review of their biological synthesis, antimicrobial activity, uptake, translocation and biotransformation in plants. J Mater Sci 53(1):185–201

Singh TB, Ali A, Prasad M, Yadav A, Shrivastav P, Goyal D, Dantu PK (2020) Role of organic fertilizers in improving soil fertility. In: Naeem M, Ansari A, Gill S (eds) Contaminants in agriculture: sources, impacts and management. Springer, Cham, pp 61–77

Singh D, Thapa S, Geat N, Mehriya ML, Rajawat MVS (2021) Biofertilizers: mechanisms and application. In: Rakshit A, Meena VS, Parihar M, Singh HB, Singh AK (eds) Biofertilizers advances in bio-inoculants. Elsevier, Amsterdam, pp 151–166

Singh P, Singh RK, Zhou Y, Wang J, Jiang Y, Shen N, Yang L, Jiang M (2022) Unlocking the strength of plant growth promoting Pseudomonas in improving crop productivity in normal and challenging environments: a review. J Plant Interact 17(1):220–238

Sivasakthi S, Usharani G, Saranraj P (2014) Biocontrol potentiality of plant growth promoting bacteria (PGPR)-*Pseudomonas fluorescens* and *Bacillus subtilis*: a review. Afr J Agric Res 9(16):1265–1277

Sizonenko TA, Dubrovskiy YA, Novakovskiy AB (2020) Changes in mycorrhizal status and type in plant communities along altitudinal and ecological gradients—a case study from the Northern Urals (Russia). Mycorrhiza 30(4):445–454

Smercina DN, Evans SE, Friesen ML, Tiemann LK (2019) To fix or not to fix: controls on free-living nitrogen fixation in the rhizosphere. Appl Environ Microbiol 85(6):e02546–ee2518

Snyder CS, Bruulsema TW, Jensen TL, Fixen PE (2009) Review of greenhouse gas emissions from crop production systems and fertilizer management effects. Agric Ecosyst Environ 133:247–266

Sood M, Kapoor D, Kumar V, Sheteiwy MS, Ramakrishnan M, Landi M, Araniti F, Sharma A (2020) Trichoderma: the "secrets" of a multitalented biocontrol agent. Plan Theory 9(6):762

Soumare A, Diedhiou AG, Thuita M, Hafidi M, Ouhdouch Y, Gopalakrishnan S, Kouisni L (2020) Exploiting biological nitrogen fixation: a route towards a sustainable agriculture. Plan Theory 9(8):1011

Soumare A, Sarr DS, Diédhiou AG (2023) Potassium sources, microorganisms, and plant nutrition: challenges and future research directions: a review. Pedosphere 33(1):105–115

Sousa JAJ, Olivares FL (2016) Plant growth promotion by Streptomycetes: ecophysiology, mechanisms and applications. Chem Biol Technol Agric 3:24

Srinivasan R, Prabhu G, Chaudhary M, Radhakrishna A, Paul S (2020) Enhancing Fodder oat (Avena sativa) production in problem soils using phosphate solubilizing fungi isolated from acid and salt affected soils of India. IGC Proceedings XXIII, New Delhi, India. 2-3-1

Sumbul A, Ansari RA, Rizvi R, Mahmood I (2020) Azotobacter: a potential bio-fertilizer for soil and plant health management. Saudi J Biol Sci 27(12):3634–3640

Swayambhu G, Bruno M, Gulick AM, Pfeifer BA (2021) Siderophore natural products as pharmaceutical agents. Curr Opin Biotechnol 69:242–251

Tahami MK, Jahan M, Khalilzadeh H, Mehdizadeh M (2017) Plant growth promoting rhizobacteria in an ecological cropping system: a study on basil (*Ocimum basilicum* L.) essential oil production. Ind Crop Prod 107:97–104

Tejani GG, Savsani VJ, Bureerat S, Patel VK (2018) Topology and size optimization of trusses with static and dynamic bounds by modified symbiotic organisms search. J Comput Civ Eng 32(2):04017085

Thomas L, Singh I (2019) Microbial biofertilizers: types and applications. In: Giri B, Prasad R, Wu QS, Varma A (eds) Biofertilizers for sustainable agriculture and environment. Springer, Cham, pp 1–19

Timsina J (2018) Can organic sources of nutrients increase crop yields to meet global food demand? Agronomy 8(10):214

Tyagi S, Naresh RK, Prakash S, Yadav G, Tiwari S, Rawat B, Tiwari S, Joshi A, Tyagi A, Sharma N (2019) Conservation agriculture, biofertilizers and biopesticides: a holistic approach for agricultural sustainability and food security: a review. Int J Chem Studies 7:3036–3046

Uwamungu JY, Shi G, Wang Y, Paliwal A, Jadhav RR, Wani AW (2022) Arbuscular mycorrhizal fungi (AMF) for sustainable soil and plant health. In: Malik JA (ed) Microbial and biotechnological interventions in bioremediation and phytoremediation. Springer, Cham, pp 135–152

Vaario LM, Matsushita N (2021) Conservation of edible ectomycorrhizal mushrooms: understanding of the ECM fungi mediated carbon and nitrogen movement within forest ecosystems. In: Ohyama T, Inubushi K (eds) Nitrogen in agriculture—physiological, agricultural and ecological aspects. IntechOpen. pp. 95399

Valadares R, Marroni F, Sillo F, Oliveira RR, Balestrini R, Perotto S (2021) A transcriptomic approach provides insights on the mycorrhizal symbiosis of the mediterranean orchid *Limodorum abortivum* in nature. Plan Theory 10(2):251

Vasanthi N, Saleena LM, Raj SA (2018) Silica solubilization potential of certain bacterial species in the presence of different silicate minerals. Silicon 10:267–275

Vessey JK (2003) Plant growth promoting rhizobacteria as biofertilizers. Plant Soil 255:571–586

Vurukonda SSKP, Giovanardi D, Stefani E (2018) Plant growth promoting and biocontrol activity of *Streptomyces* spp. as endophytes. Int J Mol Sci 19(4):952

Vurukonda SSKP, Giovanadri D, Stefani E (2021) Growth promotion and biocontrol activity of endophytic *Streptomyces* spp. Emilio 1(2021):1–55. http://hdl.handle.net/11380/1248582

Wagner SC (2011) Biological nitrogen fixation. Nat Educ Knowl 3(10):15. http://www.nature.com/scitable/knowledge/library/biological-nitrogen-fixation-23570419

Wang HY, Liu S, Zhai LM, Zhang JZ, Ren TZ, Fan BQ, Liu HB (2015) Preparation and utilization of phosphate biofertilizers using agricultural waste. J Integr Agric 14(1):158–167

Wani SA, Chand S, Wani MA, Ramzan M, Hakeem KR (2016) *Azotobacter chroococcum*—a potential biofertilizer in agriculture: an overview. In: Hakeem K, Akhtar J, Sabir M (eds) Soil science: agricultural and environmental prospectives. Springer, Cham, pp 333–348

Weyens N, van der Lelie D, Taghavi S, Newman L, Vangronsveld J (2009) Exploiting plant–microbe partnerships to improve biomass production and remediation. Trends Biotechnol 27(10):591–598

Wu Q, Jiao S, Ma M, Peng S (2020) Microbial fuel cell system: a promising technology for pollutant removal and environmental remediation. Environ Sci Pollut Res 27:6749–6764

Yadav AN (2022) Phosphate-solubilizing microorganisms for agricultural sustainability. J Appl Biol Biotechnol 10(3):1–6

Youssef MMA, Eissa MFM (2014) Biofertilizers and their role in management of plant parasitic nematodes. A review. J Biotechnol Pharm Res 5(1):1–6

Zahedi H (2016) Growth-promoting effect of potassium-solubilizing microorganisms on some crop species. In: Meena V, Maurya B, Verma J, Meena R (eds) Potassium solubilizing microorganisms for sustainable agriculture. Springer, New Delhi, pp 31–42

Zaidi A, Khan MS, Saif S, Rizvi A, Ahmed B, Shahid M (2017) Role of nitrogen-fixing plant growth-promoting rhizobacteria in sustainable production of vegetables: current perspective. In: Zaidi A, Khan M (eds) Microbial strategies for vegetable production. Springer, Cham, pp 49–79

Chapter 8
Biocontrol Agents for Soil Health Management

Shilpa Sharma, Gaurav Rana, Sonali Kalotra, Lakita Verma, and Anjali Chauhan

8.1 Introduction

Because of the increase in world population and the damage brought on by climate change, producing high-quality harvests demands adopting new effective and sustainable strategies (Poveda 2021). The agroecosystem is being harmed by the careless use of chemical fertilizers to increase the yield of agriculture (Alavaisha et al. 2019). Plant pathogens like nematodes, viruses, fungus, and bacteria attack crops grown in the field or while being stored after harvest which causes loss in the production of agricultural products and brings risk to the amount of food provided globally (Savary et al. 2019). As a result, the need for a sustainable method is alarming. The desire to choose biological control agents (BCAs) as microbial biopesticides is considered an important part of integrated pest management (IPM), as a result its application is increasing these days. Biological control is defined as the practice by which humans utilize nonpathogenic microorganisms in order to reduce infectious pathogens using an ecofriendly approach (Cook 1993). Synthetic pesticides pollute the ecosystem and an excess of chemical decreases fertility of soil and promotes soil degradation. Biocontrol is a multi-trait marvel whose success is dependent on the rhizosphere competency of microbial inoculants, how it interacts with the indigenous microbiota, ability to fight for nutrients, tolerance to changes in different environmental conditions, and increased host plant protection against pathogens (Tariq et al. 2020). BCA not only controls disease but also promotes plant

S. Sharma · A. Chauhan (✉)
Department of Soil Science and Water Management, Dr. Yashwant Singh Parmar University of Horticulture and Forestry, Solan, Himachal Pradesh, India

G. Rana · S. Kalotra · L. Verma
Department of Basic Sciences, Dr. Yashwant Singh Parmar University of Horticulture and Forestry, Solan, Himachal Pradesh, India

R. K. Bhatia, A. Walia (eds.), *Advancements in Microbial Biotechnology for Soil Health*, Microorganisms for Sustainability 50,
https://doi.org/10.1007/978-981-99-9482-3_8

growth by secreting growth hormones. Furthermore, they increase stress tolerance and elicit resistance against diseases. BCA products have the potential be used as biopesticides, plant strengtheners, and biofertilizers based on their mechanisms and effects (Khasa 2017). Plant growth-promoting rhizobacteria and fungi (PGPR and PGPF) are used as BCA, and they help plants in many ways including organic matter breakdown, increased nutrient availability, mineral solubilization, phytohormones production, and phytopathogen biocontrol (Verma et al. 2019). Hence, biological control is an effective and environmentally acceptable means of controlling phytopathogens, preserving our environment and preventing numerous health risks linked with the continual utilization of harmful chemical pesticides. In this current chapter, soil health and its characteristics, biocontrol agent and its various types, and mode of action of BCA and its applications have been discussed.

8.2 Soil Health

Soil aspects fostering and sustaining the plant development and production while preserving long-term environmental quality is referred to as "soil health" (Lehmann et al. 2020). Soil health, in its wider context, refers to the ability of soil maintaining all living forms including their growth and development within the ecosystem. In relation to sustainable agriculture, soil is considered just as vigorous living organism that functions holistically alternative than a substrate. It describes its ability to preserve environmental quality, support biological productivity, and advance the health of plants, animals, and humans (Frac et al. 2018; Tahat et al. 2020). All agricultural systems depend on good soil management, and preventing further soil deterioration is essential to maintaining current production levels. The biological, chemical, and physical characteristics required for long-term, sustainable agricultural productivity with little harm to the environment all are integrated into the soil health as offering a comprehensive picture of soil functioning (Ray et al. 2020). The biological elements of soil, namely, bacteria, fungi, actinomycetes, algae, insects, and other organisms along with plant roots and underground plant parts, all interact in the soil food web which is crucial part of soil health. Notably the management of soil-borne illness can benefit greatly from improved soil health (Bagyaraj and Ashwin 2017). High population of organisms including microorganisms that support and encourage plant growth and development is an essential element of good soil as providing vital nutrients, water, oxygen, and root support that our food-producing plants require to develop and flourish is offered by the healthy soil (Brahmaprakash et al. 2017).

8.3 Characteristics of a Healthy Soil

Although the concept of soil microorganism processing plant and animal residues and synthesizing organic matter is no unfamiliar, yet it has generally been ignored until recently with the development of molecular analytical tools. Microbial contribution in the origination of soil organic material have acquired general acceptance, which marked a dramatic change in the environment (Shahbaz et al. 2017). The well-being of the soil relies on the microorganisms existing in the soil since they support nutrient cycling, breakdown of organic materials, and the biologic control of plant pests with other processes (Kohle et al. 2019). In addition, microbial enzymes are widely acknowledged as being proximate drivers of the transformation and putrefaction of organic matter where bacteria and fungi are the key determinants of biogeochemical processes within the soil (Dotaniya et al. 2019). In agro-ecosystem, microbial enzyme activities can act as early warning signs of soil quality degradation or improvement since they represent metabolic processes. Inorganic matter is transformed and decomposed in the soil by microbial enzymes, hence, reveals the kinship between microbial function, conventional soil quality indicators, and soil organic matter. The complex habitat of the soil expanse wherever plant roots engage physically along with chemically well along the soil is known as the rhizosphere (Ahkami et al. 2017). The majority of soil microbiome that resides in the rhizosphere includes plant growth-promoting rhizobacteria that boost crop output by exceeding soil health. For instance, soil nutrients such as nitrogen and phosphorus are delivered to the plant roots via rhizosphere or via mycorhizosphere in the context of roots connected to mycorrhizal fungus (Frey 2019). Pathogens and pest populations are minimal in a healthy soil consequently; direct competition for nutrients and niche habitats occur between populations of organisms in the soil. Moreover, strong plants are better able to defend themselves from various pests; therefore, for ideal plant growth besides sustaining a healthy cycle of nutrients inside the environmental system, there ought to be sufficient plus readily available nutrients with the existence of beneficial microorganisms which builds soil healthy (Canto et al. 2020). These rhizobacteria may also be employed as biocontrol agents to guard soil and plants against disease invasion. By-products from fungi and bacteria bind soil particles form macroaggregates that improve soil stability and structure (Rawat et al. 2020).

8.4 Biological Control Agents (BCAs)

The term “antagonism” was initially used by William Roberts to describe his investigations regarding the antagonistic relationship between bacteria and *Penicillium glaucum* (Foster and Raoult 1974). In 1914, C. F. Von introduced the biological control of plant disease management (Cook 1988). Eilenberg et al. (2001) defined biocontrol as “the application of live organisms to reduce the density of population or effect of a particular pest organism, thereby making it less abundant or less toxic than it would normally be." The term "biological control agents" (BCAs) applies primarily to helpful microorganisms that have ability to control the population of soil pathogens, by eradicating or decreasing them through various mode of mechanisms (Ons et al. 2020). Biocontrol, as defined by the International Biocontrol Manufacturers Association (IBMA), is the use of methods or agents to naturally reduce the growth of pests and pathogens that affect crops.

Essentially, BCAs employ intraspecific or interspecific conflicts within the same biome to prevent infections from surviving and thriving while sustaining the soil state (Raymaekers et al. 2020; Meng et al. 2022). Additionally, BCAs also stabilize communities from toxicity; excess nutrients can cause leaching, probable ground water pollution, high drainage, and many more (Ahmad et al. 2018). These microbes resides as plant growth-promoting microorganisms that employ beneficial effects on soil and plants while colonizing the plant root environment without causing any harm. The alteration of rhizospheric microbiota is how such bacteria aid the build-out of plants as in general, having a plethora of direct and indirect processes that affect the plant growth and development (Ramakrishna et al. 2019). Direct actions involves either delivering or aiding in the acquisition of nutrients from the environment, while the secretion of antibiotics, lytic enzymes, and chelation of accessible iron at the plant-root interface are all associated to reducing the negative impacts of phytopathogens (Kohle et al. 2019; Palmieri et al. 2022).

8.5 Types of Biocontrol Agents

8.5.1 Bacteria as Biocontrol Agent

Rhizobacteria, like bacteria from the genera *Bacillus* and *Pseudomonas*, are widely recognized for their antagonistic activities and tendency to instigate induced systemic resistance (ISR) in plants (Bardin and Pugliese 2020). To prevent plant diseases, especially nematodes, beneficial bacteria used as biopesticides include *Bacillus subtilis*, *B. amyloliquefaciens*, *B. firmus*, and *B. pumilus*. Various bacterial strains are used as BCA against pathogenic microorganisms (Table 8.1). The genus *Bacillus* is capable of producing phytostimulation, namely, cytokinins, in the rhizosphere. Dihydrozeatin riboside and isopentenyl-adenosine are zeatin riboside-type cytokinins synthesized by the *B. subtilis* strain (Abd-El-Khair et al. 2020). *B. amyloliquefaciens* strains MBI 600, AH2, and D747 and the *B. subtilis* GB03 are commercially available strains that can suppress soil-borne diseases.

Lipopeptides, enzymes, volatile chemical molecules, and peptide toxins are the four groups of protectants that *Bacillus* spp. depend on for biocontrol. BT toxins are insecticide that have been commercially marketed and are the most efficient of these compounds. These toxins are produced by *Bacillus* isolates once sporulation starts and aggregate inside bacterial cells like a dormant crystalline structure (Lee et al. 2022). Then, insect larvae consumes these crystals; protease activity is required for

Table 8.1 Bacterial and fungal strains used as biocontrol agents against plant diseases

S. no.	Biocontrol agent	Target pathogen	Disease	References
Bacterial strains				
1.	*Bacillus amyloliquifaciens* Ba01	*Streptomyces scabies*	Common scab	Lin et al. (2018)
2.	*B. amyloliquefaciens* Ba13	Tomato yellow leaf curl virus spread by whiteflies (*Bemisia tabaci*)	Yellow leaf curl virus disease	Guo et al. (2019)
3.	*B. amyloliquefaciens*	*Fusarium oxysporum*	Vascular wilt and root rot	Palyzova et al. (2019)
4.	*B. cereus* SZT1	*Xanthomonas oryzae* pv. oryzae	Leaf blight	Ahmed et al. (2020)
5.	*B. subtilis* XZ18-3	*Rhizoctonia cerealis*	Sharp eyespot disease	Yi et al. (2022)
6.	*B. cereus*	*Pythium aphanidermatum*	Rhizome rot	Vinayarani and Prakash (2018)
7.	*B. subtilis* MBI600	*Fusarium oxysporum*	Crown and root rot	Samaras et al. (2021)
8.	*B. subtilis* MBI600	*Rhizoctonia solani* and *Pythium ultimum*	Damping-off	Samaras et al. (2021)
9.	*B. pumilus*	*Rhizoctonia solani*, *Fusarium solani*, and *Pythium* sp.	Black root rot	Abd-El-Kareem et al. (2021)
10.	*Bacillus firmus*	*Macrophomina phaseolina*	Dry root rot	Mawar et al. (2019)
11.	*B. thuringiensis* TbL-22	*Xanthomonas citri*	Canker disease	Islam et al. (2019)
12.	*Pseudomonas fluorescens* (Pf173)	*Fusarium oxysporum*, *Alternaria solani*, *Phytophthora infestans*, *Ralstonia solanacearum*	Wilt, early blight, late blight	Kabdwal et al. (2019)
13.	*P. putida* ASU15	*Uromyces appendiculatus*	Bean rust	Abo-Elyousr et al. (2021)
14.	*P. aeruginosa*	*Rhizoctonia solani*	Leaf blight	Vinayarani and Prakash (2018)
15.	*Pseudomonas* strain QBA5	*Botrytis cinerea*	Gray mold disease	Gao et al. (2018)
16.	*P. aeruginosa* (OG101) and *P. putida* (PP)	*Fusarium chlamydosporum*	Root rot	Palyzova et al. (2019)
Fungal strains				
17.	*Trichoderma harzianum*	*Fusarium solani*	Wilt	Awad-Allah et al. (2022)

(continued)

Table 8.1 (continued)

S. no.	Biocontrol agent	Target pathogen	Disease	References
18.	*T. viride*	*Fusarium solani, F. oxysporum, Rhizoctonia solani*	Wilt	Awad-Allah et al. (2022)
19.	*T. asperellum* strain DQ-1	*Botrytis cinerea*	Gray mold	Wang et al. (2021)
20.	*T. harzianum* (Th43)	*Alternaria solani, Phytophthora infestans, Fusarium oxysporum*, and *Ralstonia solanacearum*	Early blight, late blight, and wilts of tomato	Kabdwal et al. (2019)
21.	*T. harzianum*	*Fusarium equiseti, Colletotrichum queenslandicum*	Anthracnose	Grano-Maldonado et al. (2021)
22.	*T. harzianum*	*Fusarium oxysporum*	Vascular wilt and root rot	Palyzova et al. (2019)
23.	*T. harzianum*	*Alternaria brassicae*	Leaf spot	Jackson and Kumar (2019)
24.	*T. asperellum*	*Fusarium oxysporum*	Corm rot	Gupta et al. (2020)
25.	*T. gamsii* It-62	*Fusarium oxysporum, Fusarium solani, Macrophomina phaseolina*, and *Pythium ultimum*	Damping off	Boat et al. (2019)
26.	*T. harzianum*	*Fusarium oxysporum*	Wilt	Mawar et al. (2019)

the conversion of the crystals into an active peptide. Recognizing active peptide with the help of certain receptors on the cells of insect gut causes development of a pore, and selective permeability is lost. As a result of weakened digestion, bacteria penetrate the circulatory system and eventually kill the larvae (Fig. 8.1). The selectivity of the receptors in the gut cell membrane allows for products that target certain insects. Metabolites such as lipopeptides are often released by *B. thuringiensis* and various *Bacillus* species (Théatre et al. 2021). Many lipopeptides have potent antifungal and anti-insect properties. There are also commercially available *Bacillus*-based products that control plant pathogens like Avogreen®, *Bacillus* SPP®, Biosubtilin®, Biosafe®, Cease®, Ballad®, Companion®, and Kodiak® (Albayrak 2019).

The US Food and Drug Administration specifically defines *B. subtilis* as GRAS (generally recognized as safe) organisms for the food sectors. *Pseudomonas* is a naturally occurring biocontrol agent that thrives in soils that are disease-resistant, and it may grow quickly and successfully colonize areas where the rhizosphere is an extremely competitive environment for resources like food and space. The extensive range of antibiotics that *Pseudomonas* strains may synthesize are differentiated by their ability to control plant infections. Numerous substances, including Plt, PCA, DAPG, Prn, HCN, and bacteriocin 102, are synthesized by *Pseudomonas*. *Pseudomonas* also causes systemic plant resistance, and *P. fluorescens* (EP1 strain), *P. putida* (5–48 strain), and *P. fluorescens* are able to protect against soil-borne diseases in oak, sugarcane, and tomato plants (Abd-El-Khair 2020).

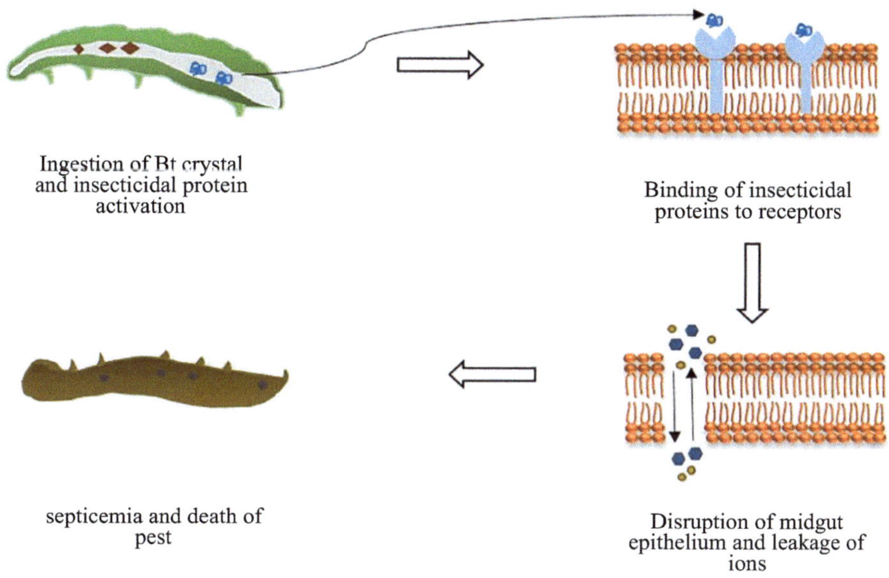

Fig. 8.1 Effect of Bt toxin on insect larvae

8.5.2 Fungi as Biocontrol Agent

Individuals from the genus *Trichoderma* have been the most thoroughly studied. *Trichoderma* strains are used to combat a variety of plant diseases, including pathogens that are carried via the soil. Certain strains have the capacity to suppress plant pathogens, especially in the soil or on plant roots, as a result of their strong antagonistic and mycoparasitic ability, hence reducing the impact of plant diseases (Abo Nouh 2019).

Trichoderma spp. are able to develop a toxin while parasitizing other fungus. Infections of hyphal coil enter the cytoplasm of host and dissolve there as part of the mycoparasitism process. Many plant diseases, such as sheath blight, *Fusarium* wilt, bacterial wilt, downy mildew, and root-knot nematode, may be controlled by *Trichoderma* spp., *Fusarium* wilt, bacterial wilt, and downy mildew (Ahmad and Zaib 2020).

Trichoderma species also decompose the polysaccharides and cell wall of fungi. *Trichoderma* spp. enhance plant tolerance against biotic and abiotic stressors, which promote plant growth and development. It enhances the growth of plant, involving root development, shoot weight, and leaf surface area (Sood et al. 2020). *Trichoderma* species, such as *T. asperellum*, *T. atroviride*, *T. harzianum*, *T. gamsii*, *T. polysporum*, and *T. viride*, have all been found effective biocontrol agents against plant pathogens, involving nematodes. Some strains that are commercially available involve *T. asperellum* ICC012 (originally *T. harzianum*), T25 (originally *T. viride*), TV1 (originally *T. harzianum*), *T. asperellum* SF 04 (URM) 5911, T211, T34, *T.*

atroviride SKT-1, IMI 206040 (formerly *T. harzianum*), T11 (originally *T. harzianum*), *T. gamsii* ICC080 (originally *T. viride*), *T. hamatum* TH382, *T. harzianum* KRL-AG2 (syn. T22), ESALQ-1306, IBL F006, ITEM 908 and T-39, and *T. virens* G-41 and GL-21 (originally *Gliocladium virens*) (Bardin and Pugliese 2020). Other fungi have been found to have antagonistic effects on a small number of fungal pathogens such as *Alternaria*, *Pythium*, *Aspergillus*, *Fusarium*, *Rhizoctonia*, *Phytophthora*, *Botrytis*, *Pyricularia*, and *Gaeumannomyces*. Various fungal strains are used as BCA against pathogenic microorganisms (Table 8.1).

In addition, *Trichoderma* species—*T. harzianum*, *T. viride*, and *T. citrinoviride*—have volatile organic compounds (VOCs) that deter pest insects, minimizing the harm they cause to plant resources. It is feasible to confirm how the termites, namely, *Odontotermes formosanus*, does not create channels in regions where *Trichoderma* is present by adding fungal spores in the soil (Xiong et al. 2018). *Trichoderma* is able to activate plant defenses against diseases and pests through root colonization. These responses are regulated by the plant hormones such as salicylic acid (SA) and jasmonic acid (JA), primarily. It can release metabolites such as cellulases, glucanases, lipases, proteases, and chitinases to aid in inserting the hyphae for intake of nutrients by pathogen, which results in the disruption of cytoplasmic contents of host cell (Amuza and Zaharia 2022).

8.5.3 Virus as Biocontrol Agent

With no conceptual objection from the research community, viruses have been included in the canon of biological control and utilized to battle pests both in direct (by pest infection caused by pathogen) and indirect (e.g., via cross-protection, i.e., "vaccination" of crops by using viral strains) way (Stenberg et al. 2021). Insect viruses are obligatory pathogens that can only replicate within their insect host. *Baculoviruses* and *Nudiviruses* have been used and are still being used on an experimental and commercial basis to control several insect species from the order *Lepidoptera*, *Hymenoptera*, and *Coleoptera* (Lacey 2017). The *Baculoviridae* family is the most common and well-studied group of insect pathogenic viruses.

Invertebrate-specific pathogens such as baculoviruses have occasionally been used as biopesticides to manage insects, notably species of the Lepidoptera (Abd-Alla et al. 2020). These viruses have circularized DNA which is double-stranded. Their genomes are shielded from potentially hazardous environmental attributes with the help of a protein matrix that is made of polyhedrin, which crystallizes to create a matrix known as polyhedral (Rohrmann 2019). Throughout their replication cycle, baculoviruses produce two distinct phenotypes: the occlusion-derived virus (ODV) that transfers the infection within larvae and the budded virus (BV) disseminate the infection inside the host. Two phenotypes of virion can be formed at two different phases of the baculovirus life cycle. Viral proteins produced during late when assembled into occlusion bodies (OB), generate a protein matrix that encases the occlusion-derived virions (ODV). Initiating the main infection of the midgut cells of host, ODV are characterized by a number of nucleocapsids in each envelope, and OB release them due to the presence of the extremely alkaline habitat of the midgut (Sosa-Gómez et al. 2020). Budded virions (BV), the second phenotype, are created during the early phases of the initial infection. BV phenotype is distinguished by the fact that these virions arise from the cytoplasmic membrane of infected cells with one nucleocapsid each envelope and are implicated within secondary or systemic infection (Rohrmann 2019). Infections due to *Baculovirus* in lepidopteran larvae causes paralysis, body becoming whiter, tree-top illness behavior, and finally death of the host, all these symptoms being easily visible. Infected insects lose their vibrant color, become less active, and the ventral side of the larvae turns pinkish white. Larvae in advanced stages develop flaccidity, brittle skin, and finally bursts. Disease larvae cling to the plants upside down. This is known as "tree top disease" or "wipfelkrankeit" (Sheik et al. 2022).

8.5.4 Nematodes as Biological Control Agent

Entomopathogenic nematodes (EPNs) are tiny roundworms belonging to the families, namely, *Heterorhabditidae* and *Steinernematidae* of the phylum Nematoda (Bhat et al. 2020). Among the different biocontrol agents that are routinely utilized, entomopathogenic nematodes (EPNs) have a reasonably broad natural host range and are efficient against a broad range of insect species, notably soil-associated insects. The usage of EPNs is regarded as the most efficient integrated pest management (IPM) strategy for controlling such insect populations (Tomar et al. 2021). Their efficacy is assessed by their ability to search prey and the outcome of their hunting ability in terms of the total number of preys caught and eliminated. The act of predation can be separated into three stages: to encounter the prey, attacking the victim, and then feeding (Moosavi 2020).

Free-living infective juveniles (IJs) aggressively migrate through the soil in search of hosts, making them a suitable biocontrol agent for insects associated with soil like *C. newmannii*. IJs invade their hosts via natural openings or via the cuticle in some circumstances. They subsequently discharge symbiotic bacteria they have in their gut, causing the insect to die quickly. The IJs mature into adults in the infected insect's haemocoel while feeding on bacteria. They mature and breed within the deceased host until all of the nutrients are depleted, at which point a new cohort of IJs departs to search for new hosts. Plant parasitic nematodes are now generally acknowledged as a possible hidden threat that might cause significant losses in a wide range of agricultural and horticultural crops (Mhatre et al. 2019). Bacterial and fungal antagonists are the most effective for biocontrol of nematodes. A wide range of microorganisms, including nematode-trapping fungi, toxin-synthesizing endoparasitic fungi, vesicular-arbuscular mycorrhiza (VAM) as well as parasites of immobile nematodes, females, and eggs make up the fungal antagonists against nematodes. The three major categories of bacterial antagonists are endophytic, epiphytic, and endoparasitic bacteria (AbdelRazek and Yaseen 2020). They use parasitism, competition, and antibiosis as biological control mechanism.

8.6 Mechanism of Action of Biocontrol Agents

Biological control can be accomplished by direct or indirect antagonism. (1) Direct mode of action involves parasitism, siderophore production, lytic enzymes, and antibiosis. (2) Indirect mode of action involves the induction of (systemic) resistance mechanism and competition (Fig. 8.2) (Akram et al. 2023).

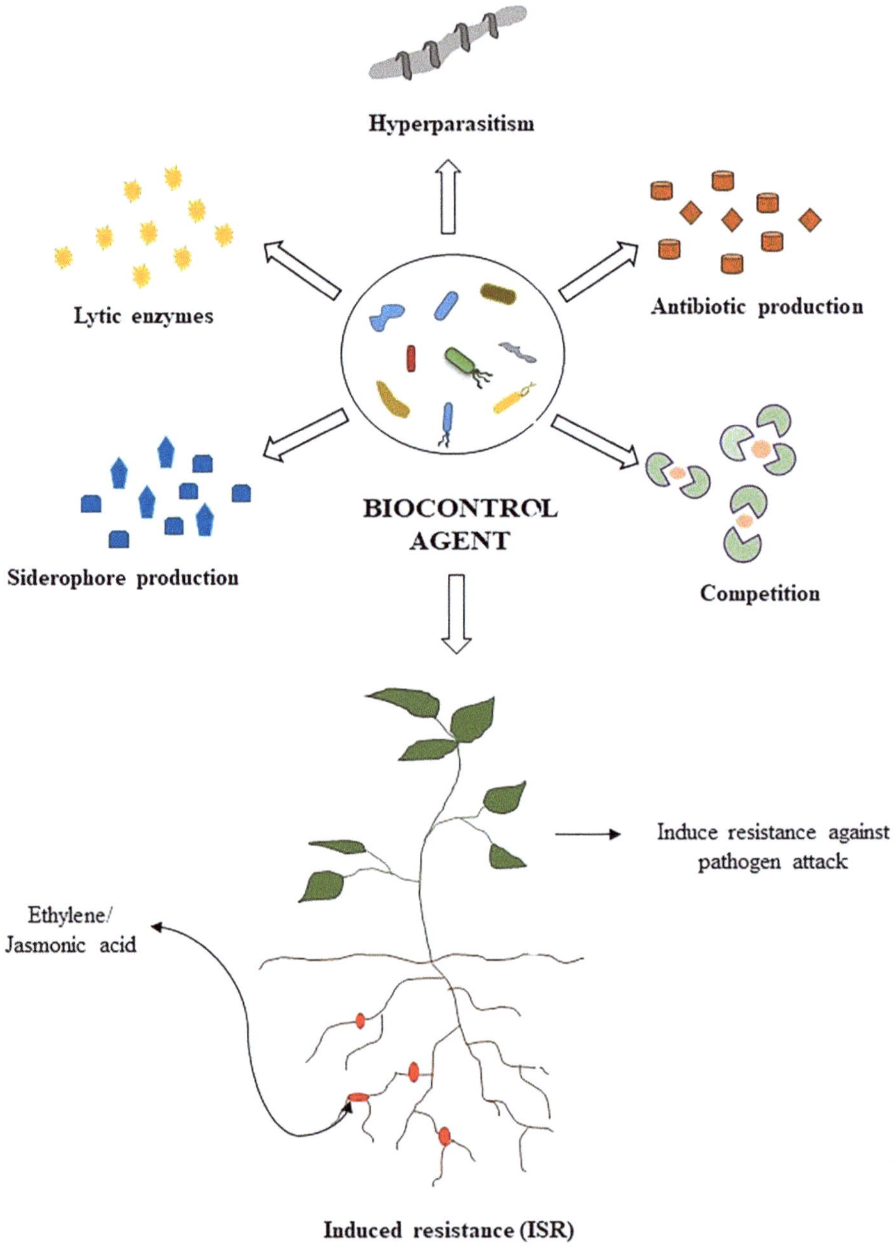

Fig. 8.2 Various mechanisms of biocontrol agent (BCA)

8.6.1 Direct Antagonism

8.6.1.1 Antibiosis

Antibiotics constitutes a broad class of organic and low-molecular-weight molecules that hinder metabolic activities or growth of other bacteria (MH Eassa et al. 2018). Antibiotics have biocidal (i.e., killing) and biostatic (i.e., slowing the growth of pathogens) effects. Antibiotics disrupt fundamental metabolic procedures, such as cell wall and protein production and DNA replication, in their hosts. It is likely that antibiosis is to be one of the most analyzed modes of action since it is simple to perform in laboratory using the dual culture technique on nutrient agar (NA). Toxic secondary metabolites are produced by the target plant pathogen as a result of the biocontrol agent. The genera *Bacillus*, *Pseudomonas*, *Streptomyces*, and *Trichoderma* are some examples of bacteria and fungus that create antimicrobial compounds.

Bacteria grow most quickly when they are in the trophophase phase of cell development, but when resources are exhausted and cell density increases dramatically, secondary metabolites such as antibiotics are created during the idiophase (Tariq et al. 2020). The sporulation, mycelial development, and/or germination of pathogens may be inhibited by these low-concentration chemicals. Antibiotics having antibacterial and antifungal functions are produced by a variety of bacterial species, including *Streptomyces*, *Burkholderia*, *Serratia*, *Pantoea*, *Lysobacter*, and *Enterobacter*. These drugs have been evaluated against fungal infections following harvest. Genetically engineered *Pseudomonas putida* WCS358r strains develop the antibiotics phenazine and 2,4-diacetylphloroglucinol (2,4-DAPG) which are efficient at preventing infections (Khasa 2017). Antibiotics with biocontrollable action are primarily produced by *Trichoderma* and *Gliocladium*. For instance, the gliotoxin that *G. virens* produces is recognized as a major element in its biocontrol effect against *P. ultimum* and *R. solani*. Biocontrol PGPR develops various antibiotics which includes 2,4-diacetylphloroglucinol, pyrrolnitrin, zwittermicin A, phenazine, tensin, and xanthobaccin that are particularly good at halting the growth of harmful fungi (Ahmad and Zaib 2020).

8.6.1.2 Hyperparasitism

Hyperparasitism is the most recognized and the direct kind antagonism. Infection occurs when a plant pathogen comes into touch with BCA. Biocontrol involves the rapid growth of a biocontrol agent in the direction of a target organism, coiling, a final assault, and the destruction of cell wall or membrane of a target pathogen with the help of enzymes (Prajapati et al. 2020). *Bdellovibrio* bacteriovorus, a hyperparasite in bacteria, feeds on gram-negative bacteria by using cytoplasm as its host (McNeely et al. 2017). Mycoparasitism is an interaction in which a fungal biocontrol agent directly attacks and feeds on the pathogen, killing it or destroying its propagules. *Ampelomyces quisqualis* is a widely recognized mycoparasitic fungus that has the ability to suppress infections which causes powdery mildew on variety of plants.

Streptomyces spp. strains are identified as hyperparasites of several phytopathogenic fungi such as *Alternaria brassicicola, Botrytis cinerea, Fusarium oxysporum, Mịcrocentro sporosarcina, Rhizoctonia solani, S. sclerotiorum, Phomopsis sclerotioides*, and *Pythium ultimum* (Palmieri et al. 2022). The basic mechanism associated with this kind of parasitism is that it can produce enzymes that dissolves cell wall, often in conjunction with several other metabolites, resulting in cell wall opening and cytoplasm disorientation. Since the cell wall of fungus is made up of chitin and glucans, the enzymes secreted include chitinases, 1,3-glucanases, and proteases, as well as cellulases in the case of oomycote fungi (Upadhyay et al. 2021). Mycoparasitism has four basic stages (Chugh et al. 2022):

1. Chemotropism is the process by which biocontrol agents are drawn by chemical stimuli produced by pathogens. For example, pathogen produces water-soluble or volatile chemicals that function as chemoattractants for its parasite.
2. Recognition: This is a process of identifying certain pathogenic fungi that are attacked by antagonistic microbes. In this, carbohydrate-surface receptors on biocontrol agents interacts with pathogen-produced lectins.
3. Attachment and disintegration: The third step involves two processes: adhesion to the cell wall via receptor-ligand interaction and then degradation.
4. Penetration: This involves the entry into the fungal cell wall of specific structures like appressoria.

8.6.1.3 Siderophore

Siderophore are low-molecular-weight chelating component with high affinity for Fe(III). The importance of siderophore formation as the process of *Erwinia carotovora* inhibition by *P. fluorescens* strains A1, TL3B1, BK1, and BK10 was originally observed by Kloepper et al. (1980). Metabolic activity, proliferation, and reproduction of plant pathogen is inhibited by microorganisms like BCAs that may produce siderophores by iron chelation. They create a ferric-siderophore complex, which sequesters iron from their surroundings and diffuses to the cell surface. Two siderophore-producing bacteria that have been studied the most are *Pseudomonas fluorescens* and *Pseudomonas aeruginosa*. They are able to produce two types of siderophores, pyochelin and pyoverdine, which may exhibit antibiotic activity (Iakimova and Sobiczewski 2022). Bacterial siderophores are classified into four categories based on the functional groups which coordinate iron: carboxylates, hydroxamates, phenol catecholates, and pyoverdines. They work as they sequester iron in the rhizosphere, depriving phytopathogens of it and preventing their proliferation.

8.6.1.4 Lytic Enzymes

One of the most often described mechanisms of biological control against the fungal diseases are lytic enzymes since they are engaged in the breakdown of cellular wall of soil pathogens by interfering in the enormous polysaccharides network causes cell lysis and death (Mhatre et al. 2019). Chitinases, cellulases, and proteases are some of most researched lytic metabolites produced by BCAs as their hydrolytic enzymes alter, perforate, and breakdown of the cell wall structure of pathogens that are found in the soil (Jadhav et al. 2017). According to reports, *Xanthomonas oryzae* responsible for leaf blight illness in rice is biologically controlled by an extracellular chitinase and cellulases that was isolated from *Pseudomonas fluorescens*. Moreover, it has been noted that *Pseudomonas* strains produced 1,3-glucanases, cellulases, and chitinases that are hostile to *Rhizoctonia solani*. Likewise, *Bacillus subtilis* as effective BCAs also demonstrated the capability of reducing growth of soil pathogens (Chen et al. 2020). Fungi also have tremendous pliability to take on many forms in poor or unfavorable conditions and therefore are particularly successful inhabitants of the soil by decomposing organic matter and producing variety of extracellular enzymes that facilitates to produce biomass and maintain an equilibrium between nutrients and carbon leading to a healthy state of soil (Peng et al. 2021).

8.6.2 Indirect Antagonism

8.6.2.1 Induced Resistance

Induced resistance is a phenomenon in which a plant creates a defense mechanism when necessary against pathogenic organisms. PGPR can decrease the activity of harmful microbes by stimulating the plant to protect itself. Van Peer et al. (1991) first identified this the mechanism as "induced systemic resistance" (ISR) which is beneficial in a variety of agronomic crops such as tobacco and beans and horticultural crops such as tomato, radish, and carnations and also *Arabidopsis thaliana*. Two basic categories into which induced resistance (ISR) is frequently classified are systemic acquired resistance (SAR) and inducible systemic resistance (ISR). The former is the plant's innate immunity which is activated either by direct interaction with biotic factors including pathogenic and nonpathogenic microorganisms or by abiotic stimuli, such as various chemical substances (Roca-Couso et al. 2021). The two major components of this process are the synthesis of salicylic acid, a chemical molecule that is frequently generated upon infection, and the accumulation of proteins associated to pathogenesis. This second reaction is comparable to a hypersensitive reaction brought on by exposure of PGP bacteria. The two phytohormones in this case that cause the reaction are ethylene and jasmonate.

It has been shown that bacterial endophytes and rhizobacteria both cause ISR. Many chemical compounds generated by PGPR strains, including lipopolysaccharides, the antibiotic DAPG, siderophores, cyclic lipopeptides, homoserine lactones, acetoin, and 2,3-butanediol, have been shown to cause systemic resistance (Sehrawat and Sindhu 2019). The ISR gets triggered due to the presence of PGPR strains within the root, and it eventually travels to the rest of the plant. Induced resistance begins with the recognition of a signal, which sets off a chain of events that may lead to systemic plant resistance. The plants signaling network comprises of the plant defense hormones such as salicylic acid, jasmonic acid, and ethylene which enables intercellular communication (Macabuhay et al. 2022). ISR has been found to be induced by a number of rhizosphere bacteria which includes *Pseudomonas*, *Bacillus*, *Serratia*, *Azospirillum*, *Trichoderma*, and mycorrhiza. In the presence of phytopathogenic fungi, viruses, or bacteria, BCAs have been found to activate ISR. The decrease in sensitivity to *Fusarium* wilt caused by *Pseudomonas* sp. on carnations is one of the earliest reports describing the ability to produce ISR (Pieterse et al. 1996). Several studies have demonstrated that a wide range of potential BCAs can cause ISR, including *Burkholderia phytofirmans* PsJN against *Botrytis cinerea* and *Verticillium dahliae* on grapevine and tomato; *Gliocladium roseum* against *Erysiphe orontii*; and *Pseudomonas* spp. (Palmieri et al. 2022).

8.6.2.2 Competition

Competition is a fundamental factor that revealed network between beneficial BCAs and pathogens in soil biome (Sood et al. 2020). Nonpathogenic microorganisms compete with pathogenic microorganisms for organic nutrients in the host plant so that they can multiply and flourish. Due to excellent compatibility with the damp soil and high level of organic manure, *Bacillus* and *Pseudomonas* bacterial strains are most frequently used as BCAs against the soil-borne plant pathogens (Dimkic et al. 2022). BCAs have a more competent nutrition absorption mechanism than phytopathogens. Pathogens and biocontrol agents typically compete for survival factors like space, nutrition, and vital micronutrients, for example, iron, manganese, and particular growth factors that promote germination. The use of strong affinity iron chelators, such as siderophores generating BCA, leads in the starving of pathogenic bacteria that require iron as a significant element for life (Chugh et al. 2022). This form of treatment is most suited for necrotic pathogen disease control. Many species of soil-borne pathogens, like *Pythium* and *Fusarium*, which spread by mycelial contact are more prone toward various soil-borne microorganisms than those that directly infects plant tissues (Upadhyay et al. 2021).

8.7 Application of BCAs

Using chemical or synthetic pesticides is one of the most widely practiced ways to eradicate pests and are successfully used, but they also carry potential menace to human health and pollute the soil environment when persistent for long time in the soil (Pirttila et al. 2021). Moreover, indiscriminately using these chemicals on a regular basis led to increased residual issues, insect resistance, and extinction of some valuable species while sacrificing the soil health. The dangers associated with chemicals application must be overcome, and an eco-friendly approach must be used for better soil health management. Using bioformulations consisting of beneficial BCAs in the crop fields can be intriguing since it provides a reliable technique for controlling soil pathogens at a same time protecting the ecosystem (Nagpal et al. 2021). Various commercially available BCA are used against plant pathogenic microorganisms. Some of them are listed below (Table 8.2).

Table 8.2 Commercialized biocontrol agents that target pathogenic microorganisms and promote plant growth

Sr. no.	BCAs	Product	Target pathogen/crop	Manufacturing industry	References
1.	*Trichoderma harzianum*	Trichopel	Fungal diseases	Agrimm Technologies Ltd., New Zealand	Babbal and Khasa (2017)
2.	*T. harzianum* and *T. polysporum*	Binab T	Wilt and wood decay	Bio-Innovation, Sweden	Savita and Sharma (2019)
3.	*T. harzianum*	T-22 and T-22HB Bio-Trek, Root Shield	*Pythium*, *Rhizoctonia*, *Fusarium*, and *Sclerotina*	BioWorks Geneva, USA	
4.	*Fusarium oxysporum*	Fusaclean	*Fusarium oxysporum*	Natural Plant Protection, France	
5.	*Gliocladium virens*	SoilGard (GlioGard)	Damping off and root infections	ThermoTrilogy, USA	
6.	*Cryptococcus albidus*	Yield Plus	*Botrytis* spp. and *Penicillin* spp.	Anchor Yeast, South Africa	
7.	*Ampelomyces quisqualis*	AQ10 Biofungicide	Powdery mildew	Ecogen Inc., USA	Babbal and Khasa (2017)
8.	*Bacillus subtilis* GB03, other *B. subtilis*, *B. licheniformis*, and *B. megaterium*	Companion	*Rhizoctonia*, *Pythium*, *Fusarium*, and *Phytophthora*	Growth Products, New York	
9.	*B. subtilis* FZB24	FZB24 li, TB, WG RhizoPlus	Vegetables (potatoes), ornamentals, turf, and woods	AbiTep, Germany	
10.	*Pseudomonas syringae* ESC10 and ESC11	Bio-Save 10LP, 110	*Penicillium* spp., *Mucor piriformis*, and *Geotrichum candidum*	EcoScience Corp. (Longwood, FL)	
11.	*Agrobacterium radiobacter* strain 84	Galtrol	*A. tumefaciens*	AgBioChem (USA)	
12.	*Bradyrhizobium japonicum*	Soil implant	Soybean	Nitragin, Argentina	
13.	*Azospirillum* spp.	Biopromoter	Millets, oilseeds, fruits (banana), and vegetables	Manidharma Biotech, India	
14.	*Streptomyces griseoviridis* K61 strain	Mycostop	Soil-borne pathogens such as *Phomopsis* spp., *Botrytis* spp., *Pythium* spp., and *Phytophthora* spp.	KemiraOyj, Finland	

(continued)

Table 8.2 (continued)

Sr. no.	BCAs	Product	Target pathogen/crop	Manufacturing industry	References
15.	Consortia of *Azotobacter* + *Azospirillum*	Multiplex Nalpak	Promotes plant growth and improves soil quality	Multiplex Bio-Tech Pvt. Ltd., Karnataka (India)	Arora et al. (2020)
16.	*Aspergillus niger*	Ambiphos	Chelates undissolved phosphorus in the soil	Ambika Biotech and Agro Services, Madhya Pradesh (India)	
1.	*Trichoderma harzianum*	Trichopel	Wide range of fungal diseases	Agrimm Technologies Ltd, New Zealand	Babbal and Khasa (2017)
2.	*T. harzianum* and *T. polysporum*	Binab T	Fungi causing wilt and wood decay	Bio-Innovation, Sweden	Savita and Sharma (2019)
3.	*T. harzianum*	T-22 and T-22HB Bio-Trek, Root Shield	*Pythium*, *Rhizoctonia*, *Fusarium*, and *Sclerotina*	BioWorks Geneva, USA	
4.	*Fusarium oxysporum*	Fusaclean	*Fusarium oxysporum*	Natural Plant Protection, France	
5.	*Gliocladium virens*	SoilGard (GlioGard)	Several plant diseases Damping off and root pathogens	ThermoTrilogy, USA	
6.	*Cryptococcus albidus*	Yield Plus	*Botrytis* spp. and *Penicillin* spp.	Anchor Yeast, South Africa	

Sr. no.	BCAs	Product	Target pathogen/crop	Manufacturing industry	References
7.	*Ampelomyces quisqualis*	AQ10 Biofungicide	Powdery mildews	Ecogen Inc., USA	Babbal and Khasa (2017)
8.	*Bacillus subtilis* GB03, other *B. subtilis*, *B. licheniformis*, and *B. megaterium*	Companion	*Rhizoctonia*, *Pythium*, *Fusarium*, and *Phytophthora*	Growth Products, New York	
9.	*B. subtilis FZB24*	FZB24 li, TB, WG RhizoPlus	Potatoes, vegetables, ornamentals, strawberries, bulbs, turf, and woods	AbiTep, Germany	
10.	*Pseudomonas syringae* ESC10 and ESC11	Bio-Save 10LP, 110	*Penicillium* spp., *Mucor piriformis*, and *Geotrichum candidum*	EcoScience Corp. (Longwood, FL)	
11.	*Agrobacterium radiobacter* strain 84	Galtrol	*A. tumefaciens*	AgBioChem (USA)	
12.	*Bradyrhizobium japonicum*	Soil implant	Soybean	Nitragin, Argentina	
13.	*Azospirillum* spp.	Biopromoter	Paddy, millets, oilseeds, fruits, vegetables, sugarcane, and banana	Manidharma Biotech, India	
14.	*Streptomyces griseoviridis* K61 strain	Mycostop	Soil borne pathogens like *Phomopsis* spp., *Botrytis* spp., *Pythium* spp., and *Phytophthora* spp.	KemiraOyj, Finland	
15.	Consortia of *Azotobacter* + *Azospirillum*	Multiplex Nalpak	Promotes plant growth and improves the quality of soil	Multiplex Bio-Tech Pvt. Ltd., Karnataka (India)	Arora et al. (2020)
16.	*Aspergillus niger*	Ambiphos	Chelates undissolved phosphorus in the soil	Ambika Biotech and Agro Services, Madhya Pradesh (India)	

In agriculture, pathogen management is a top priority to maintain plant health and in turn soil health. The use of agrochemicals represents one of the most popular approaches despite its high cost, and this problem pushes agriculture toward using efficient and long-lasting methods for controlling infection of plants (Ijaz et al. 2019). Bioformulations nowadays are present in variety such as dry formulations (dust, granules, or microgranules), wettable powder forms, and liquid formulations for blending in water like emulsions, suspension concentrates, and oil dispersions (Nagpal et al. 2021). The most important component of the preparation of microbial formulations is the carrier substrate, which needs to be able to host specific microorganisms, although, carriers are inert but degradable which still have an impact on product's effectiveness and shelf life of BCAs (Pahari et al. 2017). For bacterial and fungal-based bioformulations, peat is a preferred carrier though other substrate are also added such as charcoal, lignite, and composts. Moreover, encapsulation technique has been developed as a transport mechanism which are biodegradable and ecofriendly products that has the ability to level the use of pesticide in order to enhance BCAs delivery systems (Balla et al. 2022). The BCAs *Bacillus subtilis* and *Trichoderma harzianum* were tested in vitro in combination with graphite and silica nanoparticles against *Phytophthora infestans*, causative agent of leaf blight of tomato. The formulation delivery strategy effectively controlled the growth of phytopathogen and not hindering the efficacy of the BCAs (Djaya et al. 2021).

Since the production of Bt toxin that is effective against a wide variety of insects, *Bacillus thuringiensis* is frequently employed as a biopesticide. Plant incorporated or genetically modified crops consisting of Bt toxin offer consistent defense against pests and microbial infections. Commercial usage of Bt formulations in the form of sprays or liquids is also growing (Kumar et al. 2021). Extracellular enzymes chitinase and protease secreted by *Beauveria bassiana* and *Metarhizium anisopliae*, respectively, induced the whitefly pests (*Bemisia tabaci* and *Trialeurodes vaporariorum*) mortality rate which is responsible for wide losses of tomato yield (Gebremariam et al. 2022). Insects consuming plants such as plant-chewing insects like Lepidoptera caterpillar pests of crop plants are the main targets of virus pesticides, which act on specific targets. Among entomopathogenic viruses, *Baculovirus* has got most attention in terms of commercial biopesticide development (Rajput et al. 2020). Pests including corn earworms, cotton leaf worms, cabbage moth, beet armyworms, celery loopers, and many more are controlled with virus-based solutions. Some of commercially marketed baculovirus based products include the codling moth granulovirus, the beet armyworm multiple nucleopolyhedrovirus, and the maize earworm single-enveloped nucleopolyhedrovirus (Sabbahi et al. 2022). Several genes such as Bt toxin, juvenile hormone esterase, and wasp toxin might be expressed in nucleopolyhedrovirus to increase the effectiveness by applying recombinant technology in order to increase the pathogenicity and the action speed than wild-type counterpart.

8.8 Conclusion

All agricultural systems are dependent on good soil management and preventing soil deterioration. This can be achieved by employing biocontrol agent (BCA) that will guard soil and plants against invasion of pathogens. Bacteria, fungi, viruses, and yeasts are among the microorganisms employed as BCA. These microbes have efficiently aided in the long-term management of phytopathogens. Plant growth-promoting rhizobacteria (PGPR) has been regarded as an intriguing concept and a long-term approach of eradicating soil-borne pathogenic microorganisms and various diseases. BCA use either direct or indirect mechanisms to control soil pathogens. Direct mechanism involves antibiosis, hyperparasitism, lytic enzymes, and siderophore production, whereas indirect mechanism includes induced systematic resistance (ISR) and competition. The unfavorable consequences of frequent and heavy utilization of chemical-based pesticides includes development of soil-borne diseases, increase in pest resistance, adverse effects on humans, and ecological imbalance. These concerns have led to the restrictions on the application of chemical-based pesticides and adopting more efficient biological control techniques, leading to the development of bioformulations that involves BCAs. This approach is an eco-friendly method that is more effective than synthetic alternatives, and it also assists in restoring the soil health.

References

Abd-Alla AM, Meki IK, Demirbas-Uzel G (2020) Insect viruses as biocontrol agents: challenges and opportunities. In: Cottage industry of biocontrol agents and their applications: practical aspects to deal biologically with pests and stresses facing strategic crops. Springer, pp 277–295

Abd-El-Kareem F, Elshahawy IE, Abd-Elgawad MM (2021) Application of *Bacillus pumilus* isolates for management of black rot disease in strawberry. Egypt J Biol Pest Control 31:25

Abd-El-Khair H (2020) Biological control of phyto-pathogenic bacteria. In: Cottage industry of biocontrol agents and their applications: practical aspects to deal biologically with pests and stresses facing strategic crops, vol 59. Springer, pp 299–336

AbdelRazek GM, Yaseen R (2020) Effect of some rhizosphere bacteria on root-knot nematodes. Egypt J Biol Pest Control 30:1–11

Abo Nouh FA (2019) Endophytic fungi for sustainable agriculture. Microb Biosyst 4:31–44

Abo-Elyousr KAM, Abdel-Rahim IR, Almasoudi NM, Alghamdi SA (2021) Native endophytic *pseudomonas putida* as a biocontrol agent against common bean rust caused by *Uromyces appendiculatus*. J Fungi 7:745

Ahkami AH, White RA, Handakumbura PP, Jansson C (2017) Rhizosphere engineering: enhancing sustainable plant ecosystem productivity. Rhizosphere 3:233–243

Ahmad I, Zaib S (2020) Mighty microbes: plant growth promoting microbes in soil health and sustainable agriculture. In: Soil health. Springer, pp 243–264

Ahmad M, Pataczek L, Hilger TH, Zahir ZA, Hussain A, Rasche F, Schafleitner R, Solberg S (2018) Perspectives of microbial inoculation for sustainable development and environmental management. Front Microbiol 9:2992

Ahmed T, Shahid M, Noman M, Niazi MBK, Mahmood F, Manzoor I, Zhang Y, Li B, Yang Y, Yan C (2020) Silver nanoparticles synthesized by using Bacillus cereus SZT1 ameliorated the damage of bacterial leaf blight pathogen in rice. Pathogens 9(3):160

Akram S, Ahmed A, He P, Liu Y, Wu Y, He Y (2023) Uniting the role of endophytic fungi against plant pathogens and their interaction. J Fungi 9:1–23

Alavaisha E, Manzoni S, Lindborg R (2019) Different agricultural practices affect soil carbon, nitrogen and phosphorous in Kilombero-Tanzania. J Environ Manag 234:159–166

Albayrak CB (2019) Bacillus species as biocontrol agents for fungal plant pathogens. In: Islam M, Rahman M, Pandey P, Boehme M, Haesaert G (eds) Bacilli and agrobiotechnology: phytostimulation and biocontrol, Bacilli in climate resilient agriculture and bioprospecting. Springer, Cham, pp 239–265

Amuza AC, Zaharia R (2022) Trichoderma spp.—mechanisms of action in the control of storage pathogens—review. Sci Papers 66:781–790

Arora NK, Fatima T, Mishra I, Verma S (2020) Microbe based inoculants: role in next green revolution. In: Shukla V, Kumar N (eds) Environmental concerns and sustainable development. Springer, Singapore, pp 191–246

Awad-Allah EFA, Shams AHM, Helaly AA, Ragheb EIM (2022) Effective applications of *Trichoderma* spp. as biofertilizers and biocontrol agents mitigate tomato *fusarium* wilt disease. Agriculture 12:1950

Babbal A, Khasa YP (2017) Microbes as biocontrol agents. In: Kumar V, Kumar M, Sharma S, Prasad R (eds) Probiotics and plant health. Springer, Singapore, pp 507–552

Bagyaraj DJ, Ashwin R (2017) Soil biodiversity: role in sustainable horticulture. Biodivers Hortic Crops 5:1–18

Balla A, Silini A, Cherif-Silini H, ChenariBouket A, Alenezi FN, Belbahri L (2022) Recent advances in encapsulation techniques of plant growth-promoting microorganisms and their prospects in the sustainable agriculture. Appl Sci 12:9020

Bardin M, Pugliese M (2020) Biocontrol agents against diseases. In: Integrated pest and disease management in greenhouse crops. Springer, pp 385–407

Bhat AH, Chaubey AK, Askary TH (2020) Global distribution of entomopathogenic nematodes, *Steinernema* and *Heterorhabditis*. Egypt J Biol Pest Control 30:1–15

Boat MAB, Sameza ML, Iacomi B, Tchameni SN, Boyom FF (2019) Screening, identification and evaluation of *Trichoderma* spp. for biocontrol potential of common bean damping-off pathogens. Biocontrol Sci Tech 30:228–242

Brahmaprakash GP, Sahu PK, Lavanya G, Nair SS, Gangaraddi VK, Gupta A (2017) Microbial functions of the rhizosphere. In: Singh D, Singh H, Prabha R (eds) Plant-microbe interactions in agro-ecological perspectives. Springer, Singapore, pp 177–210

Canto CF, Simonin M, King E, Moulin L, Bennet MJ, Casatrillo G, Laplaze L (2020) An extended root phenotype: the rhizosphere its formation and its impacts on plant fitness. Plant J 103:951–964

Chen K, Tian Z, He H, Long C, Jiang F (2020) *Bacillus* species as potential biocontrol agents against citrus diseases. Biol Control 151:104419

Chugh S, Parshad J, Kuma D, Rani S (2022) Microorganisms as biocontrol agents in vegetable crops. Microbes Humanity Appl:261–288

Cook RJ (1988) Biological control and holistic plant-health care in agriculture. Am J Altern Agric 3:51–62

Cook RJ (1993) Making greater use of introduced microorganisms for biological control of plant pathogens. Annu Rev Phytopathol 31:53–80

Dimkic I, Janakiev T, Petrovic M, Degrassi G, Fira D (2022) Plant associated Bacillus and Pseudomonas antimicrobial activities in plant disease suppression via biological control mechanisms—a review. Physiol Mol Plant Pathol 117:101754

Djaya L, Tohidin Joni IM, Rahma A (2021) Biocontrol delivery system of *Bacillus subtilis* and *Trichoderma harzianum* formulated with graphite and silica nano-particles to control *Phytophthora infestans in vitro*. Adv Biol Sci Res 13:134

Dotaniya ML, Aparna K, Dotaniya CK, Singh M, Regar KL (2019) Role of soil enzymes in sustainable crop production. In: Dotaniya ML (ed) Enzymes in food biotechnology production, applications, and future prospects. Academic Press, pp 569–589

Eassa SMH, Balah AM, Afiah SA, El-Hadidy AE (2018) Allelopathic activity of Sorghum bicolor root parts and exudates on *Bipolaris sorokiniana*. J Crop Prot 7:259–271

Eilenberg J, Hajek A, Lomer C (2001) Suggestions for unifying the terminology in biological control. BioControl 46:387–400

Foster W, Raoult A (1974) Early descriptions of antibiosis. J R Coll Gen Pract 24:889–894

Frac M, Hannula SE, Belka M, Jedryczka M (2018) Fungal biodiversity and their role in soil health. Front Microbiol 9:707

Frey SD (2019) Mycorrhizal fungi as mediators of soil organic matter dynamics. Annu Rev Ecol Evol Syst 50:237–259

Gao P, Qin J, Li D, Zhou S (2018) Inhibitory effect and possible mechanism of a *Pseudomonas* strain QBA5 against gray mold on tomato leaves and fruits caused by *Botrytis cinerea*. PLoS One 13:e0190932

Gebremariam A, Chekol Y, Assefa F (2022) Extracellular enzyme activity of entomopathogenic fungi *Beauveria bassiana* and *Metarhizium anisopliae* and their pathogenicity potential as a bio-control agent against whitefly pests, *Bemisia tabaci* and *Trialeurodes vaporariorum* (Hemiptera: Aleyrodidae). BMC Res Notes 15:117

Grano-Maldonado MI, Ramos-Payan R, Rivera-Chaparro F, Aguilar-Medina M, Romero-Quintana JG, Rodríguez-Santiago A, Nieves-Soto M (2021) First molecular characterization of *Colletotrichum* sp. and *Fusarium* sp. isolated from mangrove in Mexico and the antagonist effect of *Trichoderma harzianum* as an effective biocontrol agent. J Plant Pathol 37:465–475

Guo Q, Li Y, Lou Y, Shi M, Jiang Y, Zhou J, Sun Y, Xue Q, Lai H (2019) Bacillus amyloliquefaciens Ba13 induces plant systemic resistance and improves rhizosphere microecology against tomato yellow leaf curl virus disease. Appl Soil Ecol 137:154–166

Gupta V, Kumar K, Fatima K, Razdan VK, Sharma BC, Mahajan V, Rai PK, Sharma A, Gupta V, Hassan MG, Hussain R (2020) Role of biocontrol agents in management of corm rot of saffron caused by *Fusarium oxysporum*. Agronomy 10:1398

Iakimova ET, Sobiczewski P (2022) Bacteria as biocontrol agents of infectious diseases on horticultural crops. Bulgarian J Agr Sci 28:866–875

Ijaz M, Ali Q, Ashraf S, Kamran M, Rehman A (2019) Development of future bioformulations for sustainable agriculture. In: Kumar V, Prasad R, Kumar M, Choudhary D (eds) Microbiome in plant health and disease. Springer, Singapore, pp 421–446

Islam MN, Ali MS, Choi SJ, Hyun JW, Baek KH (2019) Biocontrol of citrus canker disease caused by *Xanthomonas citri* subsp. *citri* using an endophytic *Bacillus thuringiensis*. Plant Pathol J 35:486–497

Jackson KS, Kumar AD (2019) Management of *Alternaria* leaf spot on Indian mustard through chemical and biological agents. Plant Cell Biotechnol Mol Biol 20:162–178

Jadhav HP, Shaikh SS, Sayyed RZ (2017) Role of hydrolytic enzymes in biocontrol of fungal phytopathogens: and overview. In: Rhizotrophs: plant growth promotion to bioremediation. Springer, Singapore, pp 183–203

Kabdwal BC, Sharma R, Tewari R, Tewari AK, Singh RS, Dandona JK (2019) Field efficacy of different combinations of *Trichoderma harzianum*, *Pseudomonas fluorescens*, and arbuscular mycorrhiza fungus against the major diseases of tomato in Uttarakhand (India). Egypt J Biol Pest Control 29:1

Khasa YP (2017) Microbes as biocontrol agents. In: Probiotics and plant health. Springer, pp 507–552

Kloepper JW, Leong J, Teintze M, Schroth MN (1980) *Pseudomonas* siderophores: a mechanism explaining disease-suppressive soils. Curr Microbiol 4:317–320

Kohle J, Kolnaar R, Ravensberg WJ (2019) Mode of action of microbial biological control agents against plant diseases: relevance beyond efficacy. Front Plant Sci 10:845

Kumar P, Kamle M, Borah R, Mahato DK, Sharma B (2021) *Bacillus thuringiensis* as microbial biopesticide: uses and application for sustainable agriculture. Egypt J Biol Pest Control 31:95

Lacey J (2017) Microbial control of insect and mite pests. From theory to practice. Academic Press, Elsevier, Amsterdam, pp 451–461

Lee JH, Anderson AJ, Kim YC (2022) Root-associated bacteria are biocontrol agents for multiple plant pests. Microorganisms 10:1–25

Lehmann J, Bossio DA, Kogel-Knabner I, Rilliq MC (2020) The concept and future prospects of soil health. Nat Rev Earth Environ 1:544–553

Lin C, Tsai CH, Chen PY, Wu CY, Chang YL, Yang YL, Chen YL (2018) Biological control of potato common scab by *Bacillus amyloliquefaciens* Ba01. PLoS One 13:e0196520

Macabuhay A, Arsova B, Walker R, Johnson A, Watt M, Roessner U (2022) Modulators or facilitators? Roles of lipids in plant root–microbe interactions. Trends Plant Sci 27:180–190

Mawar R, Tomar AS, Singh D (2019) Demonstration of efficacy of bio-control agents in managing soil-borne diseases of various crops in arid region of India. Indian Phytopathol 72:699–703

McNeely D, Chanyi RM, Dooley JS, Moore JE, Koval SF (2017) Biocontrol of *Burkholderia cepacia* complex bacteria and bacterial phytopathogens by *Bdellovibrio* bacteriovorus. Can J Microbiol 63:350–358

Meng J, Zhang X, Han X, Fan B (2022) Application and development of biocontrol agents in China. Pathogens 11:1120

Mhatre PH, Karthik C, Kadirvelu K, Divya KL, Venkatasalam EP, Srinivasan S, Shanmuganathan R (2019) Plant growth promoting rhizobacteria (PGPR): a potential alternative tool for nematodes bio-control. Biocatal Agric Biotechnol 17:119–128

Moosavi MR (2020) Efficacy of microbial biocontrol agents in integration with other *managing methods against phytoparasitic nematodes*. In: Ansari R, Rizvi R, Mahmood I (eds) Management of phytonematodes: recent advances and future challenges. Springer, Singapore, pp 229–258

Nagpal S, Sharma P, Kumawat KC (2021) Microbial bioformulations: revisting role in sustainable agriculture. Biofertilizers 1:329–346

Ons L, Bylemans D, Thevissen K, Cammue BPA (2020) Combining biocontrol agents with chemical fungicides for integrated plant fungal disease control. Microorganisms 8:1930

Pahari A, Pradhan S, Mishra BB (2017) Carrier based formulation of plant growth promotion *Bacillus* species and their effect on different crop plants. Int J Curr Microbiol App Sci 6:379–385

Palmieri D, Ianiri G, Del Grosso C, Barone G, De Curtis F, Castoria R, Lima G (2022) Advances and perspectives in the use of biocontrol agents against fungal plant diseases. Horticulturae 8:1–34

Palyzova A, Svobodova K, Sokolova L, Novak J, Novotny C (2019) Metabolic profiling of *Fusarium oxysporum* f. sp. *conglutinans*race 2 in dual cultures with biocontrol agents *Bacillus amyloliquefaciens*, *Pseudomonas aeruginosa*, and *Trichoderma harzianum*. Folia Microbiol 64:779–787

Peng Y, Li SJ, Yan J, Tang Y, Cheng JP, Gao AJ, Yao X, Ruan JJ, Xu BL (2021) Research progress on phytopathogenic fungi and their role as biocontrol agents. Front Microbiol 12:670135

Pieterse CM, Van Wees SC, Hoffland E, Van Pelt JA, Van Loon LC (1996) Systemic resistance in *Arabidopsis* induced by biocontrol bacteria is independent of salicylic acid accumulation and pathogenesis-related gene expression. Plant Cell 8:1225–1237

Pirttila AM, Tabas MPH, Baruah N, Koskimaki JJ (2021) Biofertilizers and biocontrol agents for agriculture: how to identify and develop new potent microbial strains and traits. Microorganisms 9:817

Poveda AJ (2021) Trichoderma as biocontrol agent against pests: new uses for a mycoparasite. Biol Control 159:1–8

Prajapati S, Kumar N, Kumar S, Maurya S (2020) Biological control a sustainable approach for plant diseases management: a review. J Pharmacogn Phytochem 9:1514–1523

Rajput VS, Jhala J, Acharya VS (2020) Biopesticides and their mode of action against insect pests: a review. Int J Chem Stud 8:2856–2862

Ramakrishna W, Yadav R, Li K (2019) Plant growth promoting bacteria in agriculture: two sides of a coin. Appl Soil Ecol 138:10–18

Rawat P, Shankhdhar D, Shankhdhar SC (2020) Plant growth-promoting rhizobacteria: a booster for ameliorating soil health and agriculture production. In: Soil health. Springer, pp 47–68

Ray P, Lakshmanan V, Labbe JL, Craven KD (2020) Microbe to microbiome: a paradigm shift in the application of microorganisms for sustainable agriculture. Front Microbiol 11:622926

Raymaekers K, Ponet L, Holtappels D, Berkmans B, Cammue BPA (2020) Screening for novel biocontrol agents applicable in plant disease management—a review. Biol Control 144:104240

Roca-Couso R, Flores-Félix JD, Rivas R (2021) Mechanisms of action of microbial biocontrol agents against *Botrytis cinerea*. J Fungi 7:1–26

Rohrmann GF (2019) Baculovirus molecular biology, 4th edn. National Centre for Biotechnology Information (US), Bethesda, MD. https://www.ncbi.nih.gov/books/NBK543458

Sabbahi R, Hock V, Azzaoui K, Saoiabi S, Hammouti B (2022) A global perspective of entomopathogens as microbial biocontrol agents of insect pests. J Agric Food Res 10:100376

Samaras A, Roumeliotis E, Ntasiou P, Karaoglanidis G (2021) *Bacillus subtilis* MBI600 promotes growth of tomato plants and induces systemic resistance contributing to the control of soilborne pathogens. Plants 10:1113

Savary S, Willocquet L, Pethybridge SJ, Esker P, McRoberts N, Nelson A (2019) The global burden of pathogens and pests on major food crops. Nat Ecol Evol 3:430–439

Savita, Sharma A (2019) Fungi as biocontrol agents. In: Giri B, Prasad R, Wu QS, Varma A (eds) Biofertilizers for sustainable agriculture and environment, Soil biology. Springer, Cham, pp 398–411

Sehrawat A, Sindhu SS (2019) Potential of biocontrol agents in plant disease control for improving food safety. Defence Life Sci J 4:220–225

Shahbaz M, Kuzyakov Y, Sanaullah M, Heitkamp F, Zelenev V, Kumar A, Blagodatskaya E (2017) Microbial decomposition of soil organic matter is mediated by quality and quantity of crop residues: mechanisms and thresholds. Biol Fertil Soils 53:287–301

Sheik R, Jee A, Prasad MD, Sharma S (2022) Role of biocontrol agents in agriculture: a review. Pharma Innov J 11:1479–1482

Sood M, Kapoor D, Kumar V, Sheteiwy MS, Ramakrishnan M, Landi M, Sharma A (2020) Trichoderma: the "secrets" of a multitalented biocontrol agent. Plants 9:1–25

Sosa-Gómez DR, Morgado FS, Corrêa RFT, Silva LA, Ardisson-Araújo DMP, Rodrigues BMP, Ribeiro BM (2020) Entomopathogenic viruses in the neotropics: current status and recently discovered species. Neotrop Entomol 49:315–331

Stenberg JA, Sundh I, Becher PG, Björkman C, Dubey M, Egan PA, Viketoft M (2021) When is it biological control? A framework of definitions, mechanisms, and classifications. J Pest Sci 94:665–676

Tahat MM, Alananbeh KM, Othman YA, Leskovar DI (2020) Soil health and sustainable agriculture. Sustainability 12:4859

Tariq M, Khan A, Asif M, Khan F, Ansari T, Shariq M, Siddiqui MA (2020) Biological control: a sustainable and practical approach for plant disease management. Acta Agric Scand B Soil Plant Sci 70:507–524

Théatre A, Cano-Prieto C, Bartolini M, Laurin Y, Deleu M, Niehren J, Fida T, Gerbinet S, Alanjary M, Medema MH (2021) The surfactin-like lipopeptides from *Bacillus* spp.: natural biodiversity and synthetic biology for a broader application range. Front Bioeng Biotechnol 9:118

Tomar P, Thakur N, Yadav AN (2021) Indigenous entomopathogenic nematode as biocontrol agents for insect pest management in hilly regions. Plant Sci Today 8:51–59

Upadhyay AK, Bandi SR, Maruthi PD (2021) BioControl agents—antagonistic magicians against soil borne pathogens: a review. Biol Forum Int J 13:232–242

Van Peer R, Niemann GF, Schippers B (1991) Induced resistance and phytoalexin accumulation in biological control of Fusarium wilt of carnation by *Pseudomonas* sp. strain WCS417r. Phytopathology 81:728–733

Verma PP, Shelake RM, Das S, Sharma P, Kim JY (2019) Plant growth-promoting rhizobacteria (PGPR) and fungi (PGPF): potential biological control agents of diseases and pests. In: Singh D, Gupta V, Prabha R (eds) Microbial interventions in agriculture and environment. Springer, Singapore, pp 281–311

Vinayarani G, Prakash HS (2018) Growth promoting rhizospheric and endophytic bacteria from *Curcuma longa* L. as biocontrol agents against rhizome rot and leaf blight diseases. Plant Pathol J 34:218–235

Wang R, Chen D, Khan RAA, Cui J, Hou J, Liu T (2021) A novel *Trichoderma asperellum* strain DQ-1 promotes tomato growth and induces resistance to gray mold caused by *Botrytis cinerea*. FEMS Microbiol Lett 368:fnab140

Xiong H, Xue K, Qin W, Chen X, Wang H, Shi X, Wang C (2018) Does soil treated with conidial formulations of *Trichoderma* spp. attract or repel subterranean termites? J Econ Entomol 111:808–816

Yi Y, Luan P, Liu S, Shan Y, Hou Z, Zhao S, Jia S, Li R (2022) Efficacy of *Bacillus subtilis* XZ18-3 as a biocontrol agent against *Rhizoctonia cerealis* on wheat. Agriculture 12:258

Chapter 9
Nutrient Recycling by Microbes for Healthy Soil

Amandeep Kaur and Pratibha Vyas

9.1 Introduction

Soil is the fundamental natural resource and a finite source of minerals and living organisms. It is an important component of ecosystem and a central interface between lithosphere, atmosphere, hydrosphere and biosphere. The soil is made up of solid mineral particles such as sand, silt and clay of various sizes, water, air and organic matter. Various forms of life such as plants, insects, annelids, invertebrates and microorganisms are the inhabitants of this resource. Microorganisms of diverse nature are abundantly present in soil, being referred to as microbiome. They play a major role in soil ecosystem and provide various services and perform many functions to maintain soil structure and sustain its quality (Carney and Matson 2005). They facilitate cycling of the nutrients in the soil and continue enriching the soil with nutrients for further growth of life forms. Microorganisms are diverse in nature phylogenetically and help in plant growth, improving soil texture and other properties by making nutrients available by recycling in different ways. Diverse soil microbes also help in the reduction of greenhouse gases and prevent leaching of nutrients (Abarenkov et al. 2010). Microorganisms present in soil belong to various phyla such as Bacteroidetes, Actinobacteria, Ascomycota, Basidiomycota, Firmicutes, Cyanobacteria, Chlorophyta, Proteobacteria, Chlorobi, Euryarchaeota, Gemmatimonadetes, Mucoromycota and Tenericutes. The type of microbiota present in soil depends on pH, moisture, particle size, depth of soil and organic matter present in it. They are present in diverse environments from acidic to alkaline soils. In soil, fertility is the major factor for the growth of crops, and nutrient recycling

A. Kaur · P. Vyas (✉)
Department of Microbiology, College of Basic Sciences and Humanities, Punjab Agricultural University, Ludhiana, Punjab, India
e-mail: vyasp2000@pau.edu

R. K. Bhatia, A. Walia (eds.), *Advancements in Microbial Biotechnology for Soil Health*, Microorganisms for Sustainability 50,
https://doi.org/10.1007/978-981-99-9482-3_9

ultimately helps to enhance the fertility by providing these nutrients to plants. There are about 16 micro- and macronutrients which are essential for plant growth and ought to present in soil. C (carbon), H (hydrogen), O (oxygen), N (nitrogen), P (phosphorus) and S (sulphur) which are abbreviated as CHONPS are the main components of soil. In fact, 95% of total biomass on this planet has these six nutrients as their main constituents. Recycling of these components happens again and again in soil (Tapia-Torres and Morón-Cruz 2020). Soil organic matter and crop residues are the main sources for recycling of nutrients. Nutrients are deposited in the form of complex organic molecules and they are broken down to simple forms to make them available to plants (Lange et al. 2015). Microorganisms help in the enrichment of soil with nutrients in two major ways—(a) by decomposition of organic matter present in soil (b) by immobilization and solubilization of minerals already present in soil in inorganic and insoluble forms. Microorganisms also help in processes like weathering of mineral ores, formation of fossils and recycling of groundwater in indirect way. Without the use of fertilizers, microbial nutrient recycling can provide enough nutrients to soil for further use of plants, thereby contributing towards sustainable agriculture. Soil microorganisms collectively play a significant role in the decomposition of soil organic matter and nutrient recycling leading to fertilization of soil and promoting the plant growth. Microorganisms form the basis of recycling by the breakdown of already available nutrients in soil which cannot be taken by plants by their own, because of their complex nature. From insoluble to soluble form, uptake by plants, microflora to microfauna, microfauna to mesofauna and then to fauna and after death to soil organic matter to microorganisms again, recycling of these nutrients in soil continues.

9.2 Decomposition of Soil Organic Matter

Soil organic matter (SOM) is one of the main components present in soil which undergo nutrient recycling by microorganisms. Soil organic matter consists of living and dead organic substances with carbon as the main constituent. Decomposition is a stepwise process which involves all the living organisms (including microorganisms) present in soil. Microorganisms decompose organic matter as it provides them the energy for their growth and secondly it is the major source for the supply of carbon for new cell formation. Soil organic matter is of two types—active and passive. Decomposition by microbes is done on active soil organic matter containing living or dead animal and plant material. Sugars and proteins present in it are digested by microorganisms. Passive organic matter, for example humus, is the biologically inert material which generally takes hundred or thousand years to fully decompose (Sierra et al. 2012). Carbon dioxide is the major component which is produced after the decomposition of soil organic matter. Carbon, hydrogen, oxygen, nitrogen, phosphorus, sulphur and potassium are the main components of soil organic matter. Organic carbon and nitrogen are inter-wined as the C:N ratio affects the process of decomposition by microbes. Bacteria helps in the digestion of new

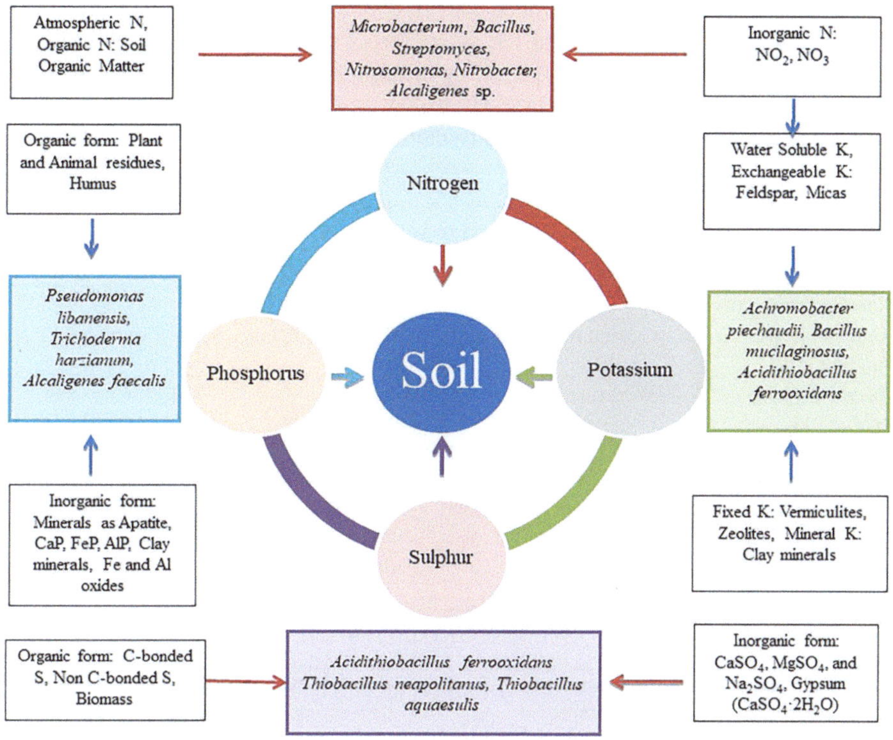

Fig. 9.1 Recycling of macronutrients by microorganisms in soil

plant and animal waste in soil and have high nitrogen content as compared to carbon. Decomposition of passive soil organic matter (including cellulose and lignin) is done by various kinds of fungi, while the transportation of mineral nutrients from soil organic matter to soil is done by mycorrhizal fungi. Protozoa generally feed on other soil microorganisms and also nematodes enter into the cycle and consume microbes. Ammonium and nitrates which are released in soil by them are further transformed in other forms with the help of microorganisms. The fraction of soil organic matter which is not broken down transformed into complex organic compounds known as humic substances. Factors such as pH, moisture, particle size, depth of soil and organic matter (its availability and type) greatly influence microbial population in soil (Hoorman 2010). In this way, the nutrients are added, consumed or decomposed; thus recycling of these nutrients in soil goes on (Fig. 9.1).

9.3 Microbes in Recycling of Macronutrients

Macronutrients are required in large quantities for the growth of plants. Carbon is one of them present as CO_2 in environment serving as the major resource for plants. Soil is one of the biggest reservoirs of carbon. Methanogenesis is done by soil microbes through root exudates and decomposition of soil organic matter. Photosynthesis done by plants is the major process involved in carbon cycling. Partial mineralization of carbon is done by microorganisms. Carbon mineralization is correlated to other nutrient cycling. Hydrogen and oxygen are present in bonding with various molecules. Geochemical dynamics of water cycling is correlated to

Table 9.1 Microorganisms involved in nutrient recycling by different activities

Microorganism	Source	Activity	Reference
Rhodospirillum rubrum	–	Selenite reduction	Sajitz et al. (1993)
Methylophaga sulfidovorans	Coastal microbial mat	Dimethyl sulphide oxidation	de Zwart et al. (1996)
Swaminathania salitolerans	Wild rice	Phosphorus solubilization	Loganathan and Nair (2004)
Thermovibrio ammonificans	Deep-sea hydrothermal vent chimney	Nitrate ammonification	Vetriani et al. (2004)
Caminibacter mediatlanticus	Deep-sea hydrothermal vent chimney	Nitrate ammonification	Voordeckers et al. (2005)
Azospirillum oryzae	Rice	Nitrogen fixation	Xie and Yokota (2005)
Azospirillum brasilense	Corn rhizosphere	Nitrogen fixation	Mehnaz et al. (2007)
Tetrathiobacter kashmirensis	Soil sample from Parlier, California	Selenite reduction	Hunter and Manter (2008)
Pandoraea thiooxydans	Sesame	Thiosulphate oxidation	Rangasamy et al. (2010)
Bacillus megaterium	Mangrove soil	Selenite reduction	Mishra et al. (2011)
Chitinophaga oryziterrae	Soil	Zinc solubilization	Chung et al. (2012)
Bacillus endoradicis	Soybean	Phosphorus solubilization	Zhang et al. (2012)
Rhizobium sp.	Soil	Zinc solubilization	Wani and Khan (2013)
Bacillus megaterium	Wheat	Phosphorus solubilization	Kumar et al. (2014)
Alcaligenes faecalis	Rice	Phosphorus solubilization	Nandini et al. (2014)
Trichoderma harzianum	Chickpea	Phosphorus solubilization	Verma et al. (2014)
Bacillus oryzicola	Paddy	Phosphorus solubilization	Chung et al. (2015)
Paenibacillus hemerocallicola	Day lily	Phosphorus solubilization	Kim et al. (2015)
Paenibacillus oenotherae	Evening primrose	Phosphorus solubilization	Kim et al. (2015)
Aspergillus niger	Soybean	Potassium solubilization	Nath et al. (2015)

(continued)

Table 9.1 (continued)

Microorganism	Source	Activity	Reference
Ensifer adhaerens	Potash mine region	Potassium solubilization	Tang et al. (2015)
Bacillus amyloliquefaciens	Soil	Zinc solubilization	Verma et al. (2015)
Aspergillus niger	Calcareous soil	Phosphorus solubilization	Yin et al. (2015)
Penicillium oxalicum	Calcareous soil	Phosphorus solubilization	Yin et al. (2015)
Spirosoma lacussanchae	Freshwater	Phosphorus solubilization	Li et al. (2017)
Mesorhizobium sp.	*Brassica napus*	Potassium solubilization	Xiao et al. (2017)
Salinicola aestuarinus	Halimione	Phosphorus solubilization	Fidalgo et al. (2019)
Enterobacter ludwigii	Rice	Phosphorus solubilization	Lee et al. (2019)
Streptomyces laurentii	Soils	Phosphorus solubilization	Kour et al. (2020)
Pseudomonas libanensis	Soils	Phosphorus solubilization	Kour et al. (2020)

nitrogen and carbon cycling. Recycling of N, P and K holds a significant place in soil ecosystem as they are required for the growth of crops (Table 9.1).

9.3.1 Nitrogen Recycling

Nitrogen is one of the major macronutrients required for the plant growth and anthropogenic life. Quality and quantity of nitrogen also affect growth in plants. ^{15}N isotope of nitrogen is abundant in nature isotope ^{14}N is light in weight and reacts readily and causes leaching to streams (Denk et al. 2017). In nitrogen cycle, firstly a molecule of dinitrogen gas is converted to ammonia, which is then assimilated into organic nitrogen. Nitrogen present in soil can be categorized as inputs and outputs. Inputs include crop residues, biological N fixation and atmospheric N whereas outputs involve gaseous N compounds resulting from nitrification, denitrification, ammonification, immobilization and leaching of nitrates (Stark and Richards 2008). About 50% of soluble N in soil is present in the form of amino acids. Microbial biomass and soil organic matter contribute these amino acids in soil pool by the breakdown of proteins and peptides present in them. Organic forms of N are mineralized to ammonium (NH_4^+) and further changed to nitrate. Nitrate is more mobile than ammonium; thus cation exchange sites absorb ammonium and further reduce its mass flow and diffusion (Miller and Cramer 2005). Microorganisms help in nitrification, denitrification, ammonification and immobilization, and mycorrhizal fungi promote the depolymerization of N (Jones et al. 2005).

9.3.1.1 Nitrogen Fixation

The nitrogen cycling depends on the biological nitrogen fixation as the first step being carried out by different symbiotic, associative or free-living bacteria. These nitrogen fixers are generally present around the roots of plants and they convert atmospheric dinitrogen to ammonia, thus making nitrogen available to plants (Elmerich and Newton 2007). Free-living N-fixing bacteria are highly sensitive to climatic conditions. Nitrogenase enzyme involved in this process is a combination of iron (Fe–) and molybdenum-iron (MoFe–) protein. These metalloproteins mediate the ATP-dependent reduction of N_2 to NH_4^+ (Hoffman et al. 2014)

9.3.1.2 Ammonification

The organic forms of nitrogen are decomposed to ammonia (NH_3). The microbial enzymes digest the polymers present in lignin and cellulose (Grayston et al. 1998). Organic nitrogen is used in microbial tissue and inorganic nitrogen is released or mineralized. Mineralization of N is one of the major events performed by soil microorganisms in N cycle. The fraction of mineralized organic N determines the C/N ratio of substrate. *Microbacterium*, *Bacillus*, *Streptomyces* and *Pseudomonas* species are the major ammonifying microorganisms. Rhizosphere of plants contain the major population of these microbes. Soil depth and other climatic conditions are associated with their growth. Mineralization of nutrients greatly depends on the C/N in the decomposed material (Wang et al. 2004).

9.3.1.3 Nitrification

NH_4^+ is the most abundant form of nitrogen present in soil. Nitrification is done in two steps: first by oxidizing ammonia to nitrite and second by oxidizing nitrite to nitrate, as ammonia formed in previous step is not a stable product. Chemolithotrophic bacteria are mainly responsible for nitrification process. *Nitrosomonas* and *Nitrosospira* species present in soil help to oxidize ammonia to nitrite. *Nitrobacter* and *Nitrospira* species further convert nitrite to nitrate. Nitrate form can easily leach into soil but microorganisms promote nitrogen retention in soil (Van Kessel et al. 2015).

9.3.1.4 Denitrification

Denitrification plays an important role in decreasing nitrate pollution by reducing it to NO^{2-}. Denitrification is done by reducing N_2O to N_2 in the presence of *Pseudomonas* and *Alcaligenes* species. It causes the depletion of soil fertility by causing the loss of ammonia to atmosphere as N_2 (Barnard et al. 2005). It is considered that all organic N is first oxidized to NH_4 and then it is taken up by soil microbes. This approach, which was substantially developed, is known as the mineralization-immobilization-turnover (MIT) route. In anaerobic circumstances, dissimilatory nitrate reduction to ammonium occurs and can be regarded an alternate route to denitrification, allowing nitrogen to be retained in soil as an accessible form for growth.

9.3.2 Phosphorus Recycling

Phosphorus is the second most important macronutrient required by plants. 50–74% of the P in mineral soils is present in the inorganic forms. They can be divided into non-labile and labile P. Inorganic form of P is present as mineral salts. In alkaline or neutral soils, calcium phosphate is present in the form of francolites and hydroxyapatite, while in acidic soils, Fe and Al oxyhydroxides or carbonates are present as major source of P (Pierzynski et al. 2005). Organic form of phosphorus in soil is

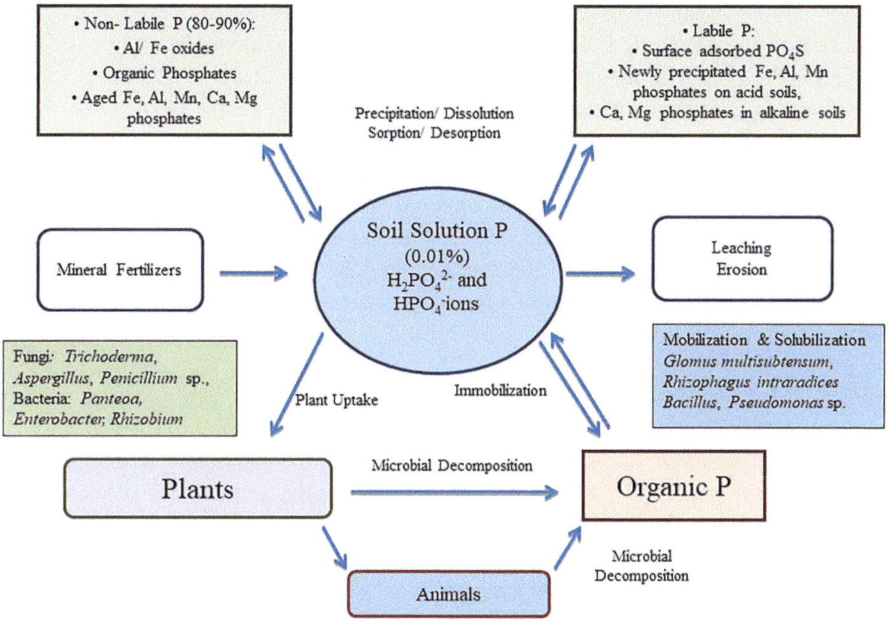

Fig. 9.2 Recycling of phosphorus in soil

present as inositol phosphate, phospholipids, teichoic acids and monophosphorylated carboxylic acids. Eighty percent of most soils have organic P present in the form of inositol phosphates. Mobilization of organic phosphorus is known as mineralization (Barea and Richardson 2015) (Fig. 9.2). Phosphorus mobilizing capacity of some microorganisms from the organic or inorganic sources helps in providing P nutrition for the plants. If the available form of phosphorus is inorganic, then the mobilization is called as solubilization.

9.3.2.1 Phosphate Solubilization

Insoluble forms of phosphate are present in the form of salts of iron, calcium or aluminium. Bacteria, such as *Bacillus*, *Pantoea*, *Enterobacter*, *Rhizobium* and fungi species of *Trichoderma*, *Penicillium* and *Aspergillus* are known to solubilize phosphates (Table 9.1). The solubilization is mainly done by the chelation of organic acids. The solubilization of soil P leading to the acidification of soil results in the production of the mineral-dissolving compounds such as hydroxyl ions, organic acids, etc. The organic acids like aspartic acid, tartaric acid, gluconic acid, oxalic acid, citric acid and lactic acid are produced by microorganisms for the solubilization of insoluble phosphates. The hydroxyl and carboxyl groups of the acid chelate the cation bound to the phosphate, thereby converting it to soluble form.

9.3.2.2 Phosphate Mineralization

Mineralization of organic P releases orthophosphate as primary (PO_4^{3-}) or secondary (HPO_4^{2-}, $H_2PO_4^-$) orthophosphates to soil. It is a hydrolytic process done by phosphatase enzymes. Microorganisms like bacteria such as *Bacillus* and *Pseudomonas* and fungi mainly *Aspergillus* and *Penicillium* are known to mineralize P. Microorganisms produce different types of enzymes categorized as specific and non-specific. Phytase is a specific enzyme which mineralize phytate and inositol phosphates to release orthophosphate. Acid and alkaline phosphatases fall under the no specific enzyme category. These enzymes cause dephosphorylation of phosphoester or phosphoanhydride bonds in soil organic matter (Nannipieri et al. 2011).

9.3.2.3 Mobilization of P

Arbuscular mycorrhiza plays an important role in the mobilization of low accessible P especially in acidic soil. AM fungi take up P from the soil pool as roots release low molecular weight organic acids (LMWOAs). Arbuscular mycorrhizal fungi helps in the mobilization of organic and inorganic forms of phosphorus (Andrino et al. 2021). Studies show that AM fungi increases the levels of phosphatase enzyme activity in mycorrhizal plants that indicates that AM fungus can uptake P from alternative sources in soil pool (Barea and Richardson 2015).

9.3.3 Potassium Recycling

Potassium is the third macronutrient essential for the growth of plants. About 90–98% of total soil potassium is unavailable. Potassium is present in three forms in soil— soluble (exchangeable), insoluble and soil solution. Insoluble potassium is present in the form of muscovite, mica, biotite and feldspar. Only 2% of potassium occurs in solution or exchangeable form available to plants for uptake (Etesami et al. 2017). Potassium is present in soil as K^+ cation, also a part of complex chemical compounds and readily adsorbed by clays and colloids in soil. Potassium solubilization is done by the excretion of organic and inorganic acids, chelation, acidolysis, polysaccharides, complexolysis and various kinds of exchanging processes. Many potassium-solubilizing microorganisms (KSMs) directly dissolve rock K, or chelation of silicon ions occurs to extract K into solution (Prajapati et al. 2013). The prominent organic acids which are produced for solubilization are citric acid, tartaric acid, succinic acid, oxalic acid, α-ketogluconic acid, gluconic acid and malic acid (Keshavarz Zarjani et al. 2013). Due to decomposition of organic materials, there is a production of ammonia and hydrogen sulphide in soil. They oxidize soil to form acids like nitric and sulphuric acid. Cation-exchange complexes in soil get disturbed as hydrogen ions displace magnesium, potassium, calcium and manganese ions (Huang et al. 2013). Microorganisms help in the decomposition of minerals like mica and illite and chelation of silicon and aluminium ions from complex minerals, thereby releasing potassium in soil solution (Zhang and Kong 2014). There are many saprophytic bacteria, actinomycetes and fungi which can solubilize K from soil. Among bacteria, *Acidithiobacillus*, *Agrobacterium*, *Arthrobacter*, *Bacillus*, *Klebsiella*, *Microbacterium*, *Myroides*, *Paenibacillus*, *Pseudomonas* and *Stenotrophomonas* have been reported as potassium-solubilizing bacteria (Rana et al. 2019).

9.3.4 Sulphur Recycling

Sulphur is the fourth essential nutrient after N, P and K in soil for the growth of plants as it plays a vital role in chlorophyll, biotin, thiamine, glutathione and coenzyme A biosynthesis. It is one of the main constituents of essential amino acids such as cysteine, methionine and methionine. It is present in soil in organic and inorganic forms. Organic S is present as heterogeneous mixture of plant and animal residues and microbes (Bloem et al. 2004). It has three fractions such as C-bonded S, non-C-bonded S and biomass (Stevenson 1985). These fractions undergo transformations from one form to another. Amino acids are part of carbon-bonded S; they are present in minor quantities as compared to lipid S (sulpholipids), soluble S and other complex forms. These complex products undergo synthesis and destruction by microorganisms repeatedly. Inorganic S is present largely in sulphate form. Gypsum ($CaSO_4{\cdot}2H_2O$) contain S in large fractions in calcareous and saline soils. $CaSO_4$,

$MgSO_4$ and Na_2SO_4 salts also contain sulphate in high amounts. Microorganisms such as green sulphur bacteria and other prokaryotes oxidize sulphur compounds to sulphur. They obligately or facultatively use sulphur as electron donors. Chemolithotrophs derive their metabolic energy from oxidation. Microorganisms mineralize S-containing cysteine and methionine present in soil organic matter within hours (Wilkinson et al. 2014). *Thiobacillus neapolitanus*, *T. thioparus*, *T. denitrificans*, *T. aquaesulis* and *Paracoccus denitrificans* are some of the examples of sulphur-oxidizing microorganisms. The sulphur cycle involves a series of oxidation-reduction potential by changing the reduction form of sulphur from −2 reduced to +6 oxidized state of organic and inorganic S complexes. Studies by Roy and Roy (2015) showed that *Acidithiobacillus ferrooxidans* and *A. thiooxidans* transform the toxic metal sulphides to less harmful sulphates.

9.4 Recycling of Micronutrients by Microorganisms

9.4.1 Zinc

Zinc is one of the important micronutrients for the growth of plant. It is present in soil in insoluble forms as zinc silicates, zinc oxide, zinc sulphide, sphalerite, smithsonite and willemite. Microorganisms secrete the metabolites which convert these insoluble forms to soluble forms. These microorganisms produce acids which lower the pH of soil and chelate the zinc and make it soluble. The efficient zinc oxide-solubilizing strains include *Pseudomonas monteilii*, *Microbacterium saperdae* (Whiting et al. 2001) and *Pseudomonas aeruginosa* (Fasim et al. 2002). Among these, *Microbacterium saperdae*, *P. monteilli* and *Enterobacter cancerogenus* have been identified as efficient in the solubilization process. Mycorrhizal fungus also helps in Zn solubilization. Ericoid mycorrhizae were also reported to solubilize insoluble Zn compounds like *Paxillus involutus*, *Hymenoscyphus ericae*, *Oidiodendron maius* and *Beauveria caledonica* (Fomina et al. 2005).

9.4.2 Selenium

Selenium is an essential trace micronutrient required for the plant growth. Selenium exists as selenide (Se^{2-}), elemental selenium (Se0), selenite (SeO_4^{2-}) and selenate (SeO_4^{2-}) in soil. Plants uptake selenium in selenite form (Gupta and Gupta 2017). Bacteria catalyse the biological reduction of soluble selenite [Se (IV) as SeO_3^{2-}] and selenate [Se(VI) as SeO_4^{2-}] to poorly soluble elemental selenium [Se(0)] (Nancharaiah and Lens 2015). Se(0) is known to react with various inorganic sulphur compounds (e.g. sulphite [SO_3^{2-}], thiosulphate [$S_2O_3^{2-}$] and sulphide [S_2−]) to form aqueous seleno-sulphur complexes that increase selenium solubility (Ball and

Milne 1995). Heterotrophic *Bacillus megaterium* strain was isolated from seleniferous soil that oxidize Se(0) to Se(IV) (Sarathchandra and Watkinson 1981). Dowdle and Oremland (1998) reported *Thiobacillus* sp. and *Leptothrix* sp. able to oxidize Se(0).

9.4.3 Iron

Iron is one of the essential micronutrients for the growth of plants. It is present in the form of hematite, goethite, lepidocrocite and ferrihydrite in soil. The cycling of iron is linked to carbon, nitrogen, phosphorus and manganese cycling. Fe is sequestered in silicate, minerals are heaviest and oxides are lightest (Wu et al. 2019). Iron exists in two forms—ferrous iron Fe(II) and ferric iron Fe(III). Fe(II) is more soluble as compared to Fe(III). Geochemical cycling of iron is the contribution of both abiotic and biotic processes. Nitrogen-reducing Fe(II)-oxidizing bacteria and Fe(III)-reducing bacteria catalyse redox reactions on the biotic part of iron. Microbial ammonium and methane oxidation are correlated to Fe(III) reduction. Fe(III) reduction plays an important role in anaerobic organic matter degradation. *Thiobacillus denitrificans* helps in reduction of nitrate. *Acidimicrobiaceae* sp. A6 is an ammonium oxidizer under Fe(III)-reducing conditions (Huang and Jaffé 2018). Phosphorus binds to Fe(III) oxides, so motility of phosphorus depends on precipitation or dissolution of iron minerals (Kappler et al. 2021). Manganese is coupled with oxidation of Fe(III). Fe(II)-oxidizing bacteria helps in fixing bicarbonate into organic carbon form. *Gallionella* and *Leptothrix* are the microaerobic oxidizers (Hu and Li 2014). Siderophore production by plants and microorganisms helps in solubilization of iron by chelating iron from its insoluble form and making them available for plant uptake and microbial cells. Soil mineral weathering is one of the consequences of siderophore production. *Suillus granulatus* produce hydromate siderophores which dissolve goethite by high efficiency (Watteau and Berthelin 1994).

9.4.4 Manganese

Manganese plays an essential role in growth and development of plants. In soil manganese is present in oxidized form as Mn^{3+}, Mn^{4+} and Mn^{6+}. Manganese oxides and hydroxides have a high capacity for sorption of heavy metal ions. Birnessite ($MnO_2 \cdot nH_2O$), pyrolusite (MnO_2), braunite ($Mn^{2+}Mn^{3+}{}_6)SiO_{12}$) and rhodochrosite ($MnCO_3$) are the major forms of Mn minerals present in soil. Plants uptake its Mn^{2+} reduced form (Rengel 2015). Bioavailability of manganese depends on Mn^{4+}–Mn^{2+} redox wheel. Sheath-forming bacteria such as *Micrococcus*, *Caulobacter*, *Leptothrix* and *Vibrio* help in oxidation of Mn. *Leptothrix discophora* plays a significant role in Mn cycling. *Bacillus polymyxa* helps to increase Mn availability to plants by using

Mn^{4+} as acceptor, and acids are generated from carbohydrates (Lal and Tabacchioni 2009).

9.5 Conclusion

Microorganisms are present in abundance in soil helping in the nutrient recycling by releasing various enzymes, secondary metabolites, etc. While carrying out these processes, the microorganisms not only decrease the bulk of the material but also provide essential nutrients required for the growth of plants and animals. Many nutrients required for plant growth are present in large quantities in soil, but they are insoluble complexes that plants cannot take up; microorganisms break them in simpler forms. Microorganisms also help to improve the structure, quality and fertility of the soil. Apart from this, they are also involved in the bioremediation of heavy metals and xenobiotics, thus decreasing the soil pollution and contributing towards healthy and fertile soils.

References

Abarenkov K, Nilsson RH, Larsson KH, Alexander IJ, Eberhardt U, Erland S, Høiland K, Kjøller R, Larsson E, Pennanen T, Sen R (2010) The UNITE database for molecular identification of fungi–recent updates and future perspectives. New Phytol 186(2):281–285

Andrino A, Guggenberger G, Kernchen S, Mikutta R, Sauheitl L, Boy J (2021) Production of organic acids by arbuscular mycorrhizal fungi and their contribution in the mobilization of phosphorus bound to iron oxides. Front Plant Sci 12:661842

Ball S, Milne J (1995) Studies on the interaction of selenite and selenium with sulfur donors. Part 3. Sulfite. Can J Chem 73(5):716–724

Barea JM, Richardson AE (2015) Phosphate mobilisation by soil microorganisms. In: Principles of plant-microbe interactions: microbes for sustainable agriculture. Springer, pp 225–234

Barnard R, Leadley PW, Hungate BA (2005) Global change, nitrification and denitrification: a review. Glob Biogeochem Cycles 19(1):1–13

Bloem E, Riemenschneider A, Volker J, Papenbrock J, Schmidt A, Salac I, Haneklaus S, Schnug E (2004) Sulphur supply and infection with Pyrenopeziza brassicae influence L-cysteine desulphydrase activity in Brassica napus L. J Exp Bot 55(406):2305–2312

Carney KM, Matson PA (2005) Plant communities, soil microorganisms, and soil carbon cycling: does altering the world belowground matter to ecosystem functioning? Ecosystems 8:928–940

Chung EJ, Park TS, Jeon CO, Chung YR (2012) *Chitinophaga oryziterrae* sp. nov., isolated from the rhizosphere soil of rice (*Oryza sativa* L.). Int J Syst Evol Microbiol 62(12):3030–3035

Chung EJ, Hossain MT, Khan A, Kim KH, Jeon CO, Chung YR (2015) *Bacillus oryzicola* sp. nov., an endophytic bacterium isolated from the roots of rice with antimicrobial, plant growth promoting, and systemic resistance inducing activities in rice. Plant Pathol J 31(2):152

de Zwart JM, Nelisse PN, Kuenen JG (1996) Isolation and characterization of *Methylophaga sulfidovorans* sp. nov.: an obligately methylotrophic, aerobic, dimethyl sulfide oxidizing bacterium from a microbial mat. FEMS Microbiol Ecol 20(4):261–270

Denk TR, Mohn J, Decock C, Lewicka-Szczebak D, Harris E, Butterbach-Bahl K, Kiese R, Wolf B (2017) The nitrogen cycle: a review of isotope effects and isotope modeling approaches. Soil Biol Biochem 105:121–137

Dowdle PR, Oremland RS (1998) Microbial oxidation of elemental selenium in soil slurries and bacterial cultures. Environ Sci Technol 32(23):3749–3755

Elmerich C, Newton WE (eds) (2007) Associative and endophytic nitrogen-fixing bacteria and cyanobacterial associations, vol 5. Springer, Dordrecht

Etesami H, Emami S, Alikhani HA (2017) Potassium solubilizing bacteria (KSB): mechanisms, promotion of plant growth, and future prospects—a review. J Soil Sci Plant Nutr 17(4):897–911

Fasim F, Ahmed N, Parsons R, Gadd GM (2002) Solubilization of zinc salts by a bacterium isolated from the air environment of a tannery. FEMS Microbiol Lett 213(1):1–6

Fidalgo C, Proença DN, Morais PV, Henriques I, Alves A (2019) The endosphere of the salt marsh plant *Halimione portulacoides* is a diversity hotspot for the genus *Salinicola*: description of five novel species *Salinicola halimionae* sp. nov., *Salinicola aestuarinus* sp. nov., *Salinicola endophyticus* sp. nov., *Salinicola halophyticus* sp. nov. and *Salinicola lusitanus* sp. nov. Int J Syst Evol Microbiol 69(1):46–62

Fomina MA, Alexander IJ, Colpaert JV, Gadd GM (2005) Solubilization of toxic metal minerals and metal tolerance of mycorrhizal fungi. Soil Biol Biochem 37(5):851–866

Grayston SJ, Wang S, Campbell CD, Edwards AC (1998) Selective influence of plant species on microbial diversity in the rhizosphere. Soil Biol Biochem 30(3):369–378

Gupta M, Gupta S (2017) An overview of selenium uptake, metabolism, and toxicity in plants. Front Plant Sci 7:2074

Hoffman BM, Lukoyanov D, Yang ZY, Dean DR, Seefeldt LC (2014) Mechanism of nitrogen fixation by nitrogenase: the next stage. Chem Rev 114(8):4041–4062

Hoorman JJ (2010) Understanding soil microbes and nutrient recycling. In: Agriculture and natural resources, pp 1–5

Hu M, Li F (2014) Soil microbe mediated iron cycling and its environmental implication. Acta Pedol Sin 51(4):683–698

Huang S, Jaffé PR (2018) Isolation and characterization of an ammonium-oxidizing iron reducer: *Acidimicrobiaceae* sp. A6. PLoS One 13(4):e0194007

Huang Z, He L, Sheng X, He Z (2013) Weathering of potash feldspar by *Bacillus* sp. L11. Wei Sheng Wu XueBao (Acta Microbiol Sin) 53(11):1172–1178

Hunter WJ, Manter DK (2008) Bio-reduction of selenite to elemental red selenium by *Tetrathiobacter kashmirensis*. Curr Microbiol 57:83–88

Jones DL, Healey JR, Willett VB, Farrar JF, Hodge A (2005) Dissolved organic nitrogen uptake by plants—an important N uptake pathway? Soil Biol Biochem 37(3):413–423

Kappler A, Bryce C, Mansor M, Lueder U, Byrne JM, Swanner ED (2021) An evolving view on biogeochemical cycling of iron. Nat Rev Microbiol 19(6):360–374

Keshavarz Zarjani J, Aliasgharzad N, Oustan S, Emadi M, Ahmadi A (2013) Isolation and characterization of potassium solubilizing bacteria in some Iranian soils. Arch Agron Soil Sci 59(12):1713–1723

Kim TS, Han JH, Joung Y, Kim SB (2015) *Paenibacillus oenotherae* sp. nov. and *Paenibacillus hemerocallicola* sp. nov., isolated from the roots of herbaceous plants. Int J Syst Evol Microbiol 65(8):2717–2725

Kour D, Rana KL, Kaur T, Sheikh I, Yadav AN, Kumar V, Dhaliwal HS, Saxena AK (2020) Microbe-mediated alleviation of drought stress and acquisition of phosphorus in great millet (*Sorghum bicolour* L.) by drought-adaptive and phosphorus-solubilizing microbes. Biocatal Agric Biotechnol 23:101501

Kumar A, Maurya BR, Raghuwanshi R (2014) Isolation and characterization of PGPR and their effect on growth, yield and nutrient content in wheat (*Triticum aestivum* L.). Biocatal Agric Biotechnol 3(4):121–128

Lal S, Tabacchioni S (2009) Ecology and biotechnological potential of *Paenibacillus polymyxa*: a minireview. Indian J Microbiol 49(1):2–10

Lange M, Eisenhauer N, Sierra CA, Bessler H, Engels C, Griffiths RI, Mellado-Vázquez PG, Malik AA, Roy J, Scheu S, Steinbeiss S (2015) Plant diversity increases soil microbial activity and soil carbon storage. Nat Commun 6(1):6707

Lee KE, Adhikari A, Kang SM, You YH, Joo GJ, Kim JH, Kim SJ, Lee IJ (2019) Isolation and characterization of the high silicate and phosphate solubilizing novel strain *Enterobacter ludwigii* GAK2 that promotes growth in rice plants. Agronomy 9(3):144

Li Y, Ai MJ, Sun Y, Zhang YQ, Zhang JQ (2017) *Spirosomala cussanchae* sp. nov., a phosphate-solubilizing bacterium isolated from a freshwater reservoir. Int J Syst Evol Microbiol 67(9):3144–3149

Loganathan P, Nair S (2004) *Swaminathania salitolerans* gen. nov., sp. nov., a salt-tolerant, nitrogen-fixing and phosphate-solubilizing bacterium from wild rice (*Porteresia coarctata Tateoka*). Int J Syst Evol Microbiol 54(4):1185–1190

Mehnaz S, Weselowski B, Lazarovits G (2007) *Azospirillum canadense* sp. nov., a nitrogen-fixing bacterium isolated from corn rhizosphere. Int J Syst Evol Microbiol 57(3):620–624

Miller AJ, Cramer MD (2005) Root nitrogen acquisition and assimilation. Plant Soil 274:1–36

Mishra RR, Prajapati S, Das J, Dangar TK, Das N, Thatoi H (2011) Reduction of selenite to red elemental selenium by moderately halotolerant bacillus megaterium strains isolated from Bhitarkanika mangrove soil and characterization of reduced product. Chemosphere 84(9):1231–1237

Nancharaiah YV, Lens PNL (2015) Ecology and biotechnology of selenium-respiring bacteria. Microbiol Mol Biol Rev 79(1):61–80

Nandini K, Preethi U, Earanna N (2014) Molecular identification of phosphate solubilizing bacterium (*Alcaligenes faecalis*) and its interaction effect with *Bradyrhizobium japonicum* on growth and yield of soybean *Glycine max* L. Afr J Biotechnol 13(31):3450–3454

Nannipieri P, Giagnoni L, Landi L, Renella G (2011) Role of phosphatase enzymes in soil. In: Phosphorus in action: biological processes in soil phosphorus cycling. Springer, pp 215–243

Nath R, Sharma GD, Barooah M (2015) Plant growth promoting endophytic fungi isolated from tea (Camellia sinensis) shrubs of Assam, India. Appl Ecol Environ Res 13(3):877–891

Pierzynski GM, McDowell RW, Thomas Sims J (2005) Chemistry, cycling, and potential movement of inorganic phosphorus in soils. In: Phosphorus agriculture and the environment, vol 46. American Society of Agronomy, pp 51–86

Prajapati K, Sharma MC, Modi HA (2013) Growth promoting effect of potassium solubilizing microorganisms on *Abelmoscus esculentus*. Int J Agric Sci 3(1):181–188

Rana KL, Kour D, Sheikh I, Yadav N, Yadav AN, Kumar V, Singh BP, Dhaliwal HS, Saxena AK (2019) Biodiversity of endophytic fungi from diverse niches and their biotechnological applications. In: Advances in endophytic fungal research: present status and future challenges. Springer, pp 105–144

Rengel Z (2015) Availability of Mn, Zn and Fe in the rhizosphere. J Soil Sci Plant Nutr 15(2):397–409

Roy S, Roy M (2015) Bioleaching of heavy metals by sulfur oxidizing bacteria: a review. Int Res J Environ Sci 4:75–79

Sajitz P, Klemme JH, Koch HG, Molitor M (1993) Isolation and properties of trimethylamine N-oxide/dimethyl sulfoxide reductase from the purple bacterium *Rhodospirillum rubrum*. Z Nat C 48(9–10):812–814

Sarathchandra SU, Watkinson JH (1981) Oxidation of elemental selenium to selenite by *bacillus megaterium*. Science 211(4482):600–601

Sierra CA, Müller M, Trumbore SE (2012) Models of soil organic matter decomposition: the SoilR package, version 1.0. Geosci Model Dev 5(4):1045–1060

Stark CH, Richards KG (2008) The continuing challenge of nitrogen loss to the environment: environmental consequences and mitigation strategies. Dyn Soil Dyn Plant 2(2):41–55

Stevenson FJ (1985) Cycles of soils: carbon, nitrogen phosphorus, sulfur, micronutrients. Wiley, New York

Tang P, Hu J, Yi L, Peng Q, Peng Q (2015) Isolation and phylogenetic analysis of potassium-solubilizing bacteria. Zhongguo Weishengtaxixue Zazhi (Chin J Microecol) 27(2):125–129
Tapia-Torres Y, Morón-Cruz A (2020) Bacteria have superpowers to recycle soil nutrients. Divers Microb World 7:44
Van Kessel MA, Speth DR, Albertsen M, Nielsen PH, Op den Camp HJ, Kartal B, Jetten MS, Lücker S (2015) Complete nitrification by a single microorganism. Nature 528(7583):555–559
Verma JP, Yadav J, Tiwari KN, Jaiswal DK (2014) Evaluation of plant growth promoting activities of microbial strains and their effect on growth and yield of chickpea (*Cicer arietinum* L.) in India. Soil Biol Biochem 70:33–37
Verma P, Yadav AN, Khannam KS, Panjiar N, Kumar S, Saxena AK, Suman A (2015) Assessment of genetic diversity and plant growth promoting attributes of psychrotolerant bacteria allied with wheat (*Triticum aestivum*) from the northern hills zone of India. Ann Microbiol 65:1885–1899
Vetriani C, Speck MD, Ellor SV, Lutz RA, Starovoytov V (2004) *Thermovibrio ammonificans* sp. nov., a thermophilic, chemolithotrophic, nitrate-ammonifying bacterium from deep-sea hydrothermal vents. Int J Syst Evol Microbiol 54(1):175–181
Voordeckers JW, Starovoytov V, Vetriani C (2005) *Caminibacter mediatlanticus* sp. nov., a thermophilic, chemolithoautotrophic, nitrate-ammonifying bacterium isolated from a deep-sea hydrothermal vent on the mid-Atlantic ridge. Int J Syst Evol Microbiol 55(2):773–779
Wang WJ, Smith CJ, Chen D (2004) Predicting soil nitrogen mineralization dynamics with a modified double exponential model. Soil Sci Soc Am J 68(4):1256–1265
Watteau F, Berthelin J (1994) Microbial dissolution of iron and aluminum from soil minerals—efficiency and specificity of hydroxamate siderophores compared to aliphatic acids. Eur J Soil Biol 30:1–9
Whiting SN, de Souza MP, Terry N (2001) Rhizosphere bacteria mobilize Zn for hyperaccumulation by *Thlaspi caerulescens*. Environ Sci Technol 35(15):3144–3150
Wilkinson A, Hill PW, Farrar JF, Jones DL, Bardgett RD (2014) Rapid microbial uptake and mineralization of amino acids and peptides along a grassland productivity gradient. Soil Biol Biochem 72:75–83
Wu B, Amelung W, Xing Y, Bol R, Berns AE (2019) Iron cycling and isotope fractionation in terrestrial ecosystems. Earth Sci Rev 190:323–352
Xiao Y, Wang X, Chen W, Huang Q (2017) Isolation and identification of three potassium-solubilizing bacteria from rape rhizospheric soil and their effects on ryegrass. Geomicrobiol J 34(10):873–880
Xie CH, Yokota A (2005) *Azospirillum oryzae* sp. nov., a nitrogen-fixing bacterium isolated from the roots of the rice plant *Oryza sativa*. Int J Syst Evol Microbiol 55(4):1435–1438
Yin Z, Shi F, Jiang H, Roberts DP, Chen S, Fan B (2015) Phosphate solubilization and promotion of maize growth by *Penicillium oxalicum* P4 and *aspergillus Niger* P85 in a calcareous soil. Can J Microbiol 61(12):913–923
Zhang C, Kong F (2014) Isolation and identification of potassium-solubilizing bacteria from tobacco rhizospheric soil and their effect on tobacco plants. Appl Soil Ecol 82:18–25
Zhang YZ, Chen WF, Li M, Sui XH, Liu HC, Zhang XX, Chen WX (2012) *Bacillus endoradicis* sp. nov., an endophytic bacterium isolated from soybean root. Int J Syst Evol Microbiol 62(2):359–363

Chapter 10
Plant-Microbe Interaction to Improve Soil Health

Nisha Dhiman, Jana Uthoff, Birgit Scharf, and Vijay Kumar

10.1 Introduction

Soil supports all living beings ranging from unicellular to multicellular forms of all classes. For example, plants derive nutrients, water, and anchorage from the soil, while it also serves as a home for numerous microbes like bacteria or fungi and a multitude of other organisms (Bardgett and Van Der Putten 2014; Banerjee and van der Heijden 2023). The health of this multipartite interaction becomes crucial for the sustainable coexistence of all organisms, which directly or indirectly rely on each other. It is the contribution of each individual partner in this interactome that keeps the soil viable (Gupta et al. 2021).

It is evident that plants are the major source of food, medicine, and other resources for all living beings including microbes, besides serving as sink for many pollutants like heavy metal(loid)s, for example, cadmium (Cd), lead (Pb), mercury (Hg), arsenic (As), pesticides, herbicides, microplastics, and nanoparticles of all types from contaminated soil (Clemens 2006; Yan et al. 2020). Their roots take up these harmful compounds and improve soil health (Yan et al. 2020; Alsafran et al. 2022). Plants are present in a continuous association with microbes in a form of an ecologically

N. Dhiman · B. Scharf
Department of Biological Sciences, Virginia Tech, Blacksburg, VA, USA

J. Uthoff
WG Fermentation and Formulation of Biologicals and Chemicals, Hochschule Bielefeld—University of Applied Sciences and Arts, Bielefeld, Germany

Department of Biochemistry and Physiology of Plants, Faculty of Biology, Bielefeld University, Bielefeld, Germany

V. Kumar (✉)
Department of Biochemistry and Physiology of Plants, Faculty of Biology, Bielefeld University, Bielefeld, Germany

R. K. Bhatia, A. Walia (eds.), *Advancements in Microbial Biotechnology for Soil Health*, Microorganisms for Sustainability 50,
https://doi.org/10.1007/978-981-99-9482-3_10

complex but stabilizing unit called holobiont and rely on these multispecies interactions for carrying out these processes (Banerjee and van der Heijden 2023). The interdependence and ecological contributions of these different life forms have been captured in well-known examples of pioneer *Lichen* species in wastelands or crucial mycorrhizal associations in nutrient-poor soils (Finlay 2008; Kallison 2021). Recent studies have also revealed the involvement of microbes in processes like phytoremediation, aiding plants in maintaining soil quality and health (Yan et al. 2020; Alsafran et al. 2022).

The network of plant-microbe interactions and involved dynamic processes are far from complete understanding (Fadiji et al. 2023). These interactions can be broadly categorized as positive, negative, and neutral associations based on respective effects on individual partners. Positive interactions constitute mutualism and commensalism, while negative interactions include parasitism. Positive plant-microbe interactions involve beneficial bacteria and fungi like in N-fixing root nodules or in mycorrhiza, respectively. On the other hand, negative interactions involve harmful organisms such as pathogenic bacteria, fungi, nematodes, and parasitic plants (Mendes et al. 2013). Some other interactions are even more complex, where beneficial fungi help plants against, for example, nematode invasion. These beneficial microbes could either stay in free-living form in the vicinity of the underground plant parts or behave as an endophyte. Endophytes, especially the beneficial ones that reside within the plant systems, play a crucial role in promoting plant growth by direct or indirect mechanisms. These mechanisms include conversion of atmospheric nitrogen to ammonia, thus providing a vital source of nitrogen; facilitating solubilization of essential mineral nutrients like phosphorus (P), zinc (Zn), or potassium (K); producing plant hormones like auxins, cytokinins, and gibberellins; producing antibiotics and siderophores; and promoting systemic resistance through activated defense systems, etc. (Woźniak et al. 2019). This has led to redirect the focus of researchers toward unravelling, identifying, and characterizing the role of the plant endophytic communities in plant growth and health (Rani et al. 2022). Endophytes reside in various plant tissues such as leaves, roots, phloem sap, and reproductive organs (Fadiji and Babalola 2020; Rani et al. 2022). Several studies have demonstrated the pivotal role of endophytic microbes in triggering production of distinct plant-specific secondary metabolites, which participate in plant defense or other essential metabolic activities. The specificity and biological significance of these biosynthetic pathways is a trending topic for investigation (Fadiji and Babalola 2020; Rani et al. 2022).

However, several pathogenic microbes affect plant health negatively by inducing serious diseases leading to severe crop losses (Nazarov et al. 2020). Crops face 27–42% yield losses due to these pathogens globally (Oerke and Dehne 2004; Noman et al. 2021). A few examples of both beneficial and pathogenic microbes involved with high-value crops have been summarized in Table 10.1. Understanding ecological, physiological, and molecular aspects of these plant-microbe interactions has become even more crucial considering current climate unpredictability. The anthropogenically enhanced global climate change, increasing landfills and wastelands, fast eroding agricultural areas due to rising saline water levels, and increased

Table 10.1 Major crops and their microbial community with positive or negative impacts

Crop	Microbe	Interaction type	Reference
Beta vulgaris	*Macrophomina phaseolina*	Negative, fungal	Duduk et al. (2023)
	Heterodera schachtii	Negative, nematodal	Oro et al. (2022)
Cicer arietinum	*Bacillus subtilis*	Positive, bacterial	Mageshwaran et al. (2022)
Glycine max	*Bradyrhizobium diazoefficiens* syn. *Bradyrhizobium japonicum*	Positive, bacterial	Ledermann et al. (2021)
	Pseudomonas syringae pv. *glycinea*	Negative, bacterial	Budde and Ullrich (2000)
	Pseudomonas savastanoi	Negative, bacterial	Tarakanov et al. (2022)
	Xanthomonas campestris	Negative, bacterial	Mingma et al. (2014)
	Septoria glycines	Negative, fungal	Lin and Mideros (2022)
Pisum sativum	*Rhizobium leguminosarum*	Positive, bacterial	Laguerre et al. (2007)
Solanum lycopersicum	*Bacillus subtilis* and *Brevibacillus brevis*	Positive, bacterial	Konappa et al. (2020)
	Pseudomonas fluorescens	Positive, bacterial	Konappa et al. (2020)
	Ralstonia solanacearum	Negative, bacterial	Yendyo et al. (2017)
	Trichoderma asperellum	Positive, fungal	Konappa et al. (2020)
Solanum tuberosum	*Phytophthora infestans*	Negative, bacterial	Ivanov et al. (2021)
	Fusarium oxysporum	Negative, fungal	Tiwari et al. (2020)
Pennisetum glaucum	*Kosakonia cowanii*, *Pantoea stewartii*, and *Pseudomonas aeruginosa*	Positive, bacterial	Kumar et al. (2021a)
Triticum aestivum	*Klebsiella pneumoniae* 342	Positive, bacterial	Raina et al. (2019)
	Puccinia graminis f. sp. *tritici*	Negative, fungal	Kangara et al. (2020)
	Ustilago tritici	Negative, fungal	Thambugala et al. (2020)
Zea mays	*Erwinia carotovora* f. sp. *zeae Sabet*	Negative, bacterial	Sinha and Prasad (1977)
	Xanthomonas vasicola pv. *vasculorum*	Negative, bacterial	Tebaldi et al. (2022)
	Pantoea stewartii	Negative, bacterial	Roper (2011)
	Ustilago maydis	Negative, fungal	Lanver et al. (2017)

(continued)

Table 10.1 (continued)

Crop	Microbe	Interaction type	Reference
Hordeum vulgare	*Ustilago hordei*	Negative, fungal	Oekmen et al. (2021)
Oryza sativa	*Azospirillum* sp. B510	Positive, bacterial	Raina et al. (2019)
	Azotobacter spp.	Positive, bacterial	Chen et al. (2018)
	Xanthomonas oryzae	Negative, bacterial	Niño-Liu et al. (2006)
	Acidovorax avenae syn. *Pseudomonas avenae*	Negative, bacterial	Xie et al. (2011)
	Acaulospora, Scutellospora	Positive, fungal	Sarkodee-Addo et al. (2020)
	Trichoderma spp.	Positive, fungal	Doni et al. (2014)
	Meloidogyne graminicola	Negative, nematodal	Mantelin et al. (2017)

soil contamination resulting from poor waste management are few examples where an immediate but sustainable solution is required (Nicholls et al. 2021; Ashraf et al. 2014; Ferronato and Torretta 2019; Kumar et al. 2021b). A better overview and understanding of naturally existing plant-microbe interactions that have positive impact on, for example, soil quality in general, can only be beneficial in achieving sustainable and economic solutions to the abovementioned problems.

For instance, plants provide food and shelter to the microbes, while microbes help plants through increased nutrient availability and therefore uptake in a more efficient way. In doing so, microbes help in the restoration of soil health and minimize the need to enrich the soil by adding synthetic fertilizers, which in most cases lead to downstream groundwater contamination (Johnson et al. 2021). Preventing eutrophication through these natural means therefore has multidimensional benefits. Thus, improved understanding of the plant microbe relations provides immense economic benefits through promoting sustainable agriculture and reducing dependency on chemical or artificial fertilizers, besides being beneficial for the soil health in general. The well-known example of evolutionarily conserved As-hyperaccumulation feature in fern *Pteris vittata* and the crucial role of the microbial partners in the process, which was discovered in this century and has been exhaustively studied in the last decade, further highlights the importance of beneficial plant-microbe interactions (Xiong et al. 2010; Yang et al. 2022). Evidence suggests a role of both epiphytic and endophytic microbes in the As-hyperaccumulation process. Some of the identified interactions are highly conserved, even to the extent of potential horizontal gene transfer events resulting in the As-hyper tolerance feature of *P. vittata* (Gu et al. 2018).

Microbes also play a crucial role in biogeochemical cycles maintaining ecosystem balance (Basu et al. 2021). Biological recycling of carbon (C), nitrogen (N), sulfur (S), and other nutrients is strictly dependent on several classes of bacteria. Their mainly heterotrophic lifestyle makes them the end acceptors of autotrophically fixed carbon. Methanogens and aerobic/anaerobic methanotrophs, green or purple bacteria, or cyanobacteria are prime examples. Further, nitrogen biofertilization by diazotrophic bacteria and archaea provides a direct entry point for fixed nitrogen into new biomolecules synthesized by plants. Also, anaerobic respiration, anoxygenic photosynthesis, and the life cycle of lithotrophic bacteria and archaea rely on reduction-oxidation reactions involving sulfates or sulfides. The subsequently improved soil nutrient enrichment increases soil fertility without leading to cross-contamination of groundwater as is the case for chemical fertilization.

This chapter provides detailed discussion of the known symbiotic associations of soil microbes with plants and crucial specific genetic factors, proteins, or biomolecules responsible for the existence of such interactions. The role of these interactions in improvement of ecological sustainability, soil quality, and therefore gains in agriculture are highlighted.

10.2 Plant-Microbe Interactions: Benefits and Losses

In this section, we present a selective and noncomprehensive view of several plant-microbe associations. This should give the reader an idea of the wide spectrum of these interactions and existing natural complexity. The idea behind the section is to highlight why extensive resource investment is required in research pertaining to all forms of plant-microbe interactions and associated molecular mechanisms.

10.2.1 Plant-Fungal Association

10.2.1.1 Mycorrhiza

More than 90% of higher plant species including forest trees, grasses, and crops exist in a natural symbiotic association with fungi wherein the fungal partner aids by increasing water and nutrient availability (e.g., inorganic phosphorus (Pi) and N), priming plants for disease resistance, or are means for biological control of natural plant pathogens. In return, fungi receive carbon from the plants for their growth and reproduction and get shelter (Bonfante and Perotto 1995; Bonfante and Genre 2010; Santoyo et al. 2021). These symbiotic fungi are classified into endo- and ectomycorrhiza, with the latter being present only in the root epidermal area and having no hyphal entry into the root cortex or intercellular spaces. Endomycorrhizal fungi include arbuscular mycorrhiza (AM) that are obligate symbionts and mostly belong to the phylum Glomeromycota (Marschner 2012). AM associate symbiotically with 80% of land plants, influence plant growth positively, and increase crop yield as validated by different studies (Morgan et al. 2005; Chen et al. 2021; Wu et al. 2022). AM are useful for the soil ecosystem as they increase the root surface area, thereby significantly enhancing plant establishment in the soil and nutrient uptake. This increases plant stress resistance and mobility of the ions in the rhizosphere, which leads to improved soil quality, and therefore allows greater plant diversity (Chen et al. 2021). These fungi, mostly belonging to the *Trichoderma* genus, are also used as biocontrol agents for suppressing fungal pathogens and are one of the most beneficial microbes for agricultural as well as horticultural crops (Sood et al. 2020; Yao et al. 2023). Furthermore, over 900 species of edible mycorrhizal fungi are known throughout the world. Along with other wild and medicinal mushrooms, they make a market share of over 60 billion USD (Pérez-Moreno et al. 2021; Royse et al. 2017). In addition, the global market share for biofertilizers (where mycorrhiza contribute majorly) has crossed 2 billion USD and is estimated to go above 5 billion USD by 2030 (Biofertilizers Market Size, Share, Growth, and Analysis 2030).

The importance and usefulness of these fungi have been reviewed widely so we keep this discussion short (see citations above). However, the cost-benefit analysis for AM fungi does not appear to be straightforward (Averill et al. 2019; Bennett and Groten 2022). From an economic standpoint, features like crop yields or stress tolerance can be quantified, but aspects like successful prevention of soil erosion, nutrient loss, or their potential for phytoremediation improvement are mostly qualitatively assessed and require new methods to evaluate their contribution to agroecosystems (Gupta and Abbott 2021). As a matter of fact, the investigated value of mycorrhizal diversity and their utility for sustainable economic gains aligns quite well with the "2030 Sustainability Goals" proposed by the UN (Pérez-Moreno et al. 2021). Quantification issues have led to the gulf between published knowledge concerning the role that native mycorrhizal communities can play as an ecofriendly method at soil restoration, sustainable agriculture, and the actual implementation in agroecosystems (Markovchick et al. 2023). Decisive initiatives and strong implementation strategies are required for safeguarding mycorrhizal diversity and truly realizing their potential benefits in agroecosystem (Markovchick et al. 2023).

10.2.1.2 Plant Pathogenic Fungi

Plant pathogenic fungi are one of the most dominant groups of fungal pathogens responsible for vast yield losses in a variety of crops (Doehlemann et al. 2017). Among plant pathogens, pathogenic fungi are ecologically and economically the most concerning menace with losses amounting to billions of USD annually

(Doehlemann et al. 2017; Shuping and Eloff 2017). Based on the pathogenesis type, they fall in two major categories, that is, "necrotrophic," which use toxins to kill the host and feed of dead plant parts, and "biotrophic," using specialized effectors to inactivate plant defenses, promote cell death, and modulate plant metabolism for their own benefit (Nazarov et al. 2020). Certain plant pathogenic fungi (hemibiotrophs) also fall in between these two categories, where after completion of the fungal reproduction cycle, host cells are damaged. Mostly obligate biotrophs show strict host specificity, while broad host range is also observed. Several different methods have evolved in fungi for attachment, penetration, and host cell metabolic regulation. Due to high genetic flexibility and broad lifestyle range, plant pathogenic fungi are rapid in developing fungicide resistance and invading new hosts through breaking plant defenses, making even more devastating impact (Doehlemann et al. 2017). Pathogenic fungi are mostly from the taxa Ascomycetes (e.g., *Fusarium*, *Alternaria*, *Uncinula*) and Basidiomycetes (e.g., *Ustilago*, *Sclerotium*), but fungi-like pathogens or pseudo fungi from the class of Oomycetes (e.g., *Phytophthora*, *Pythium*, *Albuga* genus) should also be mentioned.

Plant pathogenic fungal epidemics had severe effects like death of millions of people, mass hunger, or forced migration (Kannan et al. 2015). Also, these pathogens have had major contribution in determination of economically and socially viable crop choices in many different places throughout history. For example, coffee rust caused by *Hemileia vastatrix* destroyed coffee crops in the Ceylon region in the 1870s and changed it from one of the major coffee producers to be famous today (Sri Lanka) for its tea production (McCook and Vandermeer 2015). Even today, coffee rust poses a major threat to coffee production worldwide. Currently, farmers facing coffee rust rely on four approaches for fighting the disease, that is, bioprospecting rust-resistant plants, breeding for tolerant plants, chemical control, and improved agroecological practices. Research is focused on finding more environmentally and economically viable solutions for the problem (McCook and Vandermeer 2015). Despite better understanding and management practices for common fungal diseases, epidemics keep occurring from time to time. For example, in recent years, appearance of wheat blast from the Brazilian *Magnaporthe oryzae* (Triticum pathotype) strain in South Asia and South Africa raised eyebrows (Callaway 2016; Singh et al. 2021). Modern modeling approaches have predicted a strong effect of global climate change on, for example, tea cultivation and respective spread of three important tea fungal pathogens, namely, *Colletotrichum acutatum*, *C. camelliae*, and *Exobasidium vexans* (Tibpromma et al. 2021). Current predictions suggest a trend toward complete loss of tea survival potential in currently tea plantation-rich areas and strongly urge for innovative strategies focused on managing future cultivation of crops with a consideration of climate effects on changing diversity and host spectrum of fungal pathogens (Tibpromma et al. 2021).

A general understanding of plant cellular responses against microbial pathogen or symbiont approach exists (see Sect. 10.3). In addition, microRNAs are involved in regulation of hormonal signaling during the host establishment phase and plant immunity (Gupta et al. 2014). Various proteins have been identified to target the RNAi silencing machinery during fungal invasion (Gupta et al. 2014). What is missing is the deeper understanding and identity of molecules like secondary metabolites, which adds to the specificity of host, resistance of pathogens to fungicides, pathogen identity masking effectors, or rapid host adaptability regulating factors in pathogenic microbes, in general and specifically in fungi. For sustainable crop yields in time of the global climatic challenge, new methods need to be developed to control pathogenic fungi through better understanding of pathogenic trait evolution, engineering crop resistance, considering fungal virulence mechanisms, and identifying new strains of plant pathogenic fungi (Doehlemann et al. 2017). For example, RNAi technology has been suggested as an effective mechanism to control plant infection by obligate biotrophs (Padilla-Roji et al. 2023). Further, myconanotechnology has emerged as a green alternative to chemicals for the control of bacterial, fungal, or pest-infections (Patil et al. 2021). Utilizing modern molecular diagnostic techniques to complement traditional methods could urgently be made feasible for fast identification and control of new fungal pathogens (Aslam et al. 2017). Unlike olden times, it is required to study microbial-plant interaction in more natural settings with hugely diverse microbiomes, which certainly influences how pathogenic fungi affect its host plant. Understanding of such multipartite interactions can provide a reasonable chance for early prediction of potential epidemics and their control. Identity of such interactions will further aid in finding alternate biological control systems with preferable features like cost effectiveness and target specificity with new modes of action and biodegradability (Sivanandhan et al. 2017; Shuping and Eloff 2017; Torres-Rodriguez et al. 2022).

10.2.2 Plant-Bacterial Association

10.2.2.1 Plant Growth-Promoting Rhizospheric Bacteria (PGPRs)

Plant growth-promoting rhizospheric bacteria (PGPRs) help in better plant growth and survival (de Andrade et al. 2023). These can exist both in a free-living form and in a mutually beneficial symbiotic association within the plant roots. There are several known bacteria of agricultural importance mainly belonging to the class rhizobia, for example, *Sinorhizobium meliloti*, *Bradyrhizobium diazoefficiens*, and *Rhizobium leguminosarum* in a symbiotic association with leguminous plants alfalfa, soyabean, and pea, respectively, and many other free-living nitrogen-fixing bacteria, for example, *Azospirillum brasilense*, *Pseudomonas fluorescens*, *Azotobacter*, *Bacillus*, and *Clostridium* species. Some of the nitrogen-fixing bacteria such as *Frankia* can even form an association with nonleguminous plants (Santi et al. 2013).

A special variety of microbes called *Streptomyces* inhabits rhizospheric soil and colonize soil as well as the roots and aerial parts of the plants via its filamentous growth pattern. These are applied as biocontrol agents and biofertilizers in several crop systems due to their antagonistic behavior toward plant pathogens as well as their plant growth-promoting activity. They metabolize insoluble organic polymers present in the soil by producing lytic enzymes and promoting nutrient availability to the plants. These bacteria protect the plant from other harmful bacteria by active production of antibiotics and are also known for the typical fragrance of the forest soil due to the production of volatile organic compounds (Vurukonda et al. 2018). PGPRs can enhance plant growth by stress-priming or by imparting tolerance against biotic and abiotic stresses like pathogenic microbes or drought and salt stress, respectively (Backer et al. 2018; Kumar et al. 2021c).

10.2.2.2 Bacterial Pathogens

As per current estimates, we must roughly double the crop yield in the next 20 years to sustain the growing human population (Buttimer et al. 2017). Feeding more than 9 billion people, which is the predicted population by 2050, will be the biggest challenge, especially considering that we are losing substantial agricultural areas due to rising sea levels, anthropogenic contamination, and changing climatic conditions. In addition, microbial pathogens majorly influence net yield each year. Besides fungal pathogens as discussed above, bacterial pathogens are also highly damaging. They show a wide host range and can survive under highly diverse climatic conditions (Kannan et al. 2015). A decade ago, a group of bacteriologists came together to generate a list of the most important plant bacterial pathogens. It was hard to reach consensus on the top 10; however, a few of the most damaging or the most studied ones were part of it, for example, *Pseudomonas syringae* pathovars, *Ralstonia solanacearum*, *Agrobacterium tumefaciens*, *Xanthomonas oryzae* pv. *oryzae*, *X. campestris* pathovars, and *Erwinia amylovora* (Mansfield et al. 2012). According to a recent report of the Food and Agriculture Organization of the UN, plant pathogens are responsible for an annual loss of 220 billion USD (FAO 2019), where bacterial pathogens contribute a major part besides fungi and nematodes.

Successful pathogenesis requires bacteria to succeed not only in finding a compatible host but to also have mechanisms to thrive against adverse environmental factor like competing microbes and diverse abiotic factors. For example, bacteriocins, which are narrow-spectrum protein antibacterials produced by pathogens, rely strongly on the plant host environment to be effective (Ehau-Taumaunu and Hockett 2022). Similarly, processes deemed necessary for pathogenesis like mutations, genomic rearrangements, or even horizontal gene transfer are influenced by the environment, which in turn could influence disease emergence (Bartoli et al. 2016). A better understanding of these environmental effects could contribute toward disease control in many cases. Furthermore, detection of specificity and influence of plant immune response elements like hormones, proteins, or transcripts in pathogenesis regulation might further aid in suppressing disease emergence (Kunkel and Harper 2018; Zimaro et al. 2011; Nobori et al. 2018). Also, understanding the tolerance mechanisms in nonhost plants could provide a broad-spectrum defense strategy against pathogenic bacteria (Senthil-Kumar and Mysore 2013). Inhibitors of the very sophisticated and efficient secretory systems like the type III secretion system (T3SS), which is required by bacteria to inject effectors into the plant cell and suppress plant immunity, have been shown to be an effective management strategy for bacterial diseases (Puigvert et al. 2019). Another disease management approach with exciting outcome has been the use of bacteriophages against specific bacterial pathogens (Buttimer et al. 2017). A major advantage of this system over chemicals or antibiotics would be the capability to design a very specific cocktail of phages for a specific pathogen. A time-dependent resistance development can be avoided by modifying the phage mixture (Buttimer et al. 2017). Most of the above-discussed novel bacterial pathogenesis management strategies require more time and resource investment into the understanding of multitrophic interactions of different microbes in their natural environment.

10.2.3 Multi-microbe Interactions: Tritrophic Interaction of Plant-Fungus-Nematodes in Soil

Nematodes are one of the most abundant metazoans in the biosphere, colonize aquatic and terrestrial ecosystems worldwide (Ptatscheck and Traunspurger 2020; Yeates 1979), and can be plant parasites, omnivore-predators, bacterivore, and fungivore (Yeates et al. 1993; Song et al. 2017). Plant-parasitic nematodes (PPNs) can dramatically reduce yields resulting in substantial economic losses of agronomically important crops (Nicol et al. 2011; Jones et al. 2013; Ghahremani et al. 2022). These economic losses caused by PPNs are difficult to quantify accurately. The global estimates vary considerably, ranging from 80 billion USD (Nicol et al. 2011) to 157 billion USD per year (Abad et al. 2008). Jones et al. (2013) ranked the ten most economically catastrophic PPNs where root-knot nematodes (*Meloidogyne* spp.) and cyst nematodes (*Heterodera* and *Globodera* spp.) make the top of the list. These PPNs can either cause direct damage to their host through endo- or ectophytic feeding on plant tissues or act as vectors for plant pathogenic viruses (Nicol et al. 2011). Additionally, nematode feeding facilitates subsequent infestation by secondary pathogens, such as fungi and bacteria (Powell 1971). Some crops can be host for only a small range of PPNs while others act as host for a wider range of PPN genera and species (Nicol et al. 2011).

Root-knot nematodes (RKN) *Meloidogyne* spp. (order Tylenchida, family Heteroderidae) are one of the most damaging PPNs, infecting a wide range of crops (Carter 1985; Jones et al. 2013). *Meloidogyne* spp. utilizes the host plant by secreting effector molecules to initiate and maintain feeding sites in host roots (Abad and Williamson 2010; Fig. 10.1d). The salivary secretions of these nematodes could contain up to 400 different proteins (Abad and Williamson 2010; Fig. 10.1d). These proteins directly or indirectly assist the pathogen in establishing the feeding sites. For example, enzymes like β-1,4-endoglucanase, endo-1,4-β-xylanase, pectate lyases, polygalacturonase, and glucanases are involved in plant cell penetration, while chorismite mutase and calreticulin are involved in plant defense suppression (Abad and Williamson 2010). Apart from proteins, other small metabolites and biomolecules can play a variety of roles in plant parasitism (Mitchum et al. 2013; Shukla et al. 2016). Parthenogenetic reproduction further adds to the successful invasion by *Meloidogyne* spp. Six biological stages in this life cycle are the eggs, the juvenile stages J1–J4, and the adult. It is the infective juveniles of J2 that can invade the root elongation zone and migrate intercellularly. They establish a permanent feeding site in the vascular cylinder. J2 juveniles can induce giant root gall formations by modulating plant physiology and metabolism. For example, the enzymes from the oxidative pentose phosphate pathway, including glucose-6-phosphate dehydrogenase, ribulose-phosphate-epimerase, phosphoglycerate mutase, and diphosphoglycerate mutase play an important role during formation of giant cells. In addition, reactive oxygen species (ROS) like hydrogen peroxide (H_2O_2), resistant genes (R-genes), and aquaporin expression as well as the production of secondary metabolites like phenolic compounds, terpenoids, saponins, and glucosinolates are influenced by *Meloidogyne* spp. infection (Abad and Williamson 2010; Fig. 10.1c). Auxin conjugates and cytokinin are identified as important components of the plant machinery that are exploited in the establishment of feeding cell formation; however, the detailed mechanisms are still largely unknown (Abad and Williamson 2010; Shukla et al. 2018). After establishment inside the roots, the J2 becomes sedentary and reaches the mature adult female stage producing eggs.

10.2.3.1 Plant Defense Against Plant-Parasitic Nematodes

Plants infested with PPNs show a strong change in gene expression profiles. Shukla et al. (2018) identified in total 1827 significant differentially expressed genes in infected tomato roots under greenhouse conditions. These genes are involved in cell wall structure (e.g., plant hydrolases, endo-β-1,4-glucanases, pectin acetylesterase, b-xylosidases, xyloglucan endotransglucosylases, polygalacturonases, and expansins), development (cyclin-dependent kinases, mitotic cyclins, tubulin, and actin), primary and secondary metabolism, and defense signaling pathways (plant hormones like jasmonate and salicylic acid), depending on the infection stage (Fig. 10.1c). However, resistant tomato plants show only 25 differentially expressed genes. Additionally, some plants possess dedicated R-genes (resistance genes), which also encode for receptors to recognize specific PPN effectors. Some of the R-genes induce a rapid and intense immune defense characterized by a hypersensitive response upon perception of PPNs (Ghahremani et al. 2022). Another element

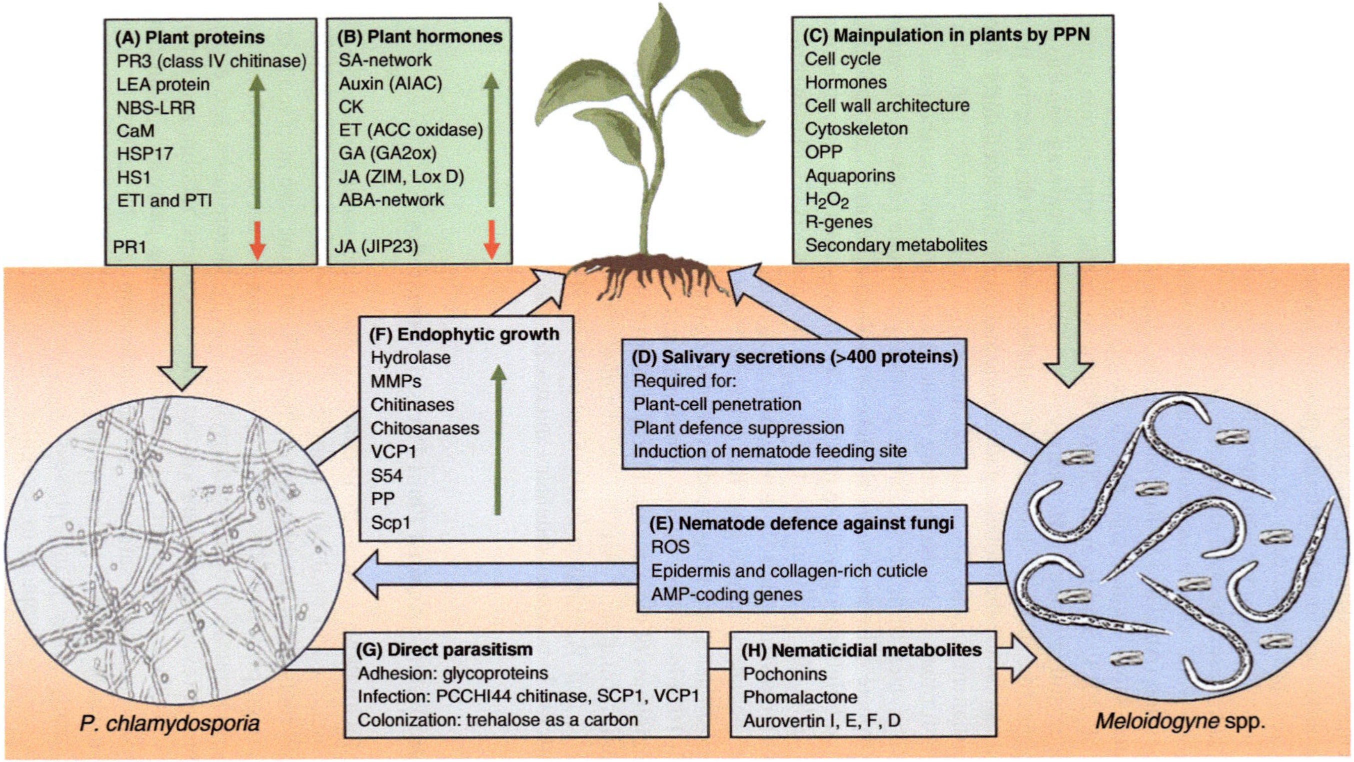

Fig. 10.1 Tritrophic interaction of plants, plant-parasitic nematodes (*Meloidogyne* spp.), and nematophagous fungus *Pochonia chlamydosporia*. (**a**, **b**) Various plant proteins and hormonal networks are influenced by endophytic growth of *P. chlamydosporia*. Up (green arrow)- or down (red arrow)-regulated proteins and metabolites are marked (*PR1/PR3* pathogenesis-related protein 1/protein 3, *LEA* late embryogenesis abundant, *NBS-LRR* nucleotide binding site-leucine-rich repeat, *CaM* calmodulins, *HSP17* heat shock protein 17, *HS1* nematode-resistance protein 1, *ETI* effector-triggered immunity, *PTI* pattern-triggered

immunity, *AIAC* auxin efflux carrier, *GA2ox* gibberellin 2-oxidase, *JIP23* jasmonate-induced protein, *SA* salicylic acid, *CK* cytokinins, *ET* ethylene, *ACC*, aminocyclopropanecarboxylate oxidase, *ZIM* zinc finger inflorescence meristem, *JA* jasmonic acid, *ABA* abscisic acid) (Tolba et al. 2021; Larriba et al. 2015). (**c**) Nematode infection also leads to changes in plant proteomics with major effects on several crucial cell maintenance activities like cell cycle, cell wall architecture, cytoskeleton rearrangements, and secondary metabolism. Cyclin-dependent kinases (CDC2a, CDC2b, and mitotic cyclins (CYC1 and CYCA2;1) are activated in nematode feeding cells (Abad and Williamson 2010). (**d**) PPNs use salivary secretions to influence plant metabolism. The salivary secretion contains >400 proteins. For induction of nematode feeding site, auxin conjugates and cytokinin are suggested to influence plant hormone balance (Abad and Williamson 2010). (**e**) Nematode defense against fungal pathogenesis include antimicrobial peptide (AMP)-coding genes, increase of reactive oxygen species (ROS) and physical barriers like epidermis and collagen-rich cuticle (Zhang et al. 2020). (**f**) For endophytic growth, *P. chlamydosporia* upregulates several proteins including hydrolases, signal transduction protein, metalloproteases (MMP, M28, M35, M36, and M43), chitinases, chitosanases, serine protease (VCP1), rhomboid protease families (S54), protein phosphatase (PP), and serine carboxypeptidase SCP1 (Larriba et al. 2014; Yi et al. 2021). (**g**) *P. chlamydosporia* also directly parasitizes on nematodes. It secretes glycoproteins for adhesion to the nematode egg. (**h**) Pochonins, phomalactone, and aurovertins are nematicidal metabolites produced by *P. chlamydosporia* (Niu et al. 2010). Pochonins are exclusive to *Pochonia* spp. (Hellwig et al. 2003; Niu 2017)

of plant immunity against PPNs is the production of small molecules with anti-nematode activity like phenolic compounds, terpenoids, saponins, or glucosinolates, either before or during nematode infection (Ghahremani et al. 2022; Fig. 10.1c).

10.2.3.2 Biological Control of Damaging Nematodes

Nematophagous fungi act as natural enemies to PPNs. A well-known example is *Pochonia chlamydosporia*, a saprophytic, endophytic, and nematophagous fungus. It is distributed worldwide and known to parasitize eggs and adults (Kerry et al. 1993) of various PPNs by using appressoria for host penetration (Larriba et al. 2014). *Heterodera* spp., *Meloidogyne* spp., *Globodera* spp. and *Nacobbus* spp. are most affected nematode taxa (Tobin et al. 2008; Manzanilla-López et al. 2011). Estimates from culture tests under greenhouse conditions suggest that *P. chlamydosporia* can reduce *Meloidogyne* spp. effectively in various crops (up to 90%; De Leij et al. 1992; De Leij et al. 1993; Vianene and Abawi 2000; Van Damme et al. 2005; Dhawan and Singh 2010). In field trails with carrots, *P. chlamydosporia* increased the marketable production by 41.7%, and the *Meloidogyne incognita* population in soil was reduced by 61.4% (Bontempo et al. 2017).

Fungal proteins like secreted chitinase (PCCHI44) and serine proteases (SCP1 and VCP1) have been described to be involved in nematode infestation, while trehalose from nematode eggs is used as carbon source by the fungi for successful colonization (Yi et al. 2021; Fig. 10.1f). *P. chlamydosporia* is available as a commercial product for biological control of PPNs in countries like Brazil (Rizotec®, Rizoflora Biotecnologia S.A., Brazil) or Cuba (KlamiC®, Centro Nacional de Sanidad Agropecuaria, Cuba). However, endophytic establishment of *P. chlamydosporia* depends on several abiotic (e.g., temperature, pH, carbon and nitrogen ratios, soil structure, soil moisture) and biotic factors (e.g., target preference, reliance on nematodes as a food source, and saprotrophic competitiveness) (Siddiqui et al. 2009; Luambano et al. 2015; Ghahremani et al. 2019, 2022). Furthermore, the interspecific interactions with plant roots and endophytic growth play an important role in the tritrophic interaction of plants, fungus, and nematodes. For example, an endophytic strain of *P. chlamydosporia* can produce indole-3-acetic acid (IAA) as observed for in vitro cultures. IAA has a critical role in the regulation of plant growth and development. An increase in IAA enhances root growth and thus possibly tolerance to nematode infection by compensating for restricted water transport (Schouten 2016). Additionally, *P. chlamydosporia* isolates can proliferate inside the roots and induce plant defense by forming papillae or through other cell wall changes associated with plant resistance in barley and tomato (Escudero and Lopez-Llorca 2012; Larriba et al. 2015; Schouten 2016). In barley, Larriba et al. (2015) described upregulated genes for chitinases and methyl esterases associated with response to pathogens during endophytic growth of *P. chlamydosporia*. Furthermore, they described other pathogen-induced genes, such as genes encoding germin A-like proteins, pathogenesis-related proteins, and late embryogenesis abundant (LEA) protein as involved in the plant response during endophytic growth (Fig. 10.1a, b). *P. chlamydosporia* root colonization in barley leads to transcriptional changes with enrichment of transcripts involved in abiotic stress response, mainly coding for heat shock proteins and proteins involved in phytohormone biosynthesis like auxin, ethylene, salicylic acid, and jasmonic acid (Larriba et al. 2015; Ghahremani et al. 2019; Tolba et al. 2021; Fig. 10.1a, b). Salicylic acid and jasmonic acid (JA) are known as signal molecules involved in systemic acquired resistance (SAR). SAR is a mechanism of plant defense that leads to systemic protection of the entire plant, after sensing an initial local pathogen signal. However, it needs to be underlined that the specific role of these altered hormonal profiles in the plant during this complex three-part interaction of plant-fungi-nematode is still unresolved. The complete resolution of the interaction networks for involved hormones like SA, JA, and others needs further research effort. Additionally, Gouveia et al. (2019) described the mechanism whereby *P. chlamydosporia* dissolves inorganic phosphate, produces extracellular phosphatases to hydrolyze organic phosphorus compounds, and synthesizes acetic, citric, and propionic acid. These endophytic fungi therefore not only promote plant growth through endophytic growth and priming plants against pathogenic nematodes, but they also improve soil nutrient status and hence benefit soil quality. Thus, they provide efficient cover against pathogens while being environment-friendly unlike chemical toxins.

10.3 Chemical Determinants of Plant-Microbe Interactions

10.3.1 Plant Signals

The phyllosphere, the aboveground surface area of a plant that provides potential habitats for all kinds of microbes, stays in a constant state of releasing phytochemicals that can function as signals for or against any potential microbial interaction and are strongly influenced by the plant age and environmental abiotic and biotic factors alike (Zhang et al. 2022). Similarly, plant roots also form the rhizosphere, a zone of high physical, biological, and biochemical activity. Most of the biochemicals in the rhizosphere originate from roots in the form of root exudates and influence plant anchorage and nutrient uptake and assimilation. This biochemical activity is also largely modified by environmental factors like water, temperature, nutrient form and availability, and biotic factors like types of symbionts or pathogens present. The root exudates correspond to 20–40% of the photosynthetically fixed carbon (Badri and Vivanco 2009; Canarini et al. 2019). These include low molecular weight compounds like amino acids, organic acids, sugars, quaternary ammonium compounds (QACs), phenolics, flavonoids, hormones, and other secondary metabolites, as well as a wide variety of high molecular weight compounds such as carbohydrates (polysaccharides), proteins, or their derivatives (Dundek et al. 2011; Canarini et al. 2019). The low molecular weight compounds are predominantly released passively against a density gradient by diffusion, while the high molecular weight compounds are actively transported out of the roots through several transport proteins (Canarini et al. 2019). The process of diffusion of root exudates occurs mostly at the root tips (Ross-Elliott et al. 2017).

The exuded compounds often play a major role either in plant-to-plant communication or plant-to-microbe communication, aiding plant processes such as nutrient absorption or stimulation of fruitful association with the symbionts (Badri and Vivanco 2009; Seitz et al. 2022). These root exudates help plants maintain a highly plant-specific but diverse population of microbes in the vicinity of their roots (Badri and Vivanco 2009). The released biomolecules can be therefore considered as the major regulators of microbial community composition in the rhizosphere (Badri and Vivanco 2009; Seitz et al. 2022). Amino acids, their derivatives, and certain organic acids are recognized to contribute as important facilitators in establishing and determining the specificity of plant-microbe interaction. They act as signaling molecules, nutrient sources, or even toxins, thereby influencing the behavior of individual microbial communities involved in the plant-microbe interactions (Badri and Vivanco 2009; Jacoby et al. 2020; Moormann et al. 2022). Another major biochemical-dependent activity that is crucial in the rhizosphere is the inter and intraspecies communication of microbes. The most common regulating factor for this type of communication is quorum sensing (QS) molecules (Singh et al. 2022). A wide variety of QS signaling networks have been studied in bacteria-to-bacteria communication relying on organic molecules like acetylated homoserine lactones (AHLs), phytohormone mimics, diketopiperazines (DKP), and other secondary metabolites and small organic compounds (Singh et al. 2022). Such biochemicals in QS are referred to as autoinducers (AIs). QS is a necessity for survival providing bacteria with a competitive edge in tough habitats. Plant root colonization is promoted by QS-dependent biofilm formation in the rhizosphere (Noirot-Gros et al. 2018). In specific cases, QS is also useful for stimulating plant growth and improving crop yields (Singh et al. 2022).

Mechanistic understanding of how plant root and microbe interaction is controlled and how specificity is determined remains poor. The directional movement of the bacteria toward a concentration gradient of specific compounds of plant origin, that is, chemotaxis is considered one of the most important phenomena in the start of this plant-microbe interaction. Several studies have been conducted to analyze the chemotactic response of different bacterial species toward different chemical compounds present in the root and seed exudates. These studies analyzed the chemotactic responses of different bacteria using capillary assays, blind well chambers, and soft agar plates (Raina et al. 2019). In case of *R. leguminosarum*, amino acids and carboxylic acids were tested using blind well chambers, while capillary assay was used for testing sugars only. On the other hand, amino acids, carboxylic acids, phenolics, and QACs were tested in case of *S. meliloti* using capillary assays. Among the tested compounds, specific amino acids such as proline, glycine, serine, and cysteine demonstrated positive chemotactic responses at a concentration range from 10^{-5} to 10^{-6}. Similarly, carboxylic acids such as succinate, malate, and fumarate exhibited positive chemotactic responses when present at concentrations between 10^{-2} and 10^{-4}. Additionally, sugars such as arabinose, ribose, xylose, glucose, and fructose showed positive chemotactic response at a concentration of 10^{-6} M concentration (Raina et al. 2019). However, the concentration of these root and seed exudate components can be different under natural conditions, that is, in the native soil than the concentrations tested in different studies under laboratory conditions. Therefore, research is being conducted in method development to analyze the metabolome of different plant species with reduced noise and improved plant species differentiation mimicking natural systems. A recent study by Williams et al. (2021) introduced a new hybrid root washing method based on three plant species from different functional groups: *Holcus lanatus*, *Rumex acetosa*, and *Trifolium repens* representing grass, forbs, and legumes, respectively. This method claims to provide efficient recovery with minimal contamination and the potential to differentiate the metabolome profiles of different plant species. Authors claim that their findings regarding improvements of the hybrid root washing method for better understanding the root exudation process are significant in an ecological context; however, further research involving a broader range of plant species could further corroborate and validate their findings (Williams et al. 2021). One of the significant challenges has been the lack of sufficient data regarding the prevailing concentrations of these microbial attractants in the native habitats of specific microbes as well as microbial sensing mechanisms for the same. These factors appear to be key components affecting bacterial chemotaxis toward specific plant species, microbial growth induction, host identification at the plant root cell surface, and successful initiation of symbiosis. Efforts are being made in this direction to address such knowledge gaps. For instance, four out of eight chemoreceptors of *S. meliloti* species have been recently characterized. These chemoreceptors are specific for a wide concentration range of amino acids, certain organic acids, and quaternary ammonium compounds (QACs) (Webb et al. 2016, 2017a, b; Compton et al. 2018; Baaziz et al. 2021).

A new class of signaling compounds, strigolactones (SLs), which regulates this mutualistic association between plants and fungi has been discovered recently (Waters et al. 2017; Mitra et al. 2021). SLs were later categorized as plant hormones due to their similarity in function to phytohormones, regulating bud growth and stem branching (Mishra et al. 2017). Most (~80%) of the plant roots are reported to release SLs. Structurally, these are terpenoid lactones with a tricyclic-lactone (ABC-ring) and hydroxymethyl butenolide (D-ring) connected through an enol ether bridge (Seto et al. 2014). The conserved nature of the D-ring across plant species reveals its importance for its biological activity (Umehara et al. 2015). They promote plant-fungal interaction by stimulating fungal spore germination and hyphal branching. They are also known to promote the parasitic association between host plant and parasitic plant partners like *Striga hermonthica* by stimulating its seed germination. Furthermore, they inhibit stem branching while promoting plant secondary growth via regulation of auxin transport (Seto et al. 2014). Although the complete biosynthesis pathway of SLs has not been elucidated, yet they are synthesized via a carotenoid pathway from their precursor molecules, that is, carlactone, which was demonstrated in vivo in rice (Seto et al. 2014). The abscisic acid (ABA) biosynthetic pathway is also involved in the regulation of SL biosynthesis in the plant roots (López-Ráez et al. 2010).

The mutualistic association facilitated by SLs is similar to the rhizobacterial association with leguminous plant roots (Carvalhais et al. 2019). However, the mycorrhizal association is believed to be more primitive. It involves the release of signals by the fungal partner, in the form of Myc factors, which are lipochito-oligosaccharides behaving like Nod factors in rhizobacteria during active symbiosis with the leguminous plants (Liang et al. 2014). Myc factors establish the symbiotic relationship of fungi with their host plants ranging from legumes to trees such as members of the family Fabaceae, Asteraceae, and Apiaceae (Sokornova et al. 2022). AM has a broad range of hosts and can successfully form its network with plant roots (Giovannetti et al. 2006; Lee et al. 2013; Figueiredo et al. 2021).

10.3.2 PAMPs (Pathogen-Associated Molecular Patterns) and Effectors

Molecular markers on the surface of pathogens, pathogen-associated molecular patterns (PAMPs), are microbial tools for host recognition and pathogenicity. However, they can also be exploited by the hosts via extracellular plant pattern recognition receptors (PRRs) to induce a suitable immune response as the basal first line of defense, that is, pattern-triggered immunity (PTI). It is worth noting that these PAMPs are not only specific for pathogens but are also displayed by all microbes (microbe-associated molecular patterns, MAMPs). In contrast, effectors are proteins and other molecules secreted specifically by pathogens into infected cells, determine host-pathogen compatibility, and induce pathogenicity. Most commonly in highly potent plant pathogens like *P. syringae* pathovars, T3SS-secretion system is used for the injection of the effector molecules (Xin et al. 2018). These effectors can be detected by the host through the second layer of intercellular plant defense proteins and induce effector-triggered immunity (ETI). From the host perspective, PAMPs are mostly conserved among all microbes and trigger a general tolerance mechanism against all classes of microbes, while effectors are species, race, and strain specific and mostly needed for pathogenic virulence (Thomma et al. 2011). Neutral or mutualistic microbes can bypass the plant immune response during pathogenic microbe detection. On the one hand, plant immune system elements play a crucial role in defining the composition of rhizosphere microbiome through MAMP detection, while on the other hand beneficial microbes like rhizobia or AMs can actively suppress plant immune response for their survival (Yu et al. 2019).

Although PTI and ETI have been understood as a well-established two-pronged strategy by plants against virulent pathogens, recent studies suggest their close interrelationship (Thomma et al. 2011; Tena 2021; Ngou et al. 2021; Yuan et al. 2021a, b). For example, in studies involving PRR mutants, it becomes evident that components downstream of PTI are crucial for the rapid and strong NLR (nucleotide binding, leucine-rich repeat receptor)-mediated plant immunity, hence feeding into ETI (Yuan et al. 2021b). Therefore, a new model is emerging where the two branches of immunity are viewed as components of one multiphasic response with each individual phase fusing with the next phase for a heightened plant immune response (Tena 2021). Deeper understanding of how the interaction of cell surface receptors and intercellular effectors-receptors potentiate the plant toward a stronger defense against pathogens has implications for sustainable crop yield (Ngou et al. 2021).

The detection of these "non-self" and "conserved" molecular signatures functions like a danger signal and subsequently activates an immune response like that generated by molecules specific for nematodes (NAMPs) or any molecular degradation products from pathogen-induced damage (DAMPs) in plants (Choi and Klessig 2016). MAMPs could include flagellin, lipopolysaccharides or lipoproteins, peptidoglycans, mannose-rich glycans, fungal chitin, and EF-Tu (elongation factor Tu) (Choi and Klessig 2016), while effectors include different virulence molecules, proteins secreted via secretory systems like T3SS, small molecule toxins, exopolysaccharides, cell wall-degrading enzymes, plant hormones, or hormone mimics as is the case in virulent *Pseudomonas* strains (Xin et al. 2018). In response, plants possess widely known PRRs and NLRs, which are differentially localized leucine rice repeat-receptor proteins (LRR). Plant PRRs are part of plasma membrane-bound multiprotein complexes and require tight regulation through phosphatases, the protein degradation machinery, and many other regulatory proteins for efficient induction of the plant immune response (Macho and Zipfel 2014). From reports involving the most studied pair of MAMPs/PRR, namely, flagellin/FLS2 and EF-Tu/EFR, it becomes evident that pathogens are equipped with effectors that can be used to inhibit PRRs directly or indirectly via the downstream signaling pathway (Boller and Felix 2009). Interestingly, the tug of war does not stop here. Plants turn to R-proteins, which function to guard the effector target proteins and hence ensure a successful immune response (Ingle et al. 2006). This protein interaction network of effector proteins and plant R proteins can be multilayered and mostly determines host specificity and success of invasion (Glowacki et al. 2011). Identification of microbial effectors and diverse plant R proteins and their specific targets or interaction partners is a work in progress. Pathogens with evolved host specificity are very potent in suppressing host plant immunity, but the existence of potentially critical R-protein networks in nonhost plants successfully protects them against disease emergence. Research focusing on the identification of such plant immunity elements could be useful in designing biological control mechanisms for pathogen susceptible crops.

P. syringae strain DC3000 infection in *Arabidopsis thaliana* has been one of the best studied pathogenesis systems (Katagiri et al. 2002; Xin et al. 2018). *P. syringae* enters plants through natural openings or wounded tissue and is recognized through PAMPs. The stimulation of basal plant immunity involves the phytohormone salicylic acid (SA), which could eventually lead to systemically acquired resistance. As a countermeasure, *P. syringae* strain is equipped with a phytohormone mimic called coronatine, which is secreted into the *Arabidopsis* cells through the T3SS along with multiple other effectors such as the proteins AvrB or HopX1. Coronatine in *Arabidopsis* activates the JA signaling response through binding to the JA-receptor coronatine-insensitive protein 1 (COI1). JA signaling thus counteracts SA signaling to inhibit the plant immune response. However, this is a simplistic description of the events. In general, the targets of several identified effectors or R-protein are still unknown. The interaction of SA-JA signaling pathways and their contribution to pathogenesis success/failure needs to be further investigated.

10.3.3 Differentiating Pathogenic, Neutral, and Mutualistic Microbes

Understanding the complex specificity pattern of plant-microbe interactions needs further investigation to properly utilize its potential for sustainable agriculture. Plant roots have a unique ability to establish symbiotic associations only with beneficial microbes while avoiding harmful ones (Yu et al. 2019). This unique ability is imparted by a diverse array of secondary metabolites that are exuded by the plant roots including phenolics and flavonoids. These metabolites act as defense molecules against pathogenic microbes, which might be one of the responsible factors contributing toward the selective recruitment of beneficial microbes (Jacoby et al. 2020). Various compounds have been identified as attractants to induce chemotactic movement of microbes toward their plant host roots. For example, in case of *Rhizobium* sp. and *Phaseolus vulgaris*, phenolics have been identified as the compounds responsible for attraction, while amino acids and QACs play a crucial role in case of the *S. meliloti* and *Medicago sativa* interaction (Dong and Song 2020; Compton and Scharf 2021). However, it is important to consider that this response can vary depending on the specific host and its microbial partners. This is further validated by a study conducted on *S. meliloti* and *M. sativa* (alfalfa) where flavonoids failed to induce a significant chemotactic response (Compton et al. 2020). Even if nonspecific microbes manage to establish contact with a nonhost plant, it will fail to carry out nodule formation or form nonfunctional nodules (Wang et al. 2018). The mechanisms underlying the specificity of these interactions and the methods by which a host plant recruits specific microbes and undergoes a symbiotic association are still under research.

10.4 Case Study: Root Nodules in Leguminous Plants

Numerous plant species like rice (*Oryza sativa*) maintain a heterogenous microbial community in their rhizosphere. Many of these bacteria also exist as an endosymbiont in the roots of rice, but they do not induce nodulation (Chen and Zhu 2013).

Nodulation is a special characteristic restricted mostly to the members of the family Fabaceae. These leguminous plants (e.g., beans, pea, clover, and alfalfa) form a symbiotic association with a specific class of bacteria called rhizobia. Among these, *S. meliloti* and its host plants *M. sativa* (alfalfa) and *M. truncatula* are the best-studied model system to understand this unique relationship that offers several benefits to both partners (Ghosh et al. 2021). Other members of Fabaceae like *Glycine max* (soybean) and *Pisum sativum* (pea) can form a symbiotic relationship with *B. diazoefficiens* and *R. leguminosarum*, respectively. Other well-studied plant symbionts include *Agrobacterium fabacearum*, *Azorhizobium caulinodans*, and *A. brasilense* (Aroney et al. 2021).

10.4.1 Mechanism of Interaction

Specialized proteins in leguminous plants and their symbionts mediate signaling between both partners to initiate the nodulation process (Fig. 10.2). Bacterial chemoreceptors sense compounds released in the rhizosphere as root exudates that enable a multistep process facilitating chemotaxis toward the roots. These chemoreceptors bind specific compounds as ligands resulting in either attraction or repulsion and play a crucial role in guiding bacterial movement. The composition and characteristics of compounds released by leguminous plants affect the diversity of the microbial population in their rhizosphere significantly as root exudates serve as carbon and nitrogen source. Moreover, certain compounds function as specific signaling molecules to regulate crucial host-microbe interactions (Feng et al. 2021a).

One requirement for a successful host-microbe interaction is the directed movement of bacterial cells toward the roots of their host plant, known as chemotaxis. The alfalfa symbiont *S. meliloti* possesses 4–6 peritrichous flagella, which rotate clockwise in a bundle and propel the cell. The random movement of the cell can be biased toward an attractant and away from a repellent (Schmitt 2002). This process is facilitated by special receptors, named methyl-accepting chemotaxis proteins (MCPs), present either in the bacterial membrane or the cytoplasm (Collins et al. 2014). *S. meliloti* possesses eight chemoreceptors, McpT–McpZ and IcpA that mediate chemotaxis (Meier et al. 2007). They are localized in a cluster at the cell pole (Meier and Scharf 2009). The function of four of these receptors is known. McpU and McpX can sense amino acids and QACs, respectively (Webb et al. 2016, 2017a, b). McpV is a sensor for short-chain carboxylic acids, such as acetic and propionic acid (Compton et al. 2018), while McpT recognizes carboxylates like glyoxylate, malate, malonate, and oxalate (Baaziz et al. 2021).

Binding of a chemoattractant to the periplasmic or cytoplasmic ligand binding domains of these specific MCPs leads to inactivation of the histidine autokinase, CheA (Meier et al. 2007; Fig. 10.2), while absence of a chemoattractant results in autophosphorylation of CheA followed by transfer of the phosphoryl group to the response regulator CheY2. Phosphorylation activates CheY2, which then interacts with the flagellar motors and regulates their rotational speed. The asynchronous

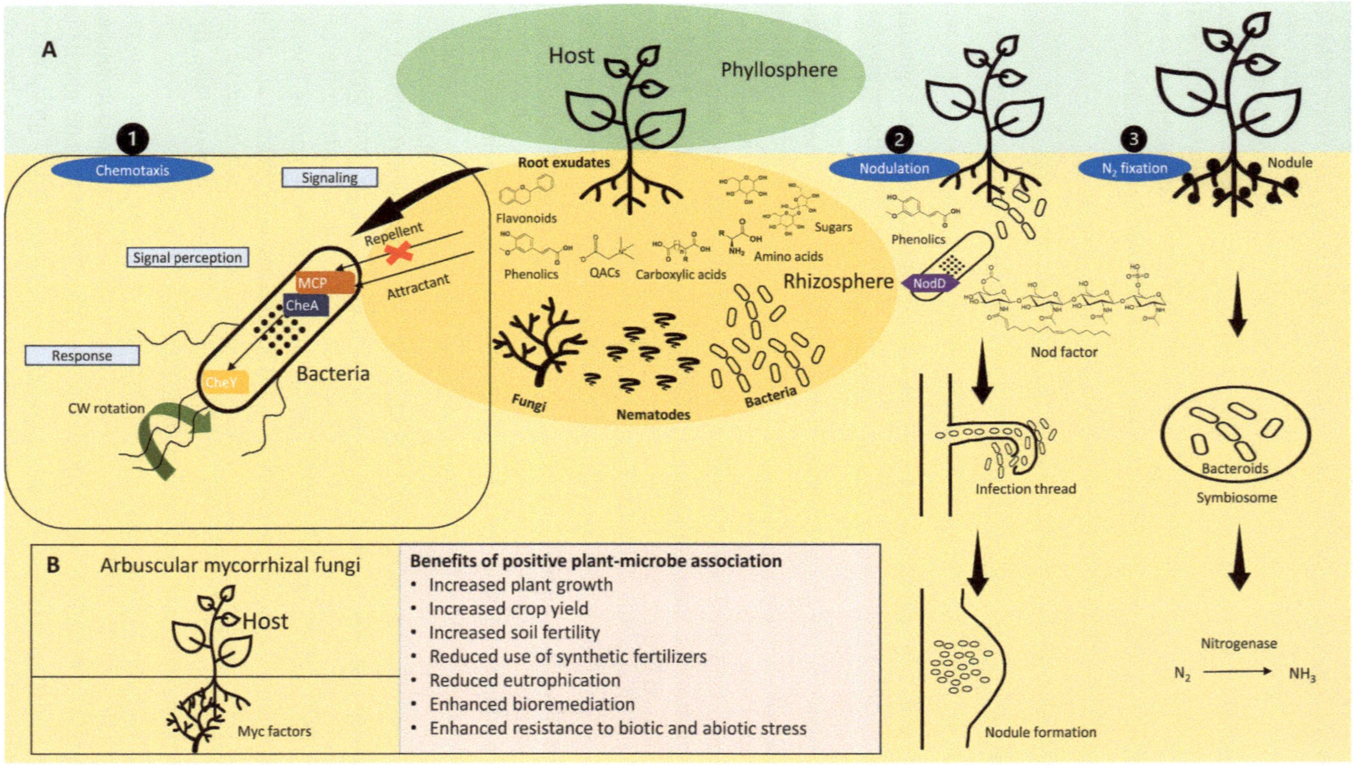

Fig. 10.2 Schematic representation of beneficial plant-microbe interactions. (**a**) Establishment of the symbiotic association in the rhizosphere; (1) Exudation of different classes of compounds (flavonoids, phenolics, carboxylic acids, amino acids, QACs, and sugars) from the roots into the rhizosphere where different microbes (fungi, nematodes, and bacteria) are present. Peritrichously flagellated legume symbionts can direct their movement toward attractants and away from

repellents through a biased random walk. Methyl-accepting chemotaxis proteins (MCPs) sense certain compounds and regulate clockwise (CW) flagellar rotation through the activity of a two-component system consisting of the histidine autokinase CheA and the response regulator CheY2. (2) Nodule formation wherein *nod* genes regulate the production of Nod factors followed by initiation of the infection thread, engulfment of bacteria by the deformed root hairs, active division of the bacteria inside the root cortex, and finally the development of nodule; (3) Bacteroides inside the symbiosome of the root nodules fix atmospheric nitrogen to ammonia (bioavailable form) through the activity of nitrogenase enzymes. (**b**) Arbuscular mycorrhizal fungi in association with its host plant leads to increased roots surface resulting in better nutrient absorption and several other benefits highlighted

speed regulation of the peritrichous flagella causes the flagellar bundle to come apart and the cell to tumble (Fig. 10.2). This biased random walk of straight runs and tumbles will direct the cell toward an attractant or away from a repellent (Scharf 2002; Attmannspacher et al. 2005).

Once bacteria reach the surface of the plant root, their exopolysaccharides play a crucial role in root invasion. Plant root exudates released under nitrogen-deficient conditions are perceived as signals, prompting bacterial cells to produce nodulation factors which are lipochitooligosaccharides. The production of these nodulation factors is regulated by the *nod* genes. A transcription factor belonging to the LysR family known as NodD regulates the *nod* genes (Dong and Song 2020). *S. meliloti* has three NodD proteins (NodD1, NodD2, and NodD3). Flavonoids like luteolin released by *M. sativa* were found to be essential for NodD1 to activate *nod* genes in *S. meliloti* (Peters et al. 1986; Peck et al. 2013), while NodD2 activates *nod* genes in the presence of betaines (Barnett and Long 2015). NodD3 does not require a specific compound of plant origin for its activity but is activated by a different transcription factor of the same LysR family (Maillet et al. 1990; Swanson et al. 1993). The complex formed by NodD and a flavonoid bind to the *nod* box ultimately lead to the production of Nod factors, which in turn are perceived by the plant roots.

Sensing these Nod factors by plant roots stimulates the deformation of root hairs (Dong and Song 2020). Once the bacterial cells reach the plant root cell surface, they induce the production of an infection thread in the deformed root hairs, followed by their entry inside the roots (Fig. 10.2). This eventually leads to their engulfment by the plant root cells and eventually the formation of nodules through rapid division inside the root cells (Dong and Song 2020). Within the nodule, the bacterial cell undergoes a transformation to a cell type that becomes metabolically inactive, which is called a bacteroid. The low oxygen concentration inside the nodules triggers a series of events leading to the activation of different regulatory elements in the bacteroid. These regulatory elements activate the expression of nitrogenase components. Bacteroides fix atmospheric nitrogen to ammonia and provide assimilable nitrogen to the plants. In return, they receive carboxylates like malate and succinate from the plant (Fig. 10.2).

Rhizobium species that participate in symbiotic relationships with the host plants also have a T3SS. The T3SS system helps the *Rhizobium* species to break host immune defenses through the injection of multiple effectors and establish their ecological niche in the root nodules of their respective hosts followed by fixing atmospheric nitrogen to be used by the host (Dong and Song 2020).

10.5 Beneficial Effects on Soil: Agricultural Sustainability

Plant-microbe interactions have several environmental health and agricultural benefits. They range from enhancement of phytoremediation potential to their role in biogeochemical cycles, promoting plant diversity and improving growth and survival through increased abiotic and biotic stress tolerance and resistance (Jalal et al. 2023).

10.5.1 Soil Health

In the present scenario, with the growing global population and increased food demand, the use of mainly chemical fertilizers has increased over time (FAO 2019). However, the continuous increased application of chemical fertilizers has resulted in several detrimental effects such as eutrophication of water sources and the deterioration of soil health. These consequences have severe implications for the marine ecosystem and the overall environmental sustainability (Krasilnikov et al. 2022). In this regard, the use of biofertilizers as an alternative to chemical fertilizers offers a promising solution. These biofertilizers can positively impact soil health by reducing the amounts of artificial sources of N, P, and K. They can also minimize the negative environmental impacts on various ecosystems. Microbes, particularly symbionts, play a crucial role in the nitrogen cycle, facilitating nitrogen fixation, and its conversion to biologically available forms for plants. Furthermore, arbuscular fungi contribute to phosphate solubilization, enabling enhanced and efficient phosphorus uptake by the plants.

Saprotrophic microbes act as decomposers and aid in releasing beneficial minerals by decomposing the organic matter in the soil and increasing its fertility (Adomako et al. 2022). The presence of specific microbes in the soil is also observed to suppress the disease incidence in crops (Enebe and Babalola 2019). The higher amounts of beneficial microbes in the rhizosphere can prevent the activity of the pathogens in two ways: (1) by producing antimicrobial compounds and (2) by stimulating the plant immune system imparting induced systemic resistance to the plants (Yu et al. 2022). Plants sometimes recruit beneficial microbes upon encountering a microbial attack by enriching and sustaining beneficial microbes. This leads to the generation of a soil memory or soilborne legacy to prepare the upcoming generations to avoid harmful pathogens. Root exudates are believed to play an important role in this process, although direct evidence for this has not yet been found.

10.5.2 Soil Enrichment

The application of PGPRs instead of chemical fertilizers helps in efficient nutrient recycling, which enriches the status of N, P, and K in the soil, creating a more suitable environment for several economically important, commercial crops. Long-term monoculture of a crop can affect microbial communities in the soil resulting in poor growth and crop yield. This is evident from a study conducted on young and old monocultured tea plantations, wherein tea plants in the older plantations showed poor growth and yield. On comparing the rhizospheric soil of tea plants from these young and old plantations, significant differences in the catechin-containing compounds as well as the pH were observed, which apparently changed the structure of microbial communities (Arafat et al. 2017). Application of Laredo soybeans green manure supported the restoration of long-term monoculture tea plantations and revealed a higher abundance of *Rhizobium* and *Bradyrhizobium* spp. in the soil (Jiang et al. 2019).

Rotation of the crop cycle involving leguminous plants can aid in the replenishment of nitrogen in the soil and further enhance crop yield without harming the soil's health (Kebede 2021). Utilization of microbial capabilities can facilitate a better and more sustainable crop rotation system that exploits biological agents for maintaining soil and plant health rather than chemical fertilizers (Ray et al. 2020). Moreover, intercropping of cereal crops and legumes can significantly enhance soil fertility followed by better crop yield. Several studies conducted on the intercropping system with maize and peanut revealed better nutrient uptake and increased plant biomass. This was accompanied by higher microbial diversity with increased accumulation of beneficial bacteria in the soil (Feng et al. 2021b).

10.5.3 Biological Hardening

Plant-microbe association is also being exploited in increasing the survival of tissue culture-raised plants during acclimatization by using PGPRs as biological hardening agents (Soumare et al. 2021). Bacterial inoculants containing *Bacillus subtilis* and *Pseudomonas corrugata* isolated from the tea rhizosphere were reported to significantly enhance the overall survival as well as growth in tissue culture-raised plants (Pandey et al. 2000). Several well-known PGPR formulations containing phosphate-solubilizing or nitrogen-fixing bacteria are available in national and international markets in the form of biofertilizer formulations. These target a broad range of high-value plants ranging from edible crops, such as cereals, maize, wheat, rice, beans, grapes, tomatoes, sunflowers, carrots, and cucumbers to spices like cumin, grasses such as alpine pinegrass, and flowers like chrysanthemum. These formulations are mostly composed of microbial inoculants such as *Bacillus* and *Pseudomonas* species either in consortia or alone targeting specific crops. *Noccaea caerulescens*, a well-known model plant involved in phytoremediation, was observed to have increased uptake of nickel from contaminated soil upon application of consortia of *Bacillus pumilus* and *Micrococcus* spp. (Aboudrar et al. 2013). Mycorrhizal inoculants are also available in the market as liquids, gels, water-soluble powder, and granules. The use of these microbial inoculants on agricultural crops has significantly enhanced crop yield by promoting its adaptability and disease resistance by providing nutrients through their action of phosphate solubilization and others.

10.5.4 Phytoremediation Efficiency Enhancement

Plant-microbe interactions enhance the plant's ability to accumulate harmful heavy metals in their roots and thus improve the soil quality through phytoremediation. The biofuel-producing crop *Helianthus tuberosus* has an enhanced ability to retain cadmium and zinc on exposure to a particular PGPR, which was isolated from a plant grown in metal-contaminated soil (Moreira et al. 2021). Plants exhibited better growth and less metal-induced stress with a significantly increased cadmium uptake in the presence of this endophytic root bacterium.

10.6 Technological Advancements: Does Modern Methods Help?

10.6.1 Technological Advancements in Understanding Plant-Microbe Interactions

A research group in Argonne laboratory in the USA has made significant advancements in studying plant-microbe interactions by developing biosensors based on advanced photon source (Noirot-Gros et al. 2020). These biosensors enable the measurement and understanding of the dynamics of compounds present in the rhizosphere. They allow for imaging of the interface between the microbe and the plant with nanometer resolution, thereby further revolutionizing the field of host-microbe interaction dynamics in real time.

10.6.2 Role of Omic Approaches to Understand the Plant-Microbe Association

High-throughput omic approaches (genomics, transcriptomics, proteomics, and metabolomics) allow the investigation of large population pools of soil bacteria as well as fungi, unravelling the regulators of the plant-microbe interactions, the effect of plant-specific exudates/allelochemicals in structuring the microbial communities in the rhizosphere, and the influence of these microbes on plant growth, yield, and survival (Vishwakarma et al. 2020). This has enhanced our understanding of the underlying mechanisms and molecular basis of these interactions leading to the development of improved crops with better resistance or symbiotic ability (Wang et al. 2018). Complete genome sequencing of the model legume *M. truncatula* has revolutionized the understanding of the changes in the plant during symbiosis and revealed the important markers involved (Kawaguchi and Minamisawa 2010). Recently, genome sequencing of *Lotus japonicus* identified the presence of a distinct symbiotic gene distribution pattern compared to *M. truncatula* (Kamal et al. 2020). This knowledge can help in the selection of superior genotypes with better symbiotic efficiency leading to enhanced yield and reduced fertilizer requirement/consumption.

RNA-seq studies have revealed that changes in the status of transcripts and their regulation pattern in symbiotically associated organisms are also specific to plant species and regions (Imam et al. 2016). The latest advances in the field of metabolomics over the past two decades have further aided in a comprehensive understanding of the composition of root exudates and the identification of novel compounds involved in plant-microbe interactions. With the advent of these modern tools and techniques, it has become possible to characterize and compare the composition of various rhizospheric microbial communities associated with specific plant species. For instance, rhizospheric soil from *Brachypodium* spp. has been analyzed with different sequencing platforms like the Roche GS FLX+ platform and Illumina MiSeq system of next-generation sequencing (NGS) technology. This analysis along with HPLC data helped to assess both rhizospheric communities and root exudates. As a result, the rhizospheric communities and root exudates from *Brachypodium* spp. revealed greater similarities to those of wheat as compared to *Arabidopsis* (Kawasaki et al. 2016).

10.7 Conclusions

The complex and continuous dynamics of plant-microbe interactions is crucial in determining soil health and productivity. This association has been exploited for successful and sustainable agriculture via traditional approaches like crop rotation and is still being exploited using modern approaches for its diverse benefits. Microbial formulations as biological control agents are being used for enhancing soil fertility and better crop productivity, but knowledge of the cellular machinery responsible for this highly specific interaction is still lacking. More efforts toward understanding the molecular mechanisms behind plant-microbe interactions, their specificity, utilization of resources, and response to the environment should be made.

Acknowledgements BS acknowledges the financial support from NSF grants MCB-1253234 and MCB-1817652. VK acknowledges the financial support from Bielefeld University.

References

Abad P, Williamson VM (2010) Plant nematode interaction: a sophisticated dialogue. In: Advances in botanical research, vol 53. Academic Press, pp 147–192

Abad P, Gouzy J, Aury JM, Castagnone-Sereno P, Danchin EG, Deleury E et al (2008) Genome sequence of the metazoan plant-parasitic nematode *Meloidogyne incognita*. Nat Biotechnol 26(8):909–915

Aboudrar W, Schwartz C, Morel JL, Boularbah A (2013) Effect of nickel-resistant rhizosphere bacteria on the uptake of nickel by the hyperaccumulator *Noccaea caerulescens* under controlled conditions. J Soils Sediments 13:501–507

Adomako MO, Roiloa S, Yu FH (2022) Potential roles of soil microorganisms in regulating the effect of soil nutrient heterogeneity on plant performance. Microorganisms 10(12):2399

Alsafran M, Usman K, Ahmed B, Rizwan M, Saleem MH, Al Jabri H (2022) Understanding the phytoremediation mechanisms of potentially toxic elements: a proteomic overview of recent advances. Front Plant Sci 13:881242

Arafat Y, Wei X, Jiang Y, Chen T, Saqib HSA, Lin S, Lin W (2017) Spatial distribution patterns of root-associated bacterial communities mediated by root exudates in different aged ratooning tea monoculture systems. Int J Mol Sci 18(8):1727

Aroney ST, Poole PS, Sánchez-Cañizares C (2021) Rhizobial chemotaxis and motility systems at work in the soil. Front Plant Sci 12:725338

Ashraf MA, Maah MJ, Yusoff I (2014) Soil contamination, risk assessment and remediation. In: Environmental risk assessment of soil contamination, vol 1. InTech, pp 3–56

Aslam S, Tahir A, Aslam MF, Alam MW, Shedayi AA, Sadia S (2017) Recent advances in molecular techniques for the identification of phytopathogenic fungi—a mini review. J Plant Interact 12(1):493–504

Attmannspacher U, Scharf B, Schmitt R (2005) Control of speed modulation (chemokinesis) in the unidirectional rotary motor of *Sinorhizobium meliloti*. Mol Microbiol 56(3):708–718

Averill C, Bhatnagar JM, Dietze MC, Pearse WD, Kivlin SN (2019) Global imprint of mycorrhizal fungi on whole-plant nutrient economics. Proc Natl Acad Sci 116(46):23163–23168

Baaziz H, Compton KK, Hildreth SB, Helm RF, Scharf BE (2021) McpT, a broad-range carboxylate chemoreceptor in *Sinorhizobium meliloti*. J Bacteriol 203(17):e00216–e00221

Backer R, Rokem JS, Ilangumaran G, Lamont J, Praslickova D, Ricci E, Subramanian S, Smith DL (2018) Plant growth-promoting rhizobacteria: context, mechanisms of action, and roadmap to commercialization of biostimulants for sustainable agriculture. Front Plant Sci 9:1473

Badri DV, Vivanco JM (2009) Regulation and function of root exudates. Plant Cell Environ 32(6):666–681

Banerjee S, van der Heijden MG (2023) Soil microbiomes and one health. Nat Rev Microbiol 21(1):6–20

Bardgett RD, Van Der Putten WH (2014) Belowground biodiversity and ecosystem functioning. Nature 515(7528):505–511

Barnett MJ, Long SR (2015) The *Sinorhizobium meliloti* SyrM regulon: effects on global gene expression are mediated by syrA and nodD3. J Bacteriol 197(10):1792–1806

Bartoli C, Roux F, Lamichhane JR (2016) Molecular mechanisms underlying the emergence of bacterial pathogens: an ecological perspective. Mol Plant Pathol 17(2):303–310

Basu S, Kumar G, Chhabra S, Prasad R (2021) Role of soil microbes in biogeochemical cycle for enhancing soil fertility. In: New and future developments in microbial biotechnology and bioengineering. Elsevier, pp 149–157

Bennett AE, Groten K (2022) The costs and benefits of plant–arbuscular mycorrhizal fungal interactions. Annu Rev Plant Biol 73:649–672

Biofertilizers Market Size, Share, Growth & Analysis (2030). https://www.fortunebusinessinsights.com/industry-reports/biofertilizers-market-100413

Boller T, Felix G (2009) A renaissance of elicitors: perception of microbe-associated molecular patterns and danger signals by pattern-recognition receptors. Annu Rev Plant Biol 60:379–406

Bonfante P, Genre A (2010) Mechanisms underlying beneficial plant–fungus interactions in mycorrhizal symbiosis. Nat Commun 1(1):48

Bonfante P, Perotto S (1995) Tansley Review No. 82. Strategies of arbuscular mycorrhizal fungi when infecting host plants. New Phytol 130(1):3–21

Bontempo AF, Lopes EA, Fernandes RH, Freitas LGD, Dallemole-Giaretta R (2017) Dose-response effect of *Pochonia chlamydosporia* against *Meloidogyne incognita* on carrot under field conditions. Revista Caatinga 30:258–262

Budde IP, Ullrich MS (2000) Interactions of *Pseudomonas syringae* pv. *glycinea* with host and nonhost plants in relation to temperature and phytotoxin synthesis. Mol Plant-Microbe Interact 13(9):951–961

Buttimer C, McAuliffe O, Ross RP, Hill C, O'Mahony J, Coffey A (2017) Bacteriophages and bacterial plant diseases. Front Microbiol 8:34

Callaway E (2016) Devastating wheat fungus appears in Asia for first time. Nature 532(7600):421–422

Canarini A, Kaiser C, Merchant A, Richter A, Wanek W (2019) Root exudation of primary metabolites: mechanisms and their roles in plant responses to environmental stimuli. Front Plant Sci 10:157

Carter CC (1985) Literature search: host range of *Meloidogyne hapla*. Int Nematol Network Newsl 2(2):16–24

Carvalhais LC, Rincon-Florez VA, Brewer PB, Beveridge CA, Dennis PG, Schenk PM (2019) The ability of plants to produce strigolactones affects rhizosphere community composition of fungi but not bacteria. Rhizosphere 9:18–26

Chen C, Zhu H (2013) Are common symbiosis genes required for endophytic rice-rhizobial interactions? Plant Signal Behav 8(9):e25453

Chen SL, Tsai MK, Huang YM, Huang CH (2018) Diversity and characterization of *Azotobacter* isolates obtained from rice rhizosphere soils in Taiwan. Ann Microbiol 68(1):17–26

Chen W, Ye T, Sun Q, Niu T, Zhang J (2021) Arbuscular mycorrhizal fungus alters root system architecture in Camellia sinensis L. as revealed by RNA-Seq analysis. Front Plant Sci 12:777357

Choi HW, Klessig DF (2016) DAMPs, MAMPs, and NAMPs in plant innate immunity. BMC Plant Biol 16:1–10

Clemens S (2006) Toxic metal accumulation, responses to exposure and mechanisms of tolerance in plants. Biochimie 88(11):1707–1719

Collins KD, Lacal J, Ottemann KM (2014) Internal sense of direction: sensing and signaling from cytoplasmic chemoreceptors. Microbiol Mol Biol Rev 78(4):672–684

Compton KK, Scharf BE (2021) Rhizobial chemoattractants, the taste and preferences of legume symbionts. Front Plant Sci 12:686465

Compton KK, Hildreth SB, Helm RF, Scharf BE (2018) *Sinorhizobium meliloti* chemoreceptor McpV senses short-chain carboxylates via direct binding. J Bacteriol 200(23):10–1128

Compton KK, Hildreth SB, Helm RF, Scharf BE (2020) An updated perspective on *Sinorhizobium meliloti* chemotaxis to alfalfa flavonoids. Front Microbiol 11:581482

de Andrade LA, Santos CHB, Frezarin ET, Sales LR, Rigobelo EC (2023) Plant growth-promoting rhizobacteria for sustainable agricultural production. Microorganisms 11(4):1088

De Leij FAAM, Kerry BR, Dennehy JA (1992) The effect of fungal application rate and nematode density on the effectiveness of *Verticillium chlamydosporium* as a biological control agent for *Meloidogyne incognita*. Nematologica 38(1–4):112–122

De Leij FAAM, Kerry BR, Dennehy JA (1993) *Verticillium chlamydosporium* as a biological control agent for *Meloidogyne incognita* and *M. hapla* in pot and micro-plot tests. Nematologica 39(1–4):115–126

Dhawan SC, Singh S (2010) Management of root-knot nematode, *Meloidogyne incognita* using *Pochonia chlamydosporia* on okra. Indian J Nematol 40(2):171–178

Doehlemann G, Ökmen B, Zhu W, Sharon A (2017) Plant pathogenic fungi. Microbiol Spectrum 5(1):5–1

Dong W, Song Y (2020) The significance of flavonoids in the process of biological nitrogen fixation. Int J Mol Sci 21(16):5926

Doni F, Isahak A, Che Mohd Zain CR, Wan Yusoff WM (2014) Physiological and growth response of rice plants (*Oryza sativa* L.) to *Trichoderma* spp. inoculants. AMB Express 4:1–7

Duduk N, Vico I, Kosovac A, Stepanović J, Ćurčić Ž, Vučković N, Rekanović E, Duduk B (2023) A biotroph sets the stage for a necrotroph to play: 'Candidatus *Phytoplasma solani*' infection of sugar beet facilitated *Macrophomina phaseolina* root rot. Front Microbiol 14:1164035

Dundek P, Holík L, Rohlík T, Hromádko L, Vranová V, Rejšek K, Formánek P (2011) Methods of plant root exudates analysis: a review. Acta Univ Agric Silvic Mendelianae Brun 59(3):241–246
Ehau-Taumaunu H, Hockett KL (2022) The plant host environment influences competitive interactions between bacterial pathogens. Environ Microbiol Rep 14(5):785–794
Enebe MC, Babalola OO (2019) The impact of microbes in the orchestration of plants' resistance to biotic stress: a disease management approach. Appl Microbiol Biotechnol 103:9–25
Escudero N, Lopez-Llorca LV (2012) Effects on plant growth and root-knot nematode infection of an endophytic GFP transformant of the nematophagous fungus *Pochonia chlamydosporia*. Symbiosis 57:33–42
Fadiji AE, Babalola OO (2020) Exploring the potentialities of beneficial endophytes for improved plant growth. Saudi J Biol Sci 27(12):3622–3633
Fadiji AE, Yadav AN, Santoyo G, Babalola OO (2023) Understanding the plant-microbe interactions in environments exposed to abiotic stresses: an overview. Microbiol Res 271:127368
FAO (2019). https://www.fao.org/state-of-food-agriculture/2019/en/
Feng H, Fu R, Hou X, Lv Y, Zhang N, Liu Y, Xu Z, Miao Y, Krell T, Shen Q, Zhang R (2021a) Chemotaxis of beneficial rhizobacteria to root exudates: the first step towards root–microbe rhizosphere interactions. Int J Mol Sci 22(13):6655
Feng C, Sun Z, Zhang L, Feng L, Zheng J, Bai W, Gu C, Wang Q, Xu Z, van der Werf W (2021b) Maize/peanut intercropping increases land productivity: a meta-analysis. Field Crop Res 270:108208
Ferronato N, Torretta V (2019) Waste mismanagement in developing countries: a review of global issues. Int J Environ Res Public Health 16(6):1060
Figueiredo AF, Boy J, Guggenberger G (2021) Common mycorrhizae network: a review of the theories and mechanisms behind underground interactions. Front Fungal Biol 2:48
Finlay RD (2008) Ecological aspects of mycorrhizal symbiosis: with special emphasis on the functional diversity of interactions involving the extraradical mycelium. J Exp Bot 59(5):1115–1126
Ghahremani Z, Escudero N, Saus E, Gabaldón T, Sorribas FJ (2019) *Pochonia chlamydosporia* induces plant-dependent systemic resistance to *Meloidogyne incognita*. Front Plant Sci 10:945
Ghahremani Z, Escudero N, Marín I, Sanz A, García S, Expósito A, Giné A, Sorribas FJ (2022) *Pochonia chlamydosporia* is the most prevalent fungal species responsible for *Meloidogyne* suppression in sustainable vegetable production systems. Sustainability 14(24):16941
Ghosh P, Adolphsen KN, Yurgel SN, Kahn ML (2021) *Sinorhizobium medicae* WSM419 genes that improve symbiosis between *Sinorhizobium meliloti* Rm1021 and *Medicago truncatula* Jemalong A17 and in other symbiosis systems. Appl Environ Microbiol 87(15):e03004–e03020
Giovannetti M, Avio L, Fortuna P, Pellegrino E, Sbrana C, Strani P (2006) At the root of the wood wide web: self recognition and nonself incompatibility in mycorrhizal networks. Plant Signal Behav 1(1):1–5
Glowacki S, Macioszek V, Kononowicz A (2011) R proteins as fundamentals of plant innate immunity. Cell Mol Biol Lett 16(1):1–24
Gouveia ADS, Monteiro TSA, Valadares SV, Sufiate BL, de Freitas LG, Ramos HJDO, de Queiroz JH (2019) Understanding how *Pochonia chlamydosporia* increases phosphorus availability. Geomicrobiol J 36(8):747–751
Gu Y, Wang Y, Sun Y, Zhao K, Xiang Q, Yu X, Zhang X, Chen Q (2018) Genetic diversity and characterization of arsenic-resistant endophytic bacteria isolated from *Pteris vittata*, an arsenic hyperaccumulator. BMC Microbiol 18(1):1–10
Gupta MM, Abbott LK (2021) Exploring economic assessment of the arbuscular mycorrhizal symbiosis. Symbiosis 83:143–152
Gupta OP, Sharma P, Gupta RK, Sharma I (2014) Current status on role of miRNAs during plant–fungus interaction. Physiol Mol Plant Pathol 85:1–7
Gupta R, Anand G, Gaur R, Yadav D (2021) Plant–microbiome interactions for sustainable agriculture: a review. Physiol Mol Biol Plants 27:165–179

Hellwig V, Mayer-Bartschmid A, Müller H, Greif G, Kleymann G, Zitzmann W, Tichy HV, Stadler M (2003) Pochonins A–F, new antiviral and antiparasitic resorcylic acid lactones from *Pochonia chlamydosporia* var. c atenulata. J Nat Prod 66(6):829–837

Imam J, Singh PK, Shukla P (2016) Plant microbe interactions in post genomic era: perspectives and applications. Front Microbiol 7:1488

Ingle RA, Carstens M, Denby KJ (2006) PAMP recognition and the plant–pathogen arms race. BioEssays 28(9):880–889

Ivanov AA, Ukladov EO, Golubeva TS (2021) *Phytophthora infestans*: an overview of methods and attempts to combat late blight. J Fungi 7(12):1071

Jacoby RP, Chen L, Schwier M, Koprivova A, Kopriva S (2020) Recent advances in the role of plant metabolites in shaping the root microbiome. F1000Research 9. https://doi.org/10.12688/f1000research.21796.1

Jalal A, Oliveira CEDS, Rosa PAL, Galindo FS, Teixeira Filho MCM (2023) Beneficial microorganisms improve agricultural sustainability under climatic extremes. Life 13(5):1102

Jiang Y, Arafat Y, Letuma P, Ali L, Tayyab M, Waqas M, Li Y, Lin W, Lin S, Lin W (2019) Restoration of long-term monoculture degraded tea orchard by green and goat manures applications system. Sustainability 11(4):1011

Johnson A, Laney K, National Academies of Sciences, Engineering, and Medicine and Environmental Health Matters Initiative (2021) Reducing the health impacts of the nitrogen problem. National Academies Press

Jones JT, Haegeman A, Danchin EG, Gaur HS, Helder J, Jones MG, Kikuchi T, Manzanilla-López R, Palomares-Rius JE, Wesemael WM, Perry RN (2013) Top 10 plant-parasitic nematodes in molecular plant pathology. Mol Plant Pathol 14(9):946–961

Kallison ER (2021) A review of the contributions by lichen to building soil. IdeaFest 5(1):1

Kamal N, Mun T, Reid D, Lin JS, Akyol TY, Sandal N, Asp T, Hirakawa H, Stougaard J, Mayer KF, Sato S (2020) Insights into the evolution of symbiosis gene copy number and distribution from a chromosome-scale *Lotus japonicus* Gifu genome sequence. DNA Res 27(3):dsaa015

Kangara N, Kurowski TJ, Radhakrishnan GV, Ghosh S, Cook NM, Yu G, Arora S, Steffenson BJ, Figueroa M, Mohareb F, Saunders DG (2020) Mutagenesis of *Puccinia graminis* f. sp. *tritici* and selection of gain-of-virulence mutants. Front Plant Sci 11:570180

Kannan VR, Bastas KK, Antony R (2015) Plant pathogenic bacteria. In: Sustainable approaches to controlling plant pathogenic bacteria. CRC Press, p 1

Katagiri F, Thilmony R, He SY (2002) The *Arabidopsis thaliana-Pseudomonas syringae* interaction. Arabidopsis Book 1:e0039

Kawaguchi M, Minamisawa K (2010) Plant–microbe communications for symbiosis. Plant Cell Physiol 51(9):1377–1380

Kawasaki A, Donn S, Ryan PR, Mathesius U, Devilla R, Jones A, Watt M (2016) Microbiome and exudates of the root and rhizosphere of *Brachypodium distachyon*, a model for wheat. PLoS One 11(10):e0164533

Kebede E (2021) Contribution, utilization, and improvement of legumes-driven biological nitrogen fixation in agricultural systems. Front Sustain Food Syst 5:767998

Kerry BR, Kirkwood IA, De Leij FAAM, Barba J, Leijdens MB, Brookes PC (1993) Growth and survival of *Verticillium chlamydosporium* Goddard, a parasite of nematodes, in soil. Biocontrol Sci Tech 3(3):355–365

Konappa N, Krishnamurthy S, Arakere UC, Chowdappa S, Ramachandrappa NS (2020) Efficacy of indigenous plant growth-promoting rhizobacteria and *Trichoderma* strains in eliciting resistance against bacterial wilt in a tomato. Egypt J Biol Pest Control 30:1–13

Krasilnikov P, Taboada MA, Amanullah (2022) Fertilizer use, soil health and agricultural sustainability. Agriculture 12(4):462

Kumar K, Verma A, Pal G, Anubha, White JF, Verma SK (2021a) Seed endophytic bacteria of pearl millet (*Pennisetum glaucum* L.) promote seedling development and defend against a fungal phytopathogen. Front Microbiol 12:774293

Kumar M, Giri VP, Pandey S, Gupta A, Patel MK, Bajpai AB, Jenkins S, Siddique KH (2021b) Plant-growth-promoting rhizobacteria emerging as an effective bioinoculant to improve the growth, production, and stress tolerance of vegetable crops. Int J Mol Sci 22(22):12245

Kumar R, Verma A, Shome A, Sinha R, Sinha S, Jha PK, Kumar R, Kumar P, Shubham D, S. and Sharma, P. (2021c) Impacts of plastic pollution on ecosystem services, sustainable development goals, and need to focus on circular economy and policy interventions. Sustainability 13(17):9963

Kunkel BN, Harper CP (2018) The roles of auxin during interactions between bacterial plant pathogens and their hosts. J Exp Bot 69(2):245–254

Laguerre G, Depret G, Bourion V, Duc G (2007) *Rhizobium leguminosarum* bv. *Viciae* genotypes interact with pea plants in developmental responses of nodules, roots and shoots. New Phytol 176(3):680–690

Lanver D, Tollot M, Schweizer G, Lo Presti L, Reissmann S, Ma LS, Schuster M, Tanaka S, Liang L, Ludwig N, Kahmann R (2017) *Ustilago maydis* effectors and their impact on virulence. Nat Rev Microbiol 15(7):409–421

Larriba E, Jaime MD, Carbonell-Caballero J, Conesa A, Dopazo J, Nislow C, Martín-Nieto J, Lopez-Llorca LV (2014) Sequencing and functional analysis of the genome of a nematode egg-parasitic fungus, *Pochonia chlamydosporia*. Fungal Genet Biol 65:69–80

Larriba E, Jaime MD, Nislow C, Martín-Nieto J, Lopez-Llorca LV (2015) Endophytic colonization of barley (*Hordeum vulgare*) roots by the nematophagous fungus *Pochonia chlamydosporia* reveals plant growth promotion and a general defense and stress transcriptomic response. J Plant Res 128:665–678

Ledermann R, Emmenegger B, Couzigou JM, Zamboni N, Kiefer P, Vorholt JA, Fischer HM (2021) *Bradyrhizobium diazoefficiens* requires chemical chaperones to cope with osmotic stress during soybean infection. MBio 12(2):10–1128

Lee EH, Eo JK, Ka KH, Eom AH (2013) Diversity of arbuscular mycorrhizal fungi and their roles in ecosystems. Mycobiology 41(3):121–125

Liang Y, Tóth K, Cao Y, Tanaka K, Espinoza C, Stacey G (2014) Lipochitooligosaccharide recognition: an ancient story. New Phytol 204(2):289–296

Lin HA, Mideros SX (2022) The effect of *Septoria glycines* and fungicide application on the soybean phyllosphere mycobiome. Phytobiomes J. https://doi.org/10.1094/PBIOMES-12-21-0075-R

López-Ráez JA, Kohlen W, Charnikhova T, Mulder P, Undas AK, Sergeant MJ, Verstappen F, Bugg TD, Thompson AJ, Ruyter-Spira C, Bouwmeester H (2010) Does abscisic acid affect strigolactone biosynthesis? New Phytol 187(2):343–354

Luambano ND, Manzanilla-López RH, Kimenju JW, Powers SJ, Narla RD, Wanjohi WJ, Kerry BR (2015) Effect of temperature, pH, carbon and nitrogen ratios on the parasitic activity of *Pochonia chlamydosporia* on Meloidogyne incognita. Biol Control 80:23–29

Macho AP, Zipfel C (2014) Plant PRRs and the activation of innate immune signaling. Mol Cell 54(2):263–272

Mageshwaran V, Gupta R, Singh S, Sahu PK, Singh UB, Chakdar H, Bagul SY, Paul S, Singh HV (2022) Endophytic *Bacillus subtilis* antagonize soil-borne fungal pathogens and suppress wilt complex disease in chickpea plants (*Cicer arietinum* L.). Front Microbiol 13:994847

Maillet F, Debellé F, Dénarié J (1990) Role of the nodD and syrM genes in the activation of the regulatory gene nodD3, and of the common and host-specific nod genes of Rhizobium meliloti. Mol Microbiol 4(11):1975–1984

Mansfield J, Genin S, Magori S, Citovsky V, Sriariyanum M, Ronald P, Dow MAX, Verdier V, Beer SV, Machado MA, Toth IAN (2012) Top 10 plant pathogenic bacteria in molecular plant pathology. Mol Plant Pathol 13(6):614–629

Mantelin S, Bellafiore S, Kyndt T (2017) *Meloidogyne graminicola*: a major threat to rice agriculture. Mol Plant Pathol 18(1):3

Manzanilla-López RH, Esteves I, Powers SJ, Kerry BR (2011) Effects of crop plants on abundance of *Pochonia chlamydosporia* and other fungal parasites of root-knot and potato cyst nematodes. Ann Appl Biol 159(1):118–129

Markovchick LM, Carrasco-Denney V, Sharma J, Querejeta JI, Gibson KS, Swaty R, Uhey DA, Belgara-Andrew A, Kovacs ZI, Johnson NC, Whitham TG (2023) The gap between mycorrhizal science and application: existence, origins, and relevance during the United Nation's Decade on Ecosystem Restoration. Restor Ecol 31(4):e13866

Marschner P (2012) Rhizosphere biology. In: Marschner's mineral nutrition of higher plants. Academic Press, pp 369–388

McCook S, Vandermeer J (2015) The big rust and the red queen: Long-term perspectives on coffee rust research. Phytopathology 105(9):1164–1173

Meier VM, Scharf BE (2009) Cellular localization of predicted transmembrane and soluble chemoreceptors in *Sinorhizobium meliloti*. J Bacteriol 191(18):5724–5733

Meier VM, Muschler P, Scharf BE (2007) Functional analysis of nine putative chemoreceptor proteins in *Sinorhizobium meliloti*. J Bacteriol 189(5):1816–1826

Mendes R, Garbeva P, Raaijmakers JM (2013) The rhizosphere microbiome: significance of plant beneficial, plant pathogenic, and human pathogenic microorganisms. FEMS Microbiol Rev 37(5):634–663

Mingma R, Pathom-aree W, Trakulnaleamsai S, Thamchaipenet A, Duangmal K (2014) Isolation of rhizospheric and roots endophytic actinomycetes from Leguminosae plant and their activities to inhibit soybean pathogen, *Xanthomonas campestris* pv. *Glycine*. World J Microbiol Biotechnol 30:271–280

Mishra S, Upadhyay S, Shukla RK (2017) The role of strigolactones and their potential cross-talk under hostile ecological conditions in plants. Front Physiol 7:691

Mitchum MG, Hussey RS, Baum TJ, Wang X, Elling AA, Wubben M, Davis EL (2013) Nematode effector proteins: an emerging paradigm of parasitism. New Phytol 199(4):879–894

Mitra D, Rad KV, Chaudhary P, Ruparelia J, Sagarika MS, Boutaj H, Mohapatra PKD, Panneerselvam P (2021) Involvement of strigolactone hormone in root development, influence and interaction with mycorrhizal fungi in plant: mini-review. Curr Res Microb Sci 2:100026

Moormann J, Heinemann B, Hildebrandt TM (2022) News about amino acid metabolism in plant–microbe interactions. Trends Biochem Sci 47:839

Moreira H, Pereira SI, Mench M, Garbisu C, Kidd P, Castro PM (2021) Phytomanagement of metal (loid)-contaminated soils: options, efficiency and value. Front Environ Sci 9:661423

Morgan JAW, Bending GD, White PJ (2005) Biological costs and benefits to plant–microbe interactions in the rhizosphere. J Exp Bot 56(417):1729–1739

Nazarov PA, Baleev DN, Ivanova MI, Sokolova LM, Karakozova MV (2020) Infectious plant diseases: etiology, current status, problems and prospects in plant protection. Acta Nat 12(3):46

Ngou BPM, Ahn HK, Ding P, Jones JD (2021) Mutual potentiation of plant immunity by cell-surface and intracellular receptors. Nature 592(7852):110–115

Nicholls RJ, Beaven RP, Stringfellow A, Monfort D, Le Cozannet G, Wahl T, Gebert J, Wadey M, Arns A, Spencer KL, Reinhart D (2021) Coastal landfills and rising sea levels: a challenge for the 21st century. Front Mar Sci 8:710342

Nicol JM, Turner SJ, Coyne DL, Nijs LD, Hockland S, Maafi ZT (2011) Current nematode threats to world agriculture. In: Genomics and molecular genetics of plant-nematode interactions. Springer, pp 21–43

Niño-liu DO, Ronald PC, Bogdanove AJ (2006) *Xanthomonas oryzae* pathovars: model pathogens of a model crop. Mol Plant Pathol 7(5):303–324

Niu XM (2017) Secondary metabolites from *Pochonia chlamydosporia* and other species of *Pochonia*. In: Perspectives in sustainable nematode management through Pochonia chlamydosporia applications for root and rhizosphere health. Springer, pp 131–168

Niu XM, Wang YL, Chu YS, Xue HX, Li N, Wei LX, Mo MH, Zhang KQ (2010) Nematodetoxic aurovertin-type metabolites from a root-knot nematode parasitic fungus *Pochonia chlamydosporia*. J Agric Food Chem 58(2):828–834

Nobori T, Velásquez AC, Wu J, Kvitko BH, Kremer JM, Wang Y, He SY, Tsuda K (2018) Transcriptome landscape of a bacterial pathogen under plant immunity. Proc Natl Acad Sci 115(13):E3055–E3064

Noirot-Gros MF, Shinde S, Larsen PE, Zerbs S, Korajczyk PJ, Kemner KM, Noirot PH (2018) Dynamics of aspen roots colonization by Pseudomonads reveals strain-specific and mycorrhizal-specific patterns of biofilm formation. Front Microbiol 9:853

Noirot-Gros MF, Shinde SV, Akins C, Johnson JL, Zerbs S, Wilton R, Kemner KM, Noirot P, Babnigg G (2020) Functional imaging of microbial interactions with tree roots using a microfluidics setup. Front Plant Sci 11:408

Noman M, Ahmed T, Ijaz U, Shahid M, Azizullah, Li D, Manzoor I, Song F (2021) Plant–microbiome crosstalk: dawning from composition and assembly of microbial community to improvement of disease resilience in plants. Int J Mol Sci 22(13):6852

Oekmen B, Schwammbach D, Bakkeren G, Neumann U, Doehlemann G (2021) The *Ustilago hordei*–barley interaction is a versatile system for characterization of fungal effectors. J Fungi 7(2):86

Oerke EC, Dehne HW (2004) Safeguarding production—losses in major crops and the role of crop protection. Crop Prot 23(4):275–285

Oro V, Trkulja N, Milosavljević A, Sečanski M, Tabaković M (2022) Sugar beet cyst nematode (*Heterodera schachtii* Schmidt): identification and antagonists. In: Sugar beet cultivation, management and processing. Singapore, Springer Nature, pp 751–776

Padilla-Roji I, Ruiz-Jiménez L, Bakhat N, Vielba-Fernández A, Pérez-García A, Fernández-Ortuño D (2023) RNAi technology: a new path for the research and management of obligate biotrophic phytopathogenic fungi. Int J Mol Sci 24(10):9082

Pandey A, Palni LMS, Bag N (2000) Biological hardening of tissue culture raised tea plants through rhizosphere bacteria. Biotechnol Lett 22:1087–1091

Patil AG, Kounaina K, Aishwarya S, Harshitha N, Satapathy P, Hudeda SP, Reddy KR, Alrafas H, Yadav AN, Raghu AV, Zameer F (2021) Myco-nanotechnology for sustainable agriculture: challenges and opportunities. In: Recent trends in mycological research, Agricultural and medical perspective, vol 1. Springer, pp 457–479

Peck MC, Fisher RF, Bliss R, Long SR (2013) Isolation and characterization of mutant *Sinorhizobium meliloti* NodD1 proteins with altered responses to luteolin. J Bacteriol 195(16):3714–3723

Pérez-Moreno J, Guerin-Laguette A, Rinaldi AC, Yu F, Verbeken A, Hernández-Santiago F, Martínez-Reyes M (2021) Edible mycorrhizal fungi of the world: what is their role in forest sustainability, food security, biocultural conservation and climate change? Plants People Planet 3(5):471–490

Peters NK, Frost JW, Long SR (1986) A plant flavone, luteolin, induces expression of *Rhizobium meliloti* nodulation genes. Science 233(4767):977–980

Powell NT (1971) Interactions between nematodes and fungi in disease complexes. Annu Rev Phytopathol 9(1):253–274

Ptatscheck C, Traunspurger W (2020) The ability to get everywhere: dispersal modes of free-living, aquatic nematodes. Hydrobiologia 847(17):3519–3547

Puigvert M, Solé M, López-Garcia B, Coll NS, Beattie KD, Davis RA, Elofsson M, Valls M (2019) Type III secretion inhibitors for the management of bacterial plant diseases. Mol Plant Pathol 20(1):20–32

Raina JB, Fernandez V, Lambert B, Stocker R, Seymour JR (2019) The role of microbial motility and chemotaxis in symbiosis. Nat Rev Microbiol 17(5):284–294

Rani S, Kumar P, Dahiya P, Maheshwari R, Dang AS, Suneja P (2022) Endophytism: a multidimensional approach to plant–prokaryotic microbe interaction. Front Microbiol 13:861235

Ray P, Lakshmanan V, Labbé JL, Craven KD (2020) Microbe to microbiome: a paradigm shift in the application of microorganisms for sustainable agriculture. Front Microbiol 11:622926

Roper MC (2011) *Pantoea stewartii* subsp. *stewartii*: lessons learned from a xylem-dwelling pathogen of sweet corn. Mol Plant Pathol 12(7):628–637

Ross-Elliott TJ, Jensen KH, Haaning KS, Wager BM, Knoblauch J, Howell AH, Mullendore DL, Monteith AG, Paultre D, Yan D, Otero S (2017) Phloem unloading in *Arabidopsis* roots is convective and regulated by the phloem-pole pericycle. elife 6:e24125

Royse DJ, Baars J, Tan Q (2017) Current overview of mushroom production in the world. In: Edible and medicinal mushrooms: technology and applications. Wiley, pp 5–13
Santi C, Bogusz D, Franche C (2013) Biological nitrogen fixation in non-legume plants. Ann Bot 111(5):743–767
Santoyo G, Gamalero E, Glick BR (2021) Mycorrhizal-bacterial amelioration of plant abiotic and biotic stress. Front Sust Food Syst 5:672881
Sarkodee-Addo E, Yasuda M, Gyu Lee C, Kanasugi M, Fujii Y, Ansong Omari R, Oppong Abebrese S, Bam R, Asuming-Brempong S, Mohammad Golam Dastogeer K, Okazaki S (2020) Arbuscular mycorrhizal fungi associated with rice (*Oryza sativa* L.) in Ghana: effect of regional locations and soil factors on diversity and community assembly. Agronomy 10(4):559
Scharf B (2002) Real-time imaging of fluorescent flagellar filaments of *Rhizobium lupini* H13-3: flagellar rotation and pH-induced polymorphic transitions. J Bacteriol 184(21):5979–5986
Schmitt R (2002) Sinorhizobial chemotaxis: a departure from the enterobacterial paradigm. Microbiology 148(3):627–631
Schouten A (2016) Mechanisms involved in nematode control by endophytic fungi. Annu Rev Phytopathol 54:121–142
Seitz VA, McGivern BB, Daly RA, Chaparro JM, Borton MA, Sheflin AM, Kresovich S, Shields L, Schipanski ME, Wrighton KC, Prenni JE (2022) Variation in root exudate composition influences soil microbiome membership and function. Appl Environ Microbiol 88(11):e00226-22
Senthil-Kumar M, Mysore KS (2013) Nonhost resistance against bacterial pathogens: retrospectives and prospects. Annu Rev Phytopathol 51:407–427
Seto Y, Sado A, Asami K, Hanada A, Umehara M, Akiyama K, Yamaguchi S (2014) Carlactone is an endogenous biosynthetic precursor for strigolactones. Proc Natl Acad Sci 111(4):1640–1645
Shukla N, Kaur P, Kumar A (2016) Molecular aspects of plant-nematode interactions. Indian J Plant Physiol 21:477–488
Shukla N, Yadav R, Kaur P, Rasmussen S, Goel S, Agarwal M, Jagannath A, Gupta R, Kumar A (2018) Transcriptome analysis of root-knot nematode (*Meloidogyne incognita*)-infected tomato (*Solanum lycopersicum*) roots reveals complex gene expression profiles and metabolic networks of both host and nematode during susceptible and resistance responses. Mol Plant Pathol 19(3):615–633
Shuping DSS, Eloff JN (2017) The use of plants to protect plants and food against fungal pathogens: a review. Afr J Tradit Complement Altern Med 14(4):120–127
Siddiqui IA, Atkins SD, Kerry BR (2009) Relationship between saprotrophic growth in soil of different biotypes of *Pochonia chlamydosporia* and the infection of nematode eggs. Ann Appl Biol 155(1):131–141
Singh PK, Gahtyari NC, Roy C, Roy KK, He X, Tembo B, Xu K, Juliana P, Sonder K, Kabir MR, Chawade A (2021) Wheat blast: a disease spreading by intercontinental jumps and its management strategies. Front Plant Sci 12:1467
Singh K, Chandra R, Purchase D (2022) Unraveling the secrets of rhizobacteria signaling in rhizosphere. Rhizosphere 21:100484
Sinha SK, Prasad M (1977) Bacterial stalk rot of maize, its symptoms and host-range. Zentralblatt für Bakteriologie, Parasitenkunde, Infektionskrankheiten und Hygiene 132(1):81–88
Sivanandhan S, Khusro A, Paulraj MG, Ignacimuthu S, Al-Dhabi NA (2017) Biocontrol properties of basidiomycetes: an overview. J Fungi 3(1):2
Sokornova S, Malygin D, Terentev A, Dolzhenko V (2022) Arbuscular mycorrhiza symbiosis as a factor of Asteraceae species invasion. Agronomy 12(12):3214
Song D, Pan K, Tariq A, Sun F, Li Z, Sun X, Zhang L, Olusanya OA, Wu X (2017) Large-scale patterns of distribution and diversity of terrestrial nematodes. Appl Soil Ecol 114:161–169
Sood M, Kapoor D, Kumar V, Sheteiwy MS, Ramakrishnan M, Landi M, Araniti F, Sharma A (2020) *Trichoderma*: the "secrets" of a multitalented biocontrol agent. Plants 9(6):762
Soumare A, Diédhiou AG, Arora NK, Tawfeeq Al-Ani LK, Ngom M, Fall S, Hafidi M, Ouhdouch Y, Kouisni L, Sy MO (2021) Potential role and utilization of plant growth promoting microbes in plant tissue culture. Front Microbiol 12:649878

Swanson JA, Mulligan JT, Long SR (1993) Regulation of syrM and nodD3 in *Rhizobium meliloti*. Genetics 134(2):435–444

Tarakanov RI, Lukianova AA, Evseev PV, Toshchakov SV, Kulikov EE, Ignatov AN, Miroshnikov KA, Dzhalilov FSU (2022) Bacteriophage control of *pseudomonas savastanoi* pv. *glycinea* in soybean. Plants 11(7):938

Tebaldi ND, Pereira JNP, Sagata E, Mota LCBM, da Hora Junior BT, Piazzoli D, Klosowski AC (2022) Occurrence and detection of *Xanthomonas vasicola* pv. *vasculorum* in naturally contaminated maize seeds in Brazil. Trop Plant Pathol 47(5):698–702

Tena G (2021) PTI and ETI are one. Nat Plants 7(12):1527–1527

Thambugala D, Menzies JG, Knox RE, Campbell HL, McCartney CA (2020) Genetic analysis of loose smut (*Ustilago tritici*) resistance in Sonop spring wheat. BMC Plant Biol 20:1–10

Thomma BP, Nürnberger T, Joosten MH (2011) Of PAMPs and effectors: the blurred PTI-ETI dichotomy. Plant Cell 23(1):4–15

Tibpromma S, Dong Y, Ranjitkar S, Schaefer DA, Karunarathna SC, Hyde KD, Jayawardena RS, Manawasinghe IS, Bebber DP, Promputtha I, Xu J (2021) Climate-fungal pathogen modeling predicts loss of up to one-third of tea growing areas. Front Cell Infect Microbiol 11:610567

Tiwari RK, Kumar R, Sharma S, Sagar V, Aggarwal R, Naga KC, Lal MK, Chourasia KN, Kumar D, Kumar M (2020) Potato dry rot disease: current status, pathogenomics and management. 3 Biotech 10:1–18

Tobin JD, Haydock PPJ, Hare MC, Woods SR, Crump DH (2008) Effect of the fungus *Pochonia chlamydosporia* and fosthiazate on the multiplication rate of potato cyst nematodes (*Globodera pallida* and *G. rostochiensis*) in potato crops grown under UK field conditions. Biol Control 46(2):194–201

Tolba SR, Rosso LC, Pentimone I, Colagiero M, Moustafa MM, Elshawaf II, Bubici G, Prigigallo MI, Ciancio A (2021) Root endophytism by *Pochonia chlamydosporia* affects defense-gene expression in leaves of monocot and dicot hosts under multiple biotic interactions. Plants 10(4):718

Torres-Rodriguez JA, Reyes-Pérez JJ, Quiñones-Aguilar EE, Hernandez-Montiel LG (2022) Actinomycete potential as biocontrol agent of phytopathogenic fungi: mechanisms, source, and applications. Plants 11(23):3201

Umehara M, Cao M, Akiyama K, Akatsu T, Seto Y, Hanada A, Li W, Takeda-Kamiya N, Morimoto Y, Yamaguchi S (2015) Structural requirements of strigolactones for shoot branching inhibition in rice and *Arabidopsis*. Plant Cell Physiol 56(6):1059–1072

Van Damme V, Hoedekie A, Viaene N (2005) Long-term efficacy of *Pochonia chlamydosporia* for management of *Meloidogyne javanica* in glasshouse crops. Nematology 7(5):727–736

Vianene NM, Abawi GS (2000) *Hirsutella rhossiliensis* and *Verticillium chlamydosporium* as biocontrol agents of the root-knot nematode *Meloidogyne hapla* on lettuce. J Nematol 32(1):85

Vishwakarma K, Kumar N, Shandilya C, Mohapatra S, Bhayana S, Varma A (2020) Revisiting plant–microbe interactions and microbial consortia application for enhancing sustainable agriculture: a review. Front Microbiol 11:560406

Vurukonda SSKP, Giovanardi D, Stefani E (2018) Plant growth promoting and biocontrol activity of *Streptomyces* spp. as endophytes. Int J Mol Sci 19(4):952

Wang Q, Liu J, Zhu H (2018) Genetic and molecular mechanisms underlying symbiotic specificity in legume-rhizobium interactions. Front Plant Sci 9:313

Waters MT, Gutjahr C, Bennett T, Nelson DC (2017) Strigolactone signaling and evolution. Annu Rev Plant Biol 68:291–322

Webb BA, Helm RF, Scharf BE (2016) Contribution of individual chemoreceptors to *Sinorhizobium meliloti* chemotaxis towards amino acids of host and nonhost seed exudates. Mol Plant-Microbe Interact 29(3):231–239

Webb BA, Compton KK, Del Campo JSM, Taylor D, Sobrado P, Scharf BE (2017a) *Sinorhizobium meliloti* chemotaxis to multiple amino acids is mediated by the chemoreceptor McpU. Mol Plant-Microbe Interact 30(10):770–777

Webb BA, Karl Compton K, Castañeda Saldaña R, Arapov TD, Keith Ray W, Helm RF, Scharf BE (2017b) *Sinorhizobium meliloti* chemotaxis to quaternary ammonium compounds is mediated by the chemoreceptor McpX. Mol Microbiol 103(2):333–346

Williams A, Langridge H, Straathof AL, Fox G, Muhammadali H, Hollywood KA, Xu Y, Goodacre R, de Vries FT (2021) Comparing root exudate collection techniques: an improved hybrid method. Soil Biol Biochem 161:108391

Woźniak M, Gałązka A, Tyśkiewicz R, Jaroszuk-Ściseł J (2019) Endophytic bacteria potentially promote plant growth by synthesizing different metabolites and their phenotypic/physiological profiles in the Biolog GEN III MicroPlateTM Test. Int J Mol Sci 20(21):5283

Wu S, Shi Z, Chen X, Gao J, Wang X (2022) Arbuscular mycorrhizal fungi increase crop yields by improving biomass under rainfed condition: a meta-analysis. PeerJ 10:e12861

Xie GL, Zhang GQ, Liu H, Lou MM, Tian WX, Li B, Zhou XP, Zhu B, Jin GL (2011) Genome sequence of the rice-pathogenic bacterium *Acidovorax avenae* subsp. *avenae* RS-1. J Bacteriol 193:5013–5014

Xin XF, Kvitko B, He SY (2018) *Pseudomonas syringae*: what it takes to be a pathogen. Nat Rev Microbiol 16(5):316–328

Xiong J, Wu L, Tu S, Van Nostrand JD, He Z, Zhou J, Wang G (2010) Microbial communities and functional genes associated with soil arsenic contamination and the rhizosphere of the arsenic-hyperaccumulating plant *Pteris vittata* L. Appl Environ Microbiol 76(21):7277–7284

Yan A, Wang Y, Tan SN, Mohd Yusof ML, Ghosh S, Chen Z (2020) Phytoremediation: a promising approach for revegetation of heavy metal-polluted land. Front Plant Sci 11:359

Yang C, Han N, Inoue C, Yang YL, Nojiri H, Ho YN, Chien MF (2022) Rhizospheric plant-microbe synergistic interactions achieve efficient arsenic phytoextraction by *Pteris vittata*. J Hazard Mater 434:128870

Yao X, Guo H, Zhang K, Zhao M, Ruan J, Chen J (2023) *Trichoderma* and its role in biological control of plant fungal and nematode disease. Front Microbiol 14:1160551

Yeates GW (1979) Soil nematodes in terrestrial ecosystems. J Nematol 11(3):213

Yeates GW, Bongers T, De Goede RG, Freckman DW, Georgieva S (1993) Feeding habits in soil nematode families and genera—an outline for soil ecologists. J Nematol 25(3):315

Yendyo S, Ramesh GC, Pandey BR (2017) Evaluation of *Trichoderma* spp., *Pseudomonas fluorescens* and *Bacillus subtilis* for biological control of *Ralstonia* wilt of tomato. F1000Research 6:2028

Yi X, Guo Y, Khan RAA, Fan Z (2021) Understanding the pathogenicity of *Pochonia chlamydosporia* to root knot nematode through omics approaches and action mechanism. Biol Control 162:104726

Yu K, Pieterse CM, Bakker PA, Berendsen RL (2019) Beneficial microbes going underground of root immunity. Plant Cell Environ 42(10):2860–2870

Yu Y, Gui Y, Li Z, Jiang C, Guo J, Niu D (2022) Induced systemic resistance for improving plant immunity by beneficial microbes. Plants 11(3):386

Yuan M, Jiang Z, Bi G, Nomura K, Liu M, Wang Y, Cai B, Zhou JM, He SY, Xin XF (2021a) Pattern-recognition receptors are required for NLR-mediated plant immunity. Nature 592(7852):105–109

Yuan M, Ngou BPM, Ding P, Xin XF (2021b) PTI-ETI crosstalk: an integrative view of plant immunity. Curr Opin Plant Biol 62:102030

Zhang Y, Li S, Li H, Wang R, Zhang KQ, Xu J (2020) Fungi–nematode interactions: diversity, ecology, and biocontrol prospects in agriculture. J Fungi 6(4):206

Zhang M, Peng C, Sun W, Dong R, Hao J (2022) Effects of variety, plant location, and season on the phyllosphere bacterial community structure of alfalfa (*Medicago sativa* L.). Microorganisms 10(10):2023

Zimaro T, Gottig N, Garavaglia BS, Gehring C, Ottado J (2011) Unraveling plant responses to bacterial pathogens through proteomics. BioMed Res Int 2011:354801

Chapter 11
Impact of Anthropogenic Activities on Microbial Diversity and Soil Health

Sangeeta Kumari, Mamta Devi, Kalpana Thakur, Bharti Minhas, Arvind Kumar Bhatt, and Neelam Kaushik

11.1 Introduction

Soil is a vital element capable of producing the resources necessary for sustaining life on Earth. Soil productivity is affected by the quantity and activity of beneficial microorganisms in the soil, which helps in the degradation of organic matter and ascertaining the availability of plant nutrients. Soil, a complex and diverse element of the terrestrial ecosystem, supports a wide range of organisms. A range of living things can be found in soil, from small mammals to small one-celled organisms. Soil microorganisms are a significant biological component of soil function, which also play an important role in biogeochemical cycles, improving soil quality, providing a variety of habitats and sustaining ecosystems (Hayat et al. 2010). Soil quality is essential for protecting the ecosystem, maintaining biodiversity and implementation of appropriate agricultural practices. The biological, biochemical and physiochemical properties of the soil are important determinants of its quality. Soil microbial populations are sensitive to anthropogenic interventions, which may have an adverse effect on soil health and the functional diversity of soil microorganisms, a crucial factor for measuring soil health (Wang et al. 2016). A number of studies demonstrated that anthropogenic activities, including urbanisation, industrialisation, climate change and agricultural intensification, have deleterious effects on soil quality to carry out a variety of essential ecosystem processes and functions (Trivedi et al. 2016). The use of inappropriate agricultural practices including intensive tillage and overuse of fertilizers and pesticides can lead to deterioration of the soil environment including microbial diversity (Yang et al. 2020). A variation in the

S. Kumari · M. Devi · K. Thakur · A. K. Bhatt · N. Kaushik (✉)
Department of Biotechnology, Himachal Pradesh University, Summer Hill, Shimla, India

B. Minhas
Department of Microbiology, Himachal Pradesh University, Summer Hill, Shimla, India

R. K. Bhatia, A. Walia (eds.), *Advancements in Microbial Biotechnology for Soil Health*, Microorganisms for Sustainability 50,
https://doi.org/10.1007/978-981-99-9482-3_11

structure and networks of soil microbial population is frequently related to changes in soil nutrient levels, such as nitrogen (N) and carbon (C) which reduce nutrient availability in the soil.

In addition to agricultural practices, urbanisation and industrialisation processes are also associated with the impact of the chemical and physical quality of the soil. Furthermore, several studies have revealed that the effects of climate change, including higher temperatures, increased CO_2 and precipitation, have an effect on the composition of the soil microbial community in some particular habitats (Yan et al. 2016). Global climate change may have a detrimental effect on ecosystems and biodiversity, leading to the extinction of some animal and plant species as well as microbes. These anthropogenic interventions pose a serious threat to agriculture productivity, soil fertility, green forest cover, ecosystem equilibrium and other factors by removing the top fertile layer. Hence, the purpose of this chapter is to examine the impact of anthropogenic activities on soil microbial diversity and soil health, as well as soil physicochemical and biological features.

11.2 Physico-chemical Properties of Soil

The quality of the soil mainly depends on the biochemical, biological and physiochemical characteristics of the soil which is impacted by urbanisation, farming habits, industrialisation and climate variability which are responsible for declined and contaminated yields. The physico-chemical analysis of soil depends on a number of variables, such as moisture, pH, texture, electrical conductivity, temperature, soil organic matter and the quantity of accessible potassium, nitrogen and phosphorus (Liu et al. 2018).

11.2.1 pH

The most important physical aspect is the soil pH. It significantly influences the soil's solute content and absorption. Many plants and soil organisms prefer an acidic (less than 6) or alkaline (greater than 8.5) habitat; hence soil pH is an important factor for crop production (Neina 2019). Mulder et al. (2005) found a significant inverse relationship between soil acidity and the presence of bacteria, fungus, nematodes and arthropods. It provides an effective way for assessing how well the nutrients are distributed in the soils. The pH of the soil is an essential factor in assuring the availability of certain plant nutrients because some of them, like Zn, Mn, Fe and Cu, are more easily accessed in acidic soil than in alkaline soil. Due to environmental changes, long-term soil acidification causes a decrease in soil microorganisms, which in turn controls the pH of the soil through ecologic and evolutionary mechanisms. The soil pH generally increases with depth. Very high and low pH leads to failure of crops because of ionic strength imbalance.

11.2.2 Temperature

The temperature of the soil is the most important characteristic, which ranges from −20 to 60 °C, because it exhibits its impact on the biological, chemical and physical processes involved in plant growth. The temperature of the soil depends up on the ratio of energy absorbed to energy lost from the soil. It fluctuates daily, affected by variation in temperature of air, season, time of the day and local climatic circumstances (Jain and Singh 2014). An increase in soil temperature speeds up chemical reactions, reduces the solubility of gases and lowers the soil pH. Furthermore, it is essential for seed germination. Temperature variations have an impact on the biomass growth (plants, wood, fuel, etc.), and microbial activity also decreases with low temperature and most activities ceases at freezing point. Temperature changes in the soil are a result of exchange activities that mostly occur at the soil surface.

11.2.3 Texture

Soil texture (loam, sandy loam and clay) is one of the most important physical aspects of the soil, which is used in both the field and the laboratory to categorise agricultural soils. The texture of the soil varies according to area and is largely influenced by the size of the soil's granules, sand, silt and clay. Sandy soils are naturally light soils with low concentration of nutrients, little capability to hold moisture, little cation exchange capacity, poor buffer capacity and a high rate of permeation (Patnaik et al. 2013). Sandy loam texture provides protection to root penetration and is less suited for crops with deep roots. In addition, clay to sandy soil reduces hydrophobicity and increases crop output. Due to their limited permeability and tendency to retain moisture for a longer period of time than soils with a lighter texture, clay soil is not ideal for crops that cannot survive prolonged soil wetness.

11.2.4 Moisture

Soil moisture is one of the most important characteristics of soil, which mainly comprises of solids (minerals), air and moisture. The soil moisture content, which also impacts the texture of the soil, greatly influences the soil capacity to absorb nutrients. Common factors influencing soil moisture include clay minerals, void ratio, organic matter, particle size and the state of the groundwater. Sandy soils have a lower water content than clay soils, which is likely to be more porous. Red and black soils have the highest water retaining capacity. Sandy soils can quickly rehydrate with soil moisture, despite the fact that they can hold just as much water as soils with heavier textures (Jain et al. 2014). The amount of moisture in the soil has a significant impact on nutrient absorption by the soil, and moisture also influences the texture of the soil.

11.2.5 Soil Organic Matter

A soil is beneficial for agricultural practices if it contains soil organic matter. The compost and animal manures are incorporated into the soil for biomass. Levelling has resulted in a decline in the organic matter content of the soil from the surface to the subsurface (Tale and Ingole 2015). The rate of organic matter decomposition tends to increase as the temperature rises, providing the best environment for plant growth. The capacity of soil organic matter to hold water and absorb cations is unparalleled, and it also provides vital nutrients. It also serves as a food source for soil microorganisms, which helps to enhance and regulate their action. The structure of the soil can be retained by its organic matter contents. It also influences the capacity of the water supply and the rate of infiltration.

11.2.6 Electrical Conductivity

The property of the soil to conduct electricity is another special feature used to assess the soil quality. With an increase in ion concentration, soil conductivity rises. Soil health can be assessed quickly, simply and affordably using electrical conductivity. Electrical conductivity determines subsurface soil characteristics like salinity, organic matter content, cation exchange capacity, drainage and soil texture. Electrical conductivity, a widely used indicator of salinity, is used to calculate the amount of soluble salts present in the soil (Wagh et al. 2013). The amount of moisture that soil particles can hold has an impact on the electrical conductivity of the soil, which is useful in monitoring the mineralisation of organic material in the soil.

11.2.7 Nitrogen

The element that plants need the most from the soil is nitrogen, which is essentially required for proper plant growth and development. In the soil system, the nitrogen cycle is vital and is regulated by biological processes. The natural green colour characteristics of good healthy plants are produced by nitrogen, which also promotes aerial growth. The rate of plant growth is proportional to the rate of nitrogen supply. However, an excess of nitrogen can damage the plants. In comparison to soil treated with manure, soil treated with inorganic fertilizer showed a higher nitrogen concentration (Tale and Ingole 2015).

11.2.8 Phosphorus

Phosphorus, which is essential for plant growth, is one of the most significant elements present in every living cell. Generally, phosphorus is the only nutrient that remains in plant nucleus and functions as a form of energy storage. Moreover, it is an integral component of the production of all oils, sugars and starch throughout the photosynthetic process. Purple stem and leaf colour as well as a low fruit yield are the symptoms of phosphorus deficiency. Phosphorus is important for both fruit and flower formations along with seed germination (Jain et al. 2014).

11.2.9 Potassium

Potassium is an essential mineral for the growth of plants and plays a vital role in different plant physiological processes. Although potassium is not a necessary component of any major plant parts, it is essential for a wide range of physiological processes, such as protein synthesis and preserving the water balance in plants. Potassium influences plant growth and division, as well as glucose synthesis, sugar transport, enzyme function and plant disease resistance (Jain et al. 2014). It also plays a key role in a variety of metabolic processes in plants, including the regulation of photosynthesis and production of plant metabolites such as sugars which serve a variety of metabolic purposes in plants and the synthesis of lignin and cellulose which are used as parts of cellular structure.

11.3 Microbial Communities Present in Soil

Microbial communities in soil constitute essential component of ecosystem because of their diversity and complexity. These communities consist of genetic divergence, community density and comparative symmetry. There are billions of microorganisms in soil, including members of almost every phylum of life forms, that is, bacteria, fungi, algae, actinomycetes, protozoa, etc. Some of the key processes that soil microbial populations regulate include the carbon cycle, N_2 fixation, ammonia oxidation, de-nitrification and ammonification. For plant growth and agricultural production, a healthy soil is one that has a wide and abundant microbial community and activity (Singh and Gupta 2018).

11.3.1 Bacteria

Soil bacteria, which make up the majority of the soil biomass, are prokaryotic and single-celled and display a wide diversity of shapes and sizes. The soil bacteria are highly diverse group with estimated number of bacterial species about 20,000 per g of soil and are able to live in different environmental conditions like high salinity or low oxygen levels. Being the most abundant microorganism in the soil, numerous metabolic processes are carried out by bacteria in the soil which directly or indirectly provide nutrients for the growth of higher plants. Soil bacteria are typically not very energy-efficient organisms; only 20% of the decomposing organic matter can turn into bacterial tissue. Bacteria have also been utilised in soil to mineralise organic contaminants, which is referred to as bioremediation of polluted soils (Zaidi et al. 2009).

11.3.2 Fungi

Fungi are eukaryotic, aerobic and heterotrophic organisms. They can produce observable macroscopic structures in soil and need an ample supply of oxygen and organic materials. There are tens of thousands of different species of fungi in the soil, making them one of the most varied groups of microbes. The cellulose, hemicellulose, starch, pectin and lignin present in the soil as organic matter are broken down or decomposed by fungi and provide food for bacteria living in the soil. Due to their metabolic activity and involvement in nutrient cycles, fungi are a key component of grassland ecosystems (Dengler et al. 2014).

11.3.3 Actinomycetes

Actinomycetes, spore-forming gram-positive bacteria characterised by substrate and aerial mycelium growth, are members of the order Actinomycetales (Bhatti et al. 2017). They lack a nuclear membrane and frequently split into spores that resemble bacterial cells. They contribute in the rotation of organic materials and make them a potential candidate to protect plants from harmful pathogens. In addition, they degrade complex polymer mixtures in decomposing plant, animal and fungal matter to provide a range of extracellular enzymes useful for crop production.

11.3.4 Algae

Eukaryotic cells called algae (photosynthetic organisms) range in size from 2 to 20 μm, and their quantity in soil normally ranges from 100 to 10,000 per g of soil. They are the main producers in the aquatic ecosystem, and through photosynthesis, they convert CO_2 from the atmosphere and minerals into organic matter in the presence of sunlight. Algal species are utilised as a biofertilizer and a soil stabiliser in various agricultural practices, and they have a significant effect on crop yields (Abdel-Raouf et al. 2012). Most algae prefer damp to wet environments to flourish, while others may live in hot or cold climate. The filamentous forms of the *Cyanophyceae*, particularly *Oscillatoria*, *Schizothrix* and *Plectonema*, were discovered to be essential for the growth of soil.

11.3.5 Protozoa

Soil protozoa are microscopic microorganisms belonging to the class ciliate/ciliophora and have short, hair-like appendages called cilia that surround their bodies and help them move. The abundance of protozoa is in the upper 15 cm layer of the soil. Due to their size and flexibility, protozoa serve as essential bacterial feeders and serve as an indicator of the soil health in the terrestrial ecosystem. The population of protozoa varies in accordance with the soil water potential, texture and structure. This group includes important soil organisms such as *Gastrostyla*, *Balantiophorus*, *Colpoda Colpidium*, *Halteria*, *Uroleptus*, *Pleurotricha* and *Vorticella*. Protozoans primarily compete with nematodes for food sources, govern bacterial and algal numbers and serve as food for nematodes, rotifers and other organisms that live in soil environment (Khan and Rao 2019).

11.4 Different Anthropogenic Activities Impacting Environment

Globally, various environmental impacts of anthropogenic activities are accountable for the significant accumulation of pollutants in the soil. Anthropogenic activities such as industrialisation, urbanisation, agricultural practices (excessive use of pesticides and fertilizers), fossil fuel combustion, climate change and fire release pollutants into the environment. Numerous dangerous chemicals including heavy metals are likely to be found in underground soil which results in soil degradation by destroying physiochemical properties of the soil and disrupting the structure of the microorganisms and decrease biodiversity. As the world economy has grown, the extent of pollution in the soil due to human activities has gradually increased, worsening the environment. The pollutants in the environment may disrupt regular

Table 11.1 Anthropogenic activities and pollutants released in the environment

Anthropogenic activity	Pollutants
Industrialisation	Different toxic gases released by industries are SO_2, SO_3, NO, NO_2, CO, CO_2, Cl_2, H_2SO_4, etc.; hydrocarbons; explosive chemicals; synthetic chemicals; organic substances, including certain heavy metals such as cadmium, arsenic, chromium, copper, lead, mercury and nickel; and radionuclides.
Urbanisation	Domestic, municipal and industrial hazardous waste, smoke, dust, gasoline and diesel release CO_2, CO, NO_2 and SO_2.
Agricultural practices	Paddy fire releases CO_2, CO and N_2O. Many pesticides are persistent organic pollutants. Cu, Zn and Cd are heavy metals that are found in fertilizers.

activities and the ecological equilibrium in the soil. Table 11.1 shows the various pollutants released in the environment due to prevalent anthropogenic activities.

11.5 Impact of Anthropogenic Activities on Microbial Diversity and Soil Health

Anthropogenic activities such as agricultural practices, urbanisation and industrialisation are continuously releasing pollutants to the soil, which is a widespread problem globally. Agrochemicals; municipal and household wastes, including wastewater; and pollutants produced as a result of industrial activities and irrigation with untreated wastewater are the main anthropogenic sources of soil contamination. All these activities produce waste that end up contaminating environment leading to loss of biodiversity and ecosystem processes. The impact of anthropogenic activities on soil health and microbial diversity is depicted in Fig. 11.1.

11.5.1 Impact of Industrialisation

Microorganisms, both beneficial and harmful, are abundant in the soil. There may be numerous consequences and an irreversible impact if groups of soil macro- and microorganisms become unbalanced. Pollutants are released into the atmosphere, water and soil by different industrial activities. The pollution comes from both point sources, such as metal smelting or mining emissions, effluents and solid discharge from industry, as well as nonpoint sources such as insecticides or the overuse of fertilizers (McGrath et al. 2001). Human activities are primarily responsible for environmental pollution, especially in the mining, metallurgy and chemical industries as well as in municipal administration and traffic emissions. In industrial areas, anthropogenic activities lead to frequent heavy metals pollution.

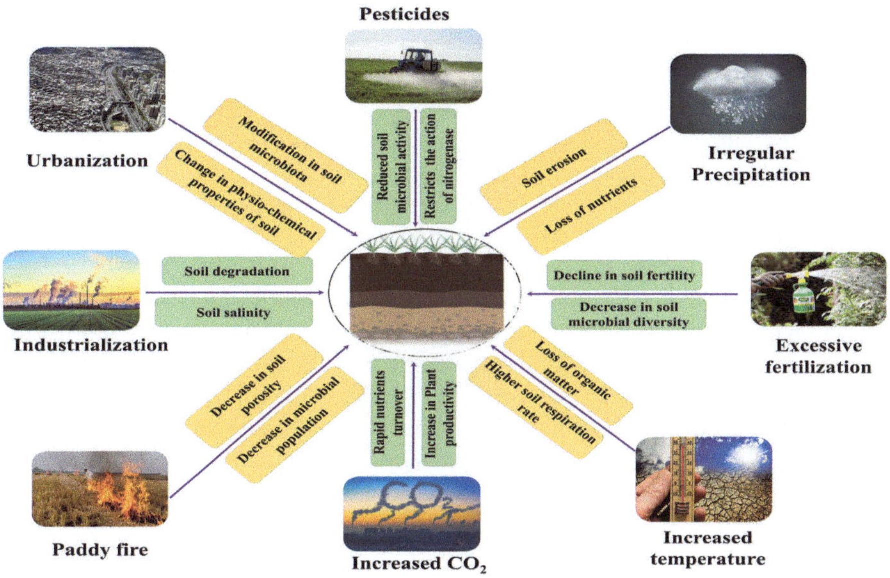

Fig. 11.1 Various anthropogenic activities and their impact on microbial diversity and soil health

Due to their persistence in the environment and inability to be eliminated, heavy metals in soils are a severe environmental hazard. Moreover, trash disposal, waste incineration, fertilizer application and long-term wastewater application in agricultural areas can all lead to heavy metal contamination of soils. The locations closest to roadways, especially in metropolitan areas, are most susceptible to pollution from vehicles. Combustion engine-powered cars release pollutants such as chromium, nickel, platinum and cadmium in addition to lead and zinc.

Microorganisms are the first members of the biota to be directly and indirectly impacted by heavy metals. In low quantities, several metals (such as Fe, Zn, Cu, Ni and Co) are crucial for many microbial processes. These metals are frequently engaged in redox and metabolic activities. However, high concentrations of heavy metals are harmful or have inhibitory effects on living beings. Detrimental effects of heavy metals on soil microorganisms result in decreased organic matter decomposition, decreased soil respiration, decreased diversity and decreased activity of certain soil enzymes. Rajapaksha et al. (2004) investigated the responses of fungi and bacteria to hazardous metals in soils, and it was concluded that the toxic quantities of heavy metals in the soil are more likely to affect bacteria than fungi. Moreover, it was observed that soil pH had a variable impact on the microbial response to soil pollution, that is, low pH in polluted soil increased the detrimental impact on bacteria and not on fungi. Furthermore, heavy metals change the structures of proteins and nucleic acids, leading to interaction between heavy metals and protein molecules further inactivating the biomolecules. As a result, the integrity of the microbial cell membrane is compromised or the entire cell is destroyed (Bong et al. 2010).

Different microbial populations may be affected differently and the outcome may also change depending on the metal or its concentration in the soil. For instance, nitrification and mineralisation of protein molecules are processes that are impacted by copper pollution of the soil. The majority of soil microbiological processes are inhibited by zinc at about 1000 ppm, which may affect nitrification processes at about 100 ppm. Microorganisms are crucial for the modification of mercury compounds in the environment. Toxic effects on soil nitrifiers may result from mercury concentrations above 10 parts per million (ppm) as some microorganisms are susceptible to excessive mercury.

Lead concentrations that are higher in the surface soil layers have a deleterious impact on the soil microbiota. On average, 0.2% of lead is absorbed by soil microorganisms, but in exceptional circumstances, this amount can exceed 40% of biomass, helping in biological remediation (Juwarkar et al. 2007). According to different studies, it has been observed that the microbial activity of the soil is negatively impacted by heavy metal pollution. The results indicated that compared to natural soil, contaminated soil had considerably lower concentrations of the microbial groups.

11.5.2 Impact of Urbanisation

The effects of urbanisation on the soil microbiota are crucial since urbanisation is expanding quickly around the world. Environmental deterioration is a result of the widespread construction of infrastructure, including housing, roads, metros and railways which are significant sources of soil pollution. Urban areas are growing at an unprecedented rate due to the large migration of individuals from rural to urban areas, which has led to physical alterations in geomorphology related to urban building construction, roads and pollution brought on by human activities (Wang et al. 2018). Due to the alteration of vegetation and rise in human activities, the habitat disturbances and alteration of land cover brought on by urbanisation have a significant impact on biodiversity, further contributing in homogenisation of biodiversity. Urbanisation causes disturbance in the physicochemical environment of the soil which ultimately changes the bacterial communities (Yi et al. 2022).

The degree to which bacterial diversity of soil and composition of community adapts to urbanisation, however, is still debatable. While some studies found that urban areas had greater soil bacterial diversity than rural ones, other research found slight to no variation among urban and rural soils (Pereira et al. 2021). Moreover, how the land is being used has a significant influence on the variety and composition of bacterial communities. Many variables, including plant diversity, pH, heavy metals, soil moisture, gradients in altitude and change in climatic factors, that is, temperature rise, precipitation and CO_2, are known to have an impact on the population of bacteria that live in the soil. Unlike other ecosystems, urban ecosystems are susceptible to a variety of external influences related to urbanisation, such as the use of technosols, soil sealing, the urban heat island effect, physical disturbance, soil compaction and city population, which can all affect soil microorganisms (Lu et al. 2020). Under various urban land cover types, these impermeable surfaces lead to variations in composition of bacterial community and a decline in bacterial diversity.

11.5.3 Impact of Agricultural Activities

With increase in world population from 5.2 billion in 1990 to 8 billion in 2022 which is expected to reach by 9.7 billion by 2050, this will increase the need to produce more food from limited resources to fulfil the food demands of expanding population (Gupta et al. 2022). As soil is the foundation of nearly all worldwide agriculture and the medium for the majority of terrestrial biological activity, therefore agriculture has an impact on soil and consequently on the qualitative and quantitative composition of soil microorganisms. The main effects of agriculture on soil microbial diversity and soil health are listed below.

11.5.3.1 Tillage

Soil tillage affects the chemical, physical and biological aspects of the soil that are essential for plant growth and optimal production. Historically, tillage systems have been utilised as a way to loosen up compacted surface soil, restricting soilborne phytopathogens, and remove plant debris but also increase the breakdown of fungal hyphal networks, erosion of soil, loss of nutrients and greenhouse gas emissions (Zuber and Villamil 2016). It is frequently required to modify the soil's physical characteristics to encourage water uptake, storage and transmission and to provide a favourable environment for seeds and roots.

In particular, conservation tillage methods such as subsoiling tillage, no-tillage and straw mulching have generated exceptional results in terms of enhancing nutrients in the soil improving soil quality and reducing erosion of soil. However, excessive soil tillage is connected to changes in the chemical and biological composition of the soil, resulting in a loss of soil stability and structure and promoting soil erosion. Furthermore, improper agricultural techniques, such as conventional and intensive tillage systems that render soil surfaces bare, have been the primary causes of agricultural soil deterioration (Blanco-Moure et al. 2012). The biological and physiochemical properties and quality of the soil are significantly influenced by both earthworms and soil microorganisms.

According to a meta-analysis study by Briones and Schmidt (2017), earthworms increase nutrient availability by releasing their casts, which contain more organic carbon (40–48%), phosphorus (84%) and nitrogen (24%), than bulk soil. In contrast, traditional tillage reduces the earthworm population up to 76–90% of the population. Using a meta-analysis, Zuber and Villamil (2016) found the detrimental effects of tillage on soil microbe biological diversity and activity of enzymes. Conservation tillage and no-tillage keep the soil surface cool and moist, whereas tillage raises the soil surface temperature and stimulates microbial development for residue breakdown (Zuber and Villamil 2016). Tillage impacted the bacterial community of actinobacteria, β-proteobacteria and γ-proteobacteria. Mbuthia et al. (2015) found that no-tillage enhanced gram-positive bacteria, *Actinomycetes*, and AMF by 5%, 6% and 17%, respectively, as compared to tillage, and showed decreased population of saprophytic fungi by 14%.

11.5.3.2 Fire

Fire is used to get rid of plant debris, stubble, straw and slash from previous harvests to make the fields ready for cultivation. Burning of crop residue is a common practice in Western Uttar Pradesh, Haryana and Punjab, but it is currently happening everywhere in the nation. Fires can alter physical, chemical, biological and mineral qualities of the soil. The heat produced by burning paddy straw penetrates the soil up to 1 cm deep, elevating the temperature to 33.8–42.2 °C. It eliminates the microbial populations and nutrients necessary for healthy soil. It is predicted that burning one ton of stubble or paddy straw results in the loss of 2.3 kg phosphorus, 5.5 kg nitrogen, 1.2 kg sulphur and 25 kg potassium, in addition to organic carbon (Sanu et al. 2023). This technique reduces soil quality by accelerating soil erosion; raising soil pH; reducing soil nutrients like carbon, nitrogen and sulphate; and lowering overall soil quality. A study by Kumar et al. (2019) found that burning paddy straw significantly reduced the population of bacteria, actinomycetes, fungi, potassium-solubilising microorganisms (KSM), phosphate-solubilising microorganisms (PSM) and cellulose degraders immediately after burning along with decrease in the microbial enzymes and other soil health factors. In addition to this, burning straw residues in open causes the release of greenhouse gases (such as N_2O and CH_4) and air pollution which has detrimental effects both on the environment and public health (Rathour et al. 2023). Therefore, other crop residue management strategies must be implemented.

11.5.3.3 Fertilizer

Any substance used to provide nutrients to the soil in order to improve soil fertility and plant growth is referred to as a fertiliser. The deficiency of these elements in the soil contributes to reduction in the crop production. Excessive fertilizer use reduces soil microbial biodiversity as well as plant biodiversity. The repeated use of these fertilizers changes soil osmolarity, carbon availability and pH and also generates toxicity due to the presence of specific ions, resulting in a decrease in Nitrospirae and Acidobacteria members and a small rise in Firmicutes and Actinobacteria. Recently Ren et al. (2020) on the basis of a 10-year field study concluded that long-term N fertilisation can remarkably impact the majority of the dominant soil bacterial species by altering the pH of the soil.

In addition to the change in microbial population, fertilizers also affect soil's organic matter and carbon composition. In order to maintain and encourage biogeochemical cycling in the soil ecosystem, organic matter is an essential component of soil's biological and physical quality (Berazneva et al. 2019). It also increases the soil's ploughing ability, improves the soil's capacity to hold water, increases the capacity for water seepage, retains and supplies fertilizer to the growing crops and withstands drought and waterlogging.

Excessive and long-term utilisation of fertilizers also affects organic matter in different soil layers, which is responsible for deterioration of agricultural soil quality. Additionally, the excessive use of chemical fertilizers poses environmental hazards by making soil harder; polluting air, water and soil; reducing soil fertility; and lowering vital nutrients and minerals in the soil.

11.5.3.4 Pesticides

Pesticides are a group of chemicals used in agriculture to manage rodents, weeds, insects and fungi. Pesticides have been shown to increase the microbial community when used in the suggested doses, but excessive use has negative effects on the microbial population. With the increased use of pesticides in modern agriculture, the problem of how these substances affect the makeup of beneficial soil microbes and the processes they control has drawn more attention. However, certain pesticides target non-specific binding sites and have an adverse impact on beneficial microorganisms (Shao and Zhang 2017). As a result, when a microbial population is disturbed, the interconnected structure of different trophic levels is disrupted, which has a variety of indirect impacts on soil microbial processes. For example, carboxylic acid fungicides by interacting with DNA topoisomerase II as the non-target-affecting site not only inhibit DNA replication and protein synthesis in some fungal pathogens but also in prokaryotic cells (Sioud et al. 2009). Similarly, glucopyranosyl antibiotics when used as pesticides are toxic to higher eukaryotic species and non-target bacteria by inhibiting the production of particular amino acids.

Most popular pesticides are insecticides (which kill insects and other arthropods), fungicides (which kill or prevent the growth of fungi and their spores) and herbicides (which kill weeds or other plants). Table 11.2 demonstrates various chemical pesticides and their impact on soil microbes. For example, glyphosate negatively impacts crops and other non-target plant because plant resistance and resilience mostly depend on plant-associated bacteria. Pesticide influence on soil microbial community is a highly complicated process that is dependent on a number of factors, including pesticide concentration and structure as well as environmental factors like soil pH, type, moisture level, salinity and temperature as well as the presence of organic substances in the soil.

Table 11.2 Impact of commonly used pesticides on soil microorganisms

Sr. no.	Pesticide	Type	Effect on soil microorganisms	References
1.	Glyphosate	Herbicide	Glyphosate negatively impacts not only rhizosphere microbes but also functionality of aerial plant endophytes, emphasising the damaging impacts of glyphosate on plant symbiotic microorganisms.	Fuchs et al. (2023)
2.	Diuron or linuron	Herbicide	The local bacterial communities were completely altered by the presence of the herbicide diuron in the soil.	Moretto et al. (2017)
3.	Atrazine	Herbicide	Decrease in bacterial, fungal and actinomycete populations along with significant reduction in percentage organic matter was observed.	Sebiomo et al. (2011)
4.	2,4-D	Herbicide	It disrupts the diversity and functionality of microbiomes and alters the bacterial population structure in contaminated soils.	de Oliveira et al. (2021)
5.	Mancozeb	Fungicides	Low concentrations of mancozeb impacted actinomycetes and fungal communities and increased quantities affected soil bacteria. The ammonification and nitrification processes were also impaired. Further, adverse impact on mineralisation of carbon and microbial biomass was observed.	Walia et al. (2014)
6.	Captan, carboxin and thiram	Fungicides	It prevents the activity of denitrification causing bacteria.	Milenkovski et al. (2010)
7.	Paclobutrazol	Fungicides	Paclobutrazol had an adverse effect on soil fertility and quality, which leads to ecological imbalance.	Kumar et al. (2021)
8.	Copper	Fungicides	Bioavailable Cu correlated positively with the relative abundances of the group Acidobacteria and negatively with the groups Bacteroidetes and proteobacteria.	Berg et al. (2012)
9.	Butachlor	Herbicide	Butachlor had decreased the activity of the protease, urease and dehydrogenase enzymes in flooded environments.	Rasool et al. (2014)
10.	Cypermethrin	Insecticides	Decline in firmicutes and an increase in Gram-negative bacteria.	Zhang et al. (2009)
11.	Carbendazim	Fungicides	Reduced the number of *Burkholderia* and *Mycobacterium* microbes.	Fang et al. (2016)

(continued)

Table 11.2 (continued)

Sr. no.	Pesticide	Type	Effect on soil microorganisms	References
12.	Chlorpyrifos	Insecticides	Chlorpyrifos considerably reduced microbial diversity, altering the organisation of microbial communities present in cotton rhizospheric soil and negatively affecting certain important bacterial species.	Wang et al. (2022)
13.	Imidacloprid and metribuzin	Insecticides and herbicides	The insecticide imidacloprid and herbicide metribuzin had an effect on the *Gibberella*, *Candida* and *Saitozyma* fungal populations.	Streletskii et al. (2022)
14.	Oxadiazon and trifluralin	Herbicide	Prevents mycorrhizae spore production and propagation.	Pasaribu et al. (2013)
15.	Carbamate	Insecticide	Restricted the action of nitrogenase in *Azospirillum* and a number of soil microorganisms.	Hussain et al. (2009)
16.	Isoproturon	Herbicide	It decreased urease activity, which is essential for the nitrogen cycle.	Chaudhary et al. (2022)
17.	Miraj	Fungicide	Reduced soil microbial activity and actinomycete, fungus and bacteria populations.	Mehjin et al. (2019)
18.	Malathion	Insecticide	At higher concentrations of malathion, the bacterial population was observed to be reduced.	Walia et al. (2018)
19.	Bensulfuron methyl, pendimethalin, 2,4-D amine, bispyribac sodium and chlorpyrifos	Herbicide and insecticides	Reduced soil bacterial diversity and abundance and induced stress in non-photosynthetic microbes.	Onwona-Kwakye et al. (2020)

11.5.3.5 Impact of Climate Change

Climate change has emerged as a major scientific and political concern worldwide. Microbial processes are dependent on environmental parameters such as moisture, temperature, enzyme activity and nutrient availability, all of which are affected directly or indirectly by climate change. Climate change has been one of the main drivers of the decline in the microbial diversity of terrestrial microorganisms (Classen et al. 2015). Global climate change is expected to have a wide range of effects on soil processes and qualities critical to restoring soil fertility and productivity. Many aspects of global climate change, such as rising carbon dioxide (CO_2) concentrations, changing temperature and precipitation patterns and an increase in the frequency of extreme weather events, are expected to have an impact on the soil microbial community. Figure 11.2 shows the key effects of global climate change on soil microbial diversity and soil health, as discussed below.

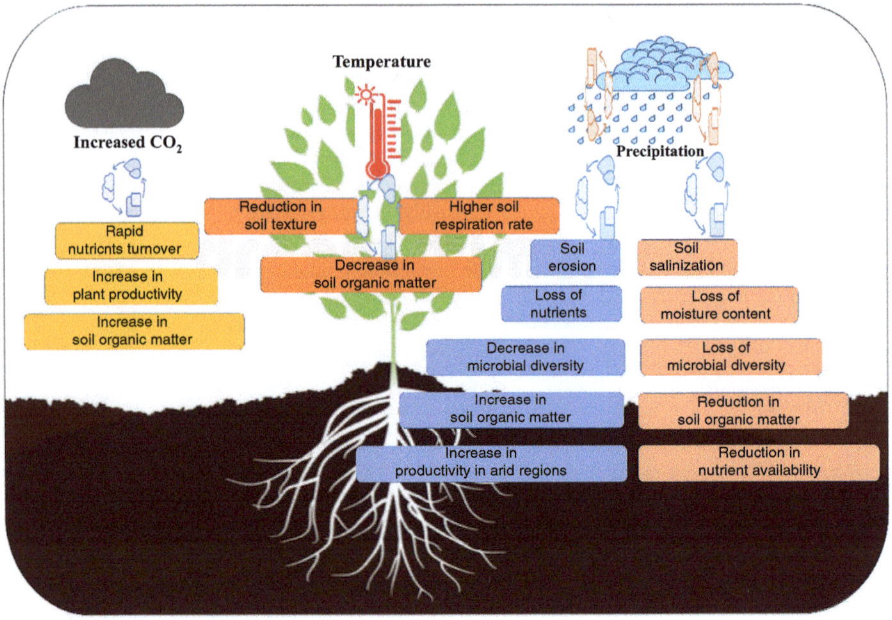

Fig. 11.2 Impact of climate change on microbial diversity and soil health

Increasing CO_2 Concentration

Globally, population growth, industrialisation, urbanisation and the use of fossil fuels are significantly contributing to increase in CO_2 concentration, which eventually leads to the increase in environmental temperature. Unexpectedly, 21% of direct CO_2 emissions originate from industry, and 24% come from agriculture, forestry and other land use, although it is widely accepted that increased CO_2 stimulates plant growth and primary production. However, the constantly increasing concentration of CO_2 in the atmosphere has significant effects on ecosystem functioning.

In the presence of increased CO_2, plants reduce the allocation of N-rich metabolites and increase the allocation of C-rich metabolites to root exudates. It causes an increase in microbial activity and consequently the CO_2 production, which has negative impact on the accumulation of organic C in the soil and consequently on potential sequestration of the soil (Reid et al. 2012). Increased CO_2 can have an impact on extracellular enzyme activities, which are frequently used for monitoring alterations in the functioning of microbial communities in soil. Yang et al. (2019) reported that elevated CO_2 reduced the activity of the nitrogen (N) fixation gene (*nifHI*) and the ammonium-oxidising gene (*amoA*) by 12.6% and 6.1%, respectively, and decreased nitrifying enzyme activity by 48.3%. Furthermore, elevated CO_2 levels effected a large number of microbial genes involved in carbon (C) breakdown. They also reported a decrease in soil N accessibility, with both nitrate (NO_3^-)

and ammonium (NH_4^+) reduction. The overall alterations in microbial communities and soil N pools caused by long-term increased CO_2 levels suggested reduced microbial N breakdown, which lowers the microbial community's maximal growth rate (Yang et al. 2019).

Increased CO_2 may also stimulate N_2O outflow from soil due to increased heterotrophic denitrification, as higher CO_2 increases root-derived carbon inputs to the soil. The effect of increased CO_2 on P mineralisation to maintain soil fertility is reported to have mixed outcomes. Some studies showed that increased CO_2 had no impact on P limits or demand. However, there are some studies that reported an increase in urease and alkaline phosphatase activities in the presence of increased CO_2, suggesting P acquisition.

Although the effects of rising CO_2 levels on plant development and soil biogeochemical cycles are well understood, the impacts of rising CO_2 levels on soil microorganisms are still debatable. Soil microbial diversity has been observed to increase, decrease or remain unchanged in response to increased CO_2 (He et al. 2012).

Due to the increased plant demand for nutrients and increased C absorption rates, the impacts of increased atmospheric CO_2 concentration on soil microbial community structure are frequently characterised by an increased mycorrhizal colonisation. This is because plant's needs for N and P will rise in along with C absorption rates, and plants will allocate additional photosynthates to roots and mycorrhizal fungi to help fulfil rising nutrient demand. As soil and microbial respiration rates are normally faster under increased CO_2, suggesting that increased plant growth under rising CO_2 increases the quantity of C that enter the soil, which in turn stimulate soil microbial activity. It is commonly considered that rising CO_2 increases soil C availability, which increases fungal biomass more than bacterial biomass because fungi have a greater C/N ratio than bacteria.

Increasing Temperature

Climate change has occurred as a result of the degree of global temperature increase. Temperature has been demonstrated to be among the most significant environmental factor influencing microorganism development and biological activity in the soil. The average global surface temperature has risen by 0.74 °C in the last 100 years and is expected to rise by 1.8–4.0 °C by 2100, which might have an adverse impact on soil carbon sequestration by potentially accelerating heterotrophic microbial activity. Additionally, higher temperatures will enhance the decomposition of organic matter, releasing CO_2 into the atmosphere and lowering carbon to nitrogen ratios (C:N). Different microorganisms have different optimal temperature requirements for the development and activity. Temperatures more than 37 °C and less than −7 °C have a negative impact on soil microbial activities (Oertel et al. 2016). As the temperature increases, the composition of the microbial population changes, and methanogenesis, fermentation and other processes speed up. Increased temperature may change the organisation of the microbial community, which in some situations may result in the loss of microbial groups that have adapted to the environment and reduces the release of soil organic carbon. Microbial biomass and enzyme both effectively break down carbon-based organic materials and release hazardous substances into the environment, making them potent tools for accelerating global warming (Jia et al. 2021).

Alteration in Precipitation Pattern

In response to global climate change, there will be changes in worldwide precipitation amounts and patterns. Due to climate change, the risks of drought due to altered precipitation pattern and flood caused by excessive precipitation and melting of glaciers are increased. Soil microbes play an important role in the decomposition of organic matter and mineralisation of nutrients; therefore variation in precipitation directly affects the soil microbes population by changing the amount of available soil water. When the rise in rainfall reaches over 40%, the number of plant species, their aerial biomass and their coverage will all be enhanced in arid and semi-arid grasslands (Zhu et al. 2022). The Intergovernmental Panel on Climate Change (IPCC) has projected a 20% increase or decrease in precipitation rate, which can affect soil moisture, terrestrial microbial community structure and the decomposition rate of soil organic carbon (Zhang et al. 2018). Therefore, one of the key factors influencing the composition of the microbial population in terrestrial ecosystems is precipitation. Temperature, pH, texture and other elements interact directly and indirectly, influencing the soil microbial community. Trivedi et al. (2013) have observed reduced plant development and microbial activity due to the lack of precipitation during the dry year.

Long dry periods may restrict microbial development and decomposition, which could have a negative feedback effect on carbon fluxes in particular ecosystems. As a result, changes in precipitation can have an impact on plant community structure and productivity as well as soil bacterial population directly by changing soil water availability. Li et al. (2022) examined the responses of microbial communities to variations in precipitation patterns on a semi-arid grassland in Northern China. According to the data, the richness of bacterial species decreased dramatically with decreased precipitation but remained constant with increased precipitation. Proteobacteria and Bacteroidetes relative abundance declined with decreasing precipitation but increased with increasing precipitation, whereas *Chloroflexi* and *Acidobacteria* relative abundance increased with decreasing precipitation but decreased with an increase in precipitation. Xu et al. (2020) performed a meta-analysis of data from 95 field experiments with drought treatments and 109 field studies with high precipitation treatments. According to the findings, microbial biomass increased by 18% during periods of increasing precipitation and declined by 17% during periods of drought.

11.6 Conclusion

The soil is essential constituent of the Earth's land which is the basis of global agriculture production. The physico-chemical, biochemical and biological characteristics of the soil play a significant role in determining soil quality. The diverse microbes present in soil play a vital role in the improvement of soil's fertility and maintains the soil health. Urbanisation, industrialisation, overuse of pesticides and fertilizers and other anthropogenic activities are releasing pollutants into the environment that may have a negative effect on the functional diversity of soil microorganisms and the health of the soil. In recent years, significant scientific evidences as discussed above have indicated that soil health is adversely impacted by anthropogenic activities by lowering contents of soil organic matter, reducing soil structure, polluting soil, reducing essential nutrients and increasing vulnerability to erosion and decrease in microbial diversity. It is suggested that soil health can be restored by using microbes that increase fertility by contributing nutrients to soil, reducing the use of synthetic chemicals, using biofertilizers, implementing agronomic techniques and other soil amendments. These approaches will help in framing strategies to increase food output and quality in order to sustainably fulfil the growing food demands of increasing global population while enhancing soil health and environmental sustainability.

References

Abdel-Raouf N, Al-Homaidan AA, Ibraheem I (2012) Microalgae and wastewater treatment. Saudi J Biol Sci 19(3):257–275

Berazneva J, Conrad JM, Guerena DT, Lehmann J, Woolf D (2019) Agricultural productivity and soil carbon dynamics: a bioeconomic model. Am J Agric Econ 101(4):1021–1046

Berg J, Brandt KK, Al-Soud WA, Holm PE, Hansen LH, Sorensen SJ, Nybroe O (2012) Selection for Cu-tolerant bacterial communities with altered composition, but unaltered richness, via long-term Cu exposure. Appl Environ Microbiol 78(20):7438–7446

Bhatti AA, Haq S, Bhat RA (2017) Actinomycetes benefaction role in soil and plant health. Microb Pathog 111:458–467

Blanco-Moure N, Moret-Fernandez D, Lopez MV (2012) Dynamics of aggregate destabilization by water in soils under long-term conservation tillage in semiarid Spain. Catena 99:34–41

Bong CW, Malfatti F, Azam F, Obayashi Y, Suzuki S (2010) The effect of zinc exposure on the bacteria abundance and proteolytic activity in seawater. In: Hamamura N, Suzuki NS, Mendo S, Barroso CM, Iwata H, Tanabe S (eds) Interdisciplinary studies on environmental chemistry-biological responses to contaminants. Terrapub, Tokyo, pp 57–63

Briones MJI, Schmidt O (2017) Conventional tillage decreases the abundance and biomass of earthworms and alters their community structure in a global meta-analysis. Glob Chang Biol 23(10):4396–4419

Chaudhary P, Singh S, Chaudhary A, Agri U, Bhandari G (2022) Agrochemicals and their effects on soil microbial population. In: Plant protection: from chemicals to biologicals. De Gruyter, pp 45–58

Classen AT, Sundqvist MK, Henning JA, Newman GS, Moore JA, Cregger MA, Moorhead LC, Patterson CM (2015) Direct and indirect effects of climate change on soil microbial and soil microbial-plant interactions: what lies ahead. Ecosphere 6(8):1–21

de Oliveira EP, Rovida AFDS, Martins JG, Pileggi SAV, Schemczssen-Graeff Z, Pileggi M (2021) Tolerance of pseudomonas strain to the 2, 4-D herbicide through a peroxidase system. PLoS One 16(12):e0257263

Dengler J, Janisova M, Torok P, Wellstein C (2014) Biodiversity of Palaearctic grasslands: a synthesis. Agric Ecosyst Environ 182:1–14

Fang H, Han L, Cui Y, Xue Y, Cai L, Yu Y (2016) Changes in soil microbial community structure and function associated with degradation and resistance of carbendazim and chlortetracycline during repeated treatments. Sci Total Environ 572:1203–1212

Fuchs B, Saikkonen K, Damerau A, Yang B, Helander M (2023) Herbicide residues in soil decrease microbe mediated plant protection. Plant Biol 25(4):571–587

Gupta A, Singh UB, Sahu PK, Paul S, Kumar A, Malviya D, Saxena AK (2022) Linking soil microbial diversity to modern agriculture practices: a review. Int J Environ Res Public Health 19(5):3141

Hayat R, Ali S, Amara U, Khalid R, Ahmed I (2010) Soil beneficial bacteria and their role in plant growth promotion: a review. Ann Microbiol 60:579–598

He Z, Piceno Y, Deng Y, Xu M, Lu Z, DeSantis T, Andersen G, Hobbie SE, Reich PB, Zhou J (2012) The phylogenetic composition and structure of soil microbial communities shifts in response to elevated carbon dioxide. ISME J 6(2):259–272

Hussain S, Siddique T, Saleem M, Arshad M, Khalid A (2009) Impact of pesticides on soil microbial diversity, enzymes, and biochemical reactions. Adv Agron 102:159–200

Jain P, Singh D (2014) Analysis the physic-chemical and microbial diversity of different variety of soil collected from Madhya Pradesh, India. Scholarly J Agric Sci 4(2):103–108

Jain SA, Jagtap MS, Patel KP (2014) Physico-chemical characterization of farmland soil used in some villages of Lunawada Taluka, Dist: Mahisagar (Gujarat) India. Int J Sci Res Publi 4(3):1–5

Jia Q, Zhang H, Wang J, Xiao X, Chang S, Zhang C, Liu Y, Hou F (2021) Planting practices and mulching materials improve maize net ecosystem C budget, global warming potential and production in semi-arid regions. Soil Tillage Res 207:104850

Juwarkar AA, Nair A, Dubey KV, Singh SK, Devotta S (2007) Biosurfactant technology for remediation of cadmium and lead contaminated soils. Chemosphere 68:1996–2002

Khan A, Rao TS (2019) Molecular evolution of xenobiotic degrading genes and mobile DNA elements in soil bacteria. In: Microbial diversity in the genomic era. Academic Press, pp 657–678

Kumar A, Kushwaha KK, Singh S, Shivay YS, Meena MC, Nain L (2019) Effect of paddy straw burning on soil microbial dynamics in sandy loam soil of indo-Gangetic plains. Environ Technol Innov 16:100469

Kumar G, Lal S, Maurya SK, Bhattacherjee AK, Chaudhary P, Gangola S, Rajan S (2021) Exploration of Klebsiella pneumoniae M6 for paclobutrazol degradation, plant growth attributes, and biocontrol action under subtropical ecosystem. PLoS One 16(12):e0261338

Li J, Benti G, Wang D, Yang Z, Xiao R (2022) Effect of alteration in precipitation amount on soil microbial community in a semi-arid grassland. Front Microbiol 13:842446

Liu D, Huang Y, An S, Sun H, Bhople P, Chen Z (2018) Soil physicochemical and microbial characteristics of contrasting land-use types along soil depth gradients. Catena 162:345–353

Lu C, Kotze DJ, Setala HM (2020) Soil sealing causes substantial losses in C and N storage in urban soils under cool climate. Sci Total Environ 725:138369

Mbuthia LW, Acosta-Martínez V, DeBruyn J, Schaeffer S, Tyler D, Odoi E, Mpheshea M, Walker F, Eash N (2015) Long term tillage, cover crop, and fertilization effects on microbial community structure, activity: implications for soil quality. Soil Biol Biochem 89:24–34

McGrath SP, Zhao FJ, Lombi E (2001) Plant and rhizosphere process involved in phytoremediation of metal-contaminated soils. Plant Soil 232:207–214

Mehjin AMA, Hmoshi RM, Kannan IA, Thanoon AA (2019) Effect of pesticides on soil microorganisms. J Phys Conf Ser 1294:072007

Milenkovski S, Baath E, Lindgren PE, Berglund O (2010) Toxicity of fungicides to natural bacterial communities in wetland water and sediment measured using leucine incorporation and potential denitrification. Ecotoxicology 19:285–294

Moretto JAS, Altarugio LM, Andrade PA, Fachin AL, Andreote FD, Stehling EG (2017) Changes in bacterial community after application of three different herbicides. FEMS Microbiol Lett 364(13). https://doi.org/10.1093/femsle/fnx113

Mulder C, Cohen JE, Setaa H, Bloem J, Breure AM (2005) Bacterial traits, organism mass, and numerical abundance in the detrital soil food web of Dutch agricultural grasslands. Ecol Lett 8:80–90

Neina D (2019) The role of soil pH in plant nutrition and soil remediation. Appl Environ Soil Sci 2019:1–9

Oertel C, Matschullat J, Zurba K, Zimmermann F, Erasmi S (2016) Greenhouse gas emissions from soil. Geochemistry 76(3):327–352

Onwona-Kwakye M, Plants-Paris K, Keita K, Lee J, Brink PJVD, Hogarh JN, Darkoh C (2020) Pesticides decrease bacterial diversity and abundance of irrigated rice fields. Microorganisms 8(3):318

Pasaribu A, Mohamad RB, Hashim A, Rahman ZA, Omar D, Morshed MM, Selangor DE (2013) Effect of herbicide on sporulation and infectivity of vesicular arbuscular mycorrhizal (*Glomus mosseae*) symbiosis with peanut plant. J Anim Plant Sci 23(3):1671–1678

Patnaik L, Raut D, Behera L, Nayak A, Misha S, Swain S (2013) Physico-chemical and heavy metal characterization of soil from industrial belt of Cuttack, Orissa. Asian J Exp Biol Sci 4(2):219–225

Pereira MC, O'Riordan R, Stevens C (2021) Urban soil microbial community and microbial-related carbon storage are severely limited by sealing. J Soils Sediments 21:1455–1465

Rajapaksha RMCP, Tabor-Kapłon MA, Baath E (2004) Metal toxicity affects fungal and bacterial activities in soil differently. Appl Environ Microbiol 70:2966–2973

Rasool N, Reshi ZA, Shah MA (2014) Effect of butachlor (G) on soil enzyme activity. Eur J Soil Biol 61:94–100

Rathour RK, Devi M, Dahiya P, Sharma N, Kaushik N, Kumari D, Kumar P, Baadhe RR, Walia A, Bhatt AK, Bhatia RK (2023) Recent trends, opportunities and challenges in sustainable management of rice straw waste biomass for green biorefinery. Energies 16(3):1429

Reid JP, Adair EC, Hobbie SE, Reich PB (2012) Biodiversity, nitrogen deposition, and CO_2 affect grassland soil carbon cycling but not storage. Ecosystems 15:580–590

Ren N, Wang Y, Ye Y, Zhao Y, Huang Y, Fu W, Chu X (2020) Effects of continuous nitrogen fertilizer application on the diversity and composition of rhizosphere soil bacteria. Front Microbiol 11:1948

Sanu K, Kumari A, Kumar A, Sharma S, Kumar A (2023) Parali burning: a threat to the environment. Vigyan Varta 4(2):1–4

Sebiomo A, Ogundero VW, Bankole SA (2011) Effect of four herbicides on microbial population, soil organic matter and dehydrogenase activity. Afr J Biotechnol 10(5):770–778

Shao H, Zhang Y (2017) Non-target effects on soil microbial parameters of the synthetic pesticide carbendazim with the biopesticides cantharidin and norcantharidin. Sci Rep 7(1):5521

Singh JS, Gupta VK (2018) Soil microbial biomass: a key soil driver in management of ecosystem functioning. Sci Total Environ 634:497–500

Sioud M, Boudabous A, Cekaite L (2009) Transcriptional responses of *Bacillus subtilis* and *thuringiensis* to antibiotics and anti-tumour drugs. Int J Mol Med 23(1):33–39

Streletskii R, Astaykina A, Krasnov G, Gorbatov V (2022) Changes in bacterial and fungal community of soil under treatment of pesticides. Agronomy 12(1):124

Tale KS, Ingole S (2015) A review on role of physico-chemical properties in soil quality. Chem Sci Rev Lett 4(13):57–66

Trivedi P, Anderson IC, Singh BK (2013) Microbial modulators of soil carbon storage: integrating genomic and metabolic knowledge for global prediction. Trends Microbiol 21:641–651
Trivedi P, Delgado-Baquerizo M, Anderson IC, Singh BK (2016) Response of soil properties and microbial communities to agriculture: implications for primary productivity and soil health indicators. Front Plant Sci 7:990
Wagh GS, Chavhan DM, MRG S (2013) Physicochemical analysis of soil from eastern part of Pune city. Univers J Environ Res Technol 3(1):93–99
Walia A, Mehta P, Guleria S, Chauhan A, Shirkot CK (2014) Impact of fungicide mancozeb at different application rates on soil microbial populations, soil biological processes, and enzyme activities in soil. Sci World J 2014:702909
Walia A, Sumal K, Kumari S (2018) Effect of chlorpyrifos and malathion on soil microbial population and enzyme activity. Acta Sci Microbiol 1(4):14–22
Wang W, Wang H, Feng Y, Wang L, Xiao X, Xi Y, Luo X, Sun R, Ye X, Huang Y, Zhang Z (2016) Consistent responses of the microbial community structure to organic farming along the middle and lower reaches of the Yangtze River. Sci Rep 6(1):1–11
Wang H, Cheng M, Dsouza M, Weisenhorn P, Zheng T (2018) Soil bacterial diversity is associated with human population density in urban greenspaces. Environ Sci Technol 52:5115–5124
Wang X, Wang J, Wang Y, Zhu X, Cheng J, Wang W (2022) Changes in microbial diversity, soil function, and plant biomass of cotton rhizosphere soil under the influence of Chlorpyrifos. Curr Microbiol 79(11):323
Xu S, Geng W, Sayer EJ, Zhou G, Zhou P, Liu C (2020) Soil microbial biomass and community responses to experimental precipitation change: a meta-analysis. Soil Ecol Lett 2:93–103
Yan B, Li J, Xiao N, Qi Y, Fu G, Liu G, Qiao M (2016) Urban-development-induced changes in the diversity and composition of the soil bacterial community in Beijing. Sci Rep 6:38811
Yang Y, Roderick ML, Zhang S, McVicar TR, Donohue RJ (2019) Hydrologic implications of vegetation response to elevated CO_2 in climate projections. Nat Clim Chang 9(1):44–48
Yang T, Siddique KH, Liu K (2020) Cropping systems in agriculture and their impact on soil health—a review. Glob Ecol Conserv 23:e01118
Yi X, Ning C, Feng S, Gao H, Zhao J, Liao J, Peng Y, Zhao S, Liu S (2022) Urbanization-induced environmental changes strongly affect wetland soil bacterial community composition and diversity. Environ Res Lett 17:014027
Zaidi A, Khan MS, Ahemad M, Oves M, Wani PA (2009) Recent advances in plant growth promotion by phosphate-solubilizing microbes. In: Khan MS, Zaidi A, Ahemad M, Oves M, Wani PA (eds) Microbial strategies for crop improvement. Springer-Verlag, Berlin, pp 23–50
Zhang B, Bai Z, Hoefel D, Tang L, Wang X, Li B, Li Z, Zhuang G (2009) The impacts of cypermethrin pesticide application on the non-target microbial community of the pepper plant phyllosphere. Sci Total Environ 407(6):1915–1922
Zhang W, Villarini G, Vecchi GA, Smith JA (2018) Urbanization exacerbated the rainfall and flooding caused by hurricane Harvey in Houston. Nature 563(7731):384–388
Zhu Y, Shen H, Akinyemi DS, Zhang P, Feng Y, Zhao M, Kang J, Zhao X, Hu H, Fang J (2022) Increased precipitation attenuates shrub encroachment by facilitating herbaceous growth in a Mongolian grassland. Funct Ecol 36(9):2356–2366
Zuber SM, Villamil MB (2016) Meta-analysis approach to assess effect of tillage on microbial biomass and enzyme activities. Soil Biol Biochem 97:176–187

Chapter 12
Organic and Natural Farming to Boost Soil Immunity

Pooja Kumari, Shweta Sagar, Akriti, Partiksha Rana, Rishika Mahajan, Jyoti Bala, Rakesh Chauhan, Rameshwar Kumar, and Abhishek Walia

12.1 Introduction

Our country's half population totally depends upon agriculture for their subsistence and for that they completely rely upon conventional farming methods. Continuous and overuse of chemical fertilizers and agrochemicals lead to environmental pollution causing global warming, eutrophication of water systems, and soil degradation. It has an impact on soil micro- and macro-fauna, which has a direct impact on the C–N ratio and nitrification activity (Shaikh and Gachande 2015). The soil is a crucial part of our ecosystem because it controls processes including the breakdown of organic matter, the cycling of nutrients, and the adsorption and decomposition of xenobiotics, all of which are important for plant growth (Doran et al. 2000). The quality of soil is related to sustainable crop production. Organic and natural farming are two sustainable agricultural models which greatly affects the soil health and fertility and conserve soil resources. Natural farming is a new concept with an aim of sustainable development as it replaces the use of chemical fertilizers and low-cost preparations made from desi cow. Without any investment, it is a natural way of farming. Natural farming is becoming more popular because it preserves soil health

P. Kumari
Department of Soil Science, College of Agriculture, CSK Himachal Pradesh Agricultural University, Palampur, Himachal Pradesh, India

S. Sagar · Akriti · P. Rana · R. Mahajan · A. Walia (✉)
Department of Microbiology, College of Basic Sciences, CSK Himachal Pradesh Agricultural University, Palampur, Himachal Pradesh, India
e-mail: abhishek@hillagric.ac.in

J. Bala · R. Chauhan · R. Kumar
Department of Organic Agriculture and Natural Farming, College of Agriculture, CSK Himachal Pradesh Agricultural University, Palampur, Himachal Pradesh, India

R. K. Bhatia, A. Walia (eds.), *Advancements in Microbial Biotechnology for Soil Health*, Microorganisms for Sustainability 50,
https://doi.org/10.1007/978-981-99-9482-3_12

for sustained crop production through microbial growth, diversification, nutrient recycling, and interaction of beneficial microbes (Khadse and Rosset 2019), while organic farming is an agriculture production system which rely upon farm agronomic, mechanical and biological methods such as green manure, crop rotation, compost, crop residues, and nutrient mobilization which enhances agroecosystem health, soil biological activity, and biodiversity. Research studies showed that organic manures, organic nutrients, and microbial biofertilizers and biopesticides are used in organic agriculture (Prasad 2015; Prasad et al. 2019). Organic farming strategies not only ensure food safety but they also improve biodiversity of soil (Megali et al. 2013). In organic agriculture, soil fertilization focuses upon organic materials in order to improve the supply of nutrients and conserve the handling of fields (Araujo et al. 2008). The bioinoculants cause no adverse effect to the ecosystem, increase crop productivity, and improve soil health (Sahoo et al. 2014). The presence of various kinds of micro- and macronutrients in organic fertilizers such as vermicompost keeps the soil environment rich via biodegradation of organic matter, release of plant growth-promoting hormones, nitrogen fixation, and phosphate solubilization or mineralization by beneficial microbes (Sinha et al. 2010). Biofertilizers multiply and take part in nutrient cycling and aid in the productivity of crop when utilized as soil inoculants (Singh et al. 2011).

12.2 Principles, Components, Objectives, and Adoption of Organic and Natural Farming

12.2.1 Historical Background

Natural farming was introduced by Masanobu Fukuoka (1913–2008) who is a Japanese farmer and philosopher. He described it in his famous book *The One-Straw Revolution*. Natural farming is also known as "the Fukuoka method," "the natural way of farming," or "do-nothing farming." Shri Subhash Palekar introduced the idea of natural farming in India, for which he was conferred the Padma Shri award in 2016. According to Subhash Palekar, the components of natural farming have more microbial load which increases the soil microflora that mineralize the soil micro- and macronutrients and make them available for growth of plant. Early farmers engaged in organic farming thousands of years ago. Rudolf Steiner started organic revolution in the mid-1920s in Central Europe. His biodynamic agriculture system is an old version of organic agriculture system. Sir Albert Howard in 1940s developed organic agriculture in England. He is known as "father of organic farming" for his post-Industrial Revolution organic movement. The production of organic products since then has shifted from small to huge farms, and products have been put up for sale under a unique organic label.

12.2.2 *Principles of Organic and Natural Farming*

12.2.2.1 Principles of Organic Farming

The principles of organic farming increase the biological cycles in the agriculture system which in turn enhance the soil fertility, reduce pollution, eradicate the use of synthetic fertilizers, and produce sufficient quantity of high-quality food.

1. **The principle of health**: Every community's and person's health depends on a healthy ecosystem and soil that yields healthy crops. Organic agriculture improves the soil health and ecosystem as it abstains from the use of synthetic fertilizers that have adverse health effects.
2. **The principle of ecology**: Organic agriculture is based on ecological cycles and systems. It maintains the ecological balance by designing farming system and genetic diversity and establishing agriculture.
3. **The principle of care**: Through organic agriculture, the health of the current and future generations and the environment is safeguarded by dealing with in a precautionary and responsible manner.
4. **The principle of fairness**: In case of principle of fairness, organic agriculture is based on alliance that ensures fairness with respect to the opportunities of life and common environment. Resources such as natural or environmental for consumption and production should be utilized in a way that is socially and economically held in trust for upcoming generations.

12.2.2.2 Principles of Natural Farming

(a) **Intercropping**: To supply nitrogen, phosphorus, potassium, and other micronutrients to the main crop, intercrops are grown along with the main crops in natural farming. The main crop is also protected from insect pest attack as intercrops trap them and also root nodules of intercrops help in nitrogen fixation.

(b) **Cow**: Natural farming is completely based on indigenous (*desi*) cow. As compared to foreign breeds of cow, 1 g dung of indigenous cow holds 300–500 crores of beneficial microorganisms. The soil which is treated with urine and dung of indigenous cow has more activity of earthworms which in turn increases the soil fertility. Sixteen major nutrients are present in the dung of *desi* cow which are needed for proper growth and development of the plants.

(c) **Contours and bunds**: Contours and bunds are made in order to collect rainwater. These keep the field moist and make water available to plants for longer period of time.

(d) **Ploughing**: Deep ploughing causes the increase in the temperature of soil. Due to this, carbon from the soil starts vaporizing decreasing the fertility level of the soil.

(e) **Water management**: In natural farming, water management system increases the crop production. Irrigation is done in such a way that only 10% water is used and remaining 90% water is saved.
(f) **Plant direction**: To entrap sunlight for longer duration, the plants are grown in north-south direction. Greater plant to plant spacing is done which triggers better growth by enhanced availability of sunlight.
(g) **Capillary action**: Through capillary action, the plants tap nutrients from deeper layers of soil which also increases the activities of microorganisms. The plants are thus able to take adequate nutrients from the soil and increase the productivity.
(h) **Microclimate**: Mulching creates an appropriate microclimate by providing 25–32 °C temperature, humus, 65–72% moisture, warmth, and darkness in the soil.

12.2.3 Components of Organic and Natural Farming

12.2.3.1 Components of Organic Farming

Crop Rotation

In crop rotation, different types of crops are grown in the same area in a sequential way. Crop is rotated year after year according to the need of nutrients and water but the land is fixed. It improves the fertility of soil, prevents soil erosion, and increases income.

Composting

It is an organic matter which is rich in nutrients used for manuring crops. Compost is made by decomposing farm waste, food and vegetable waste, and crop residues to a certain temperature and humidity under controlled conditions for a certain time period.

Green Manure

Green manure is an undecomposed green material which is used as manure. It increases the soil fertility and water-holding capacity and decreases soil erosion. It also reduces weed proliferation and helps in reclamation of alkaline soils.

Mulching

Mulching covers the soil by crop residues or plant material. It provides microclimate for the best growth of plants and acts as organic matter after decomposition.

Biofertilizers

Biofertilizers are living microorganisms which when applied to seed or soil provide nutrients to plants. They increase the availability of nutrients to the plants and make good produce from crop. Examples of green manure are dhaincha, sun hemp, cluster beans, and *Sesbania rostrata*.

Vermicompost

Various species of earthworms especially red earthworms are used in the production of vermicompost. Vermicompost is the compost product using earthworms and decomposed vegetation. Earthworms are known as "farmer's friend." Earthworms improve soil fertility by maintaining soil texture and good aeration.

Weed Management

Unwanted and undesirable plants that grow and compete for nutrients, space, water, and light with crops are known as weeds. Organic farming aims to reduce the population of weeds by the action of living organisms without using chemical herbicides. Schonbeck in 2010 classified physical and mechanical weed control procedures used on organic farms as follows:

- **Tillage**—to integrate the remains of crop, rotate the soil among crops, and eliminate accessible weed development.
- **Cultivation**—after seeding upsetting the soil.
- **Thermal weeding and flame weeding**—to kill weeds, heating should be done.
- **Cutting and mowing**—removing apex enlargement of weeds.

Livestock Production

Livestock plays an important role in organic farming and increases the sustainability of the farm. Organic material like dung which is rich in nutrients and urine which is a rich source of nitrogen are very useful in making compost manure.

Soil Management

The soul of organic farming is soil. It is very important to maintain the fertility of soil by manuring or composting as after taking one crop, the fertility of soil decreases and loses most of its nutrients.

12.2.3.2 Components of Natural Farming

The main pillars or components of natural farming are as follows:

Beejamrit

Beejamrit is an ancient formulation used for the treatment of seed prior to sowing. *Beejamrit* contains many beneficial microorganisms that have antimicrobial properties. Studies revealed that seeds treated with *Beejamrit* produce IAA (indole-3-acetic acid) and GA (gibberellic acid) as well as shielding crops from hazardous soilborne diseases (Table 12.1) (Sreenivasa et al. 2010; Mukherjee et al. 2022; Kumhar et al. 2022).

Jeevamrit

It is a fermented microbial culture that enhances the activity of beneficial soil microorganisms and earthworms in the soil. According to research studies, *Jeevamrit* boosts the microbial activity in soil and improves soil fertility (Joshi 2012; Imran et al. 2022). This formulation is a source of crores of beneficial microorganisms that helps in providing nonavailable vital nutrients present in soil available to plants (Kumari et al. 2022). *Jeevamrit* also increases the organic carbon quantity in the soil, and it also has antifungal and antibacterial properties. According to Palekar, *Jeevamrit* should be prepared from dung, and urine of *desi* cow and urine taken from one cow are sufficient for organic agriculture of 12 ha (Palekar 2009) (Table 12.2). It provides nutrients to crops which enhance their growth and development. In a study by Kaur et al. (2021) on the enhancement of productivity through the doses and application time of *Jeevamrit*, it was found that although 20%

Table 12.1 Materials required for *Beejamrit* preparation

Sr. no.	Material	Quantity
1.	Dung of *desi* cow	5 kg
2.	Urine of *desi* cow	5 L
3.	Burnt lime	50 g
4.	Water	20 L
5.	Fertile soil	100 g

Table 12.2 Materials required for *Jeevamrit* preparation

Sr. no.	Material	Quantity
1.	Dung of *desi* cow	10 kg
2.	Urine of *desi* cow	10 L
3.	Gur	2 kg
4.	Gram flour	2 kg
5.	Water	200 L
6.	Fertile soil	100 g

Jeevamrit after every 2 weeks, 10% at 2–3-week interval and 20% at 3-week interval have resulted in higher yield with respect to wheat but by comparing lower cost of cultivation and gaining higher net return, *Jeevamrit* at 20% after every 3 weeks was noticed to be the most effective and economical.

Mulching (*Acchadana*)

Mulching refers to covering of upper layer of soil to reduce evaporation. It provides micro-climate in soil; increases humus, microbial population, and availability of nutrients; and preserves the soil moisture.

Following are the different types of mulching materials:

1. Straw mulch: Crop residues or any dead plant materials are used as straw mulch. It prevents the emergence of weed plants and improves soil water retention in the field.
2. Soil mulch: The top layer of soil is protected by tilling operations during cultivation. It improves aeration and soil moisture during cultivation.

Whapasa (Moisture)

The condition when both air and water molecules are available in the soil is known as *Whapasa*. It enables better water management in the soil and makes faster humus production. Irrigation should be done in alternate furrows at noon to make water and air molecules remain in soil according to Palekar.

12.2.4 Objectives of Organic and Natural Farming

12.2.4.1 Objectives of Organic Farming

1. To work with natural system in collaboration rather than dominating them
2. To conserve and enhance the long-term soil fertility

3. To accelerate the biological cycles within the farming system which includes soil flora and fauna, microorganisms, plants, and animals
4. To supply nutritious food in sufficient quantities
5. To avoid all sorts of pollution that may result from farming practices
6. To conserve the genetic diversity of the agricultural system and its surroundings
7. To contemplate wider and environmental impact of the farming system (Deshmukh 2010)
8. To assert the ecological balance
9. To make use of renewable resources to a greater extent in an organized agricultural system
10. To work in a closed organized agricultural system with respect to organic matter and nutrient elements

12.2.4.2 Objectives of Natural Farming

1. To improve crop productivity
2. To ensure better health and reduce health risks by eliminating the use of synthetic chemicals
3. To improve soil health and agro-diversity
4. To increase farmer's income by adopting natural farming
5. To reduce water consumption by covering the soil to prevent unnecessary water loss
6. To generate employment for small and marginal farmers
7. To minimize cost of production by encouraging farmers to prepare essential bio-formulations using natural resources
8. To rejuvenates soil health by enhancing activity of microbes and earthworms present in soil
9. To integrate livestock sustainability in the farming system and helps to restore the ecosystem
10. To ensure food security

12.2.5 Adoption of Organic Agriculture

Under organic agriculture, the global area has increased significantly from 15 million ha in 2000 to 51 million ha in 2015. However, only 1% agricultural land is considered under organic agriculture out of total around worldwide in 2015. The largest certified organic area is under Australia and majority of the area is extensively managed grassland. Other such countries are the United States, Spain, and Argentina. The developed countries account for the majority of the global organic area. About 88% of the organic farmers are living in developing countries in Latin America, Africa, and Asia. The countries having the most organic growers include

India, Ethiopia, and Mexico. Developing countries' organic farmers mainly produce traditional crops such as tea and coffee (Willer and Lernoud 2017). Subsidies from government for organic agriculture play an very important role that influence farmers to convert from conventional agriculture to organic agriculture (Brenes-Munoz et al. 2016). Another major factor influencing adoption is high-value certified marketplaces, where buyers are willing to pay a premium for organic products (Lapple 2010; Wollni and Andersson 2014). Organic farming adoption in the small farm sector of developing countries depends on developmental initiatives which provides training and marketing assistance to farmers (Bolwig et al. 2009). With the increase in organic land area, the demand for organic products has also increased. In terms of per-capita expenditure on organic products, Sweden, Denmark, and Switzerland are the leading countries (Willer and Lernoud 2017). Due to significantly higher prices, poor people cannot afford organic products. Organic products have 50% higher costs than conventional products in their processing, production, and distribution.

12.3 What Is the Need to Prefer Organic Farming?

Food is a primary source of human energy. There is a significant demand for consumption of goods due to the growing population. Although a sufficient amount of food is produced using the conventional farming system, heavy use of artificial pesticides and fertilizer results in diseases, eventually, resulting in a shorter average lifespan for people. Therefore, organic food production is necessary. Organic farming is one that maintains the agro-ecosystem, recovers soil fertility, and produces food with more nutrients than food growing by other farming methods. According to estimates, there will be 9 billion people by the year 2050, and the demand for food will rise sharply (Offermann and Nieberg 2000). By using conventional farming methods, we have many varieties and hybrids, but these are not organic. Although conventional farming produces high yield of food, these are harmful to human health due to the use of chemical fertilizers during growth of plants and leads to reduction in human life expectancy (Korav et al. 2020). So it is necessary to breed for farming in organic. Organic farming may be defined as production method that improves well-being of the agro-ecosystem, where artificial products are not used, that is, pesticides and chemical fertilizers, and help to prevent the destruction to natural resources, thus contributing to the restoration of biological diversity. Zero tillage, intercropping, multiple rotation of crop, integrated management of pest, and biological control are involved in organic farming. Organic agriculture helps protect the environment and increases productivity and fertility of soil, maintains the quality of the agro-ecosystem, and produces foods that are nutritious and healthier than foods produced by using organic production methods. India ranks second with 1 ha under organic agriculture followed by China (1.6 million ha). In general, the Asian continent has 9% (2.9 million ha) of land which is under organic farming as compared to the world (Lichtfouse et al. 2009).

- **Efficiency of nutrient uptake**: Organic agriculture increases the microorganism's activity in the soil and also assists plant root growth whereas in conventional agriculture it decreases. Organic farming restricts the accessibility of nutrients such as nitrogen; when manures decompose, nutrients are released slowly. Bacteria such as arbuscular mycorrhiza and plant growth promoter rhizobia help in nutrient uptake and protect the plant root by defending soilborne pathogens. For instance, compared to the modern wheat varieties, mycorrhizal wheat produces greater yield in low phosphoric soil than in phosphorus-rich soil. Apart from these factors, moisture content, intensity of light, and temperature play an indispensable role in efficiency of absorption of nutrient and application (Ramesh et al. 2005).
- **Seed/soilborne disease resistance**: As mentioned above, activity of microorganisms plays a necessary role in the development of root and microbes such as PGPR bacteria and mycorrhizae that help protect against diseases such as *Rhizoctonia* and *F. culmorum*. A recent study shows that soil microbes and wheat genotypes have a significant relationship (Lori et al. 2017).
- **Management of weed**: Increasing competition among weed with crops for uptake of water and nutrients ultimately leads to huge loss in yield. For example, wild oats reduce 17–62% yield of wheat. In organic agriculture, mechanical methods are used to control weeds rather than spraying herbicides and the weed residue served as manure (Reddy 2010). We can also control weeds by summer ploughing, and it can also aid in killing disease causing microorganisms by exposing the microbes to the outside environment. Allelopathy has been used as a suppressant in recent years and can be seen as competitive advantage gained by plants through chemical processes within them. For instance, wheat have allelopathic compound which inhibit the *Bromus japonica* and *Chenopodium album* growth.
- **Pest/insect control**: Insect repellant, mass trapping, beetle tank establishment, and pheromone traps are used to control pest because insecticides are prohibited in organic farming. Pest attack can be avoided by early sowing. For instance, *Brassica* vegetable crops have been sown in early season to avoid attack by cabbage flies. Biocontrol methods such as BT and Spinosad are used for controlling DBM (diamond back moth) pest (Trewavas 2004).
- **High nutrition value**: In comparison to agricultural products from conventional farming, organically produced goods are safer and healthier. Comparing organic and conventional products, Worthington (2021) found that organic products have excessive levels of magnesium, phosphorus, vitamin C, iron, and low level of nitrates (Table 12.3).

Table 12.3 Conventional farming and organic farming product's nutrition comparison

Farming methods	Magnesium (mg)	Iron (mg)	Vitamin C (mg)	Phosphorus (mg)
Conventional	68.7	3.0	67.9	111.9
Organic	80.1	3.6	89.3	124.1

The varieties that are suitable for conventional agriculture may or may not be satisfactory for organic farming. Utility of genetically modified plants or crops is currently prohibited in organic agriculture due to naturalness standards. Organic farming must be practiced in order to supply current demand of healthy food to human consumption, to conserve natural resources, and to preserve the agro-ecosystem. New varieties should be created through selection, etc., to increase yield. With regard to organic farming.

12.3.1 Environmental Benefits of Organic Farming

People in developed countries prefer organic farming because it is an environmentally friendly form of agriculture. Because it does not contain synthetic pesticides, it offers healthier organic food. The structure of soil in organic plantation is much better than others, which results in less pollution by nitrate and much healthier crops. Farming by organic agriculture is also more desirable for environment than other methods and is chemical-free. Organic farming offers consumers healthy, dependable food while taking into account the natural life cycle processes. Customers who purchase organic goods benefit from them in terms of their health as well as the planet's environment. Organic farming preserves biodiversity and decreases environmental pollution, including that of the air, soil, and water. There are many environmental benefits of organic farming, for example, it preserves biodiversity; minimizes air, water, and soil pollution; and resists climate change (Duddigan et al. 2022) (Table 12.4).

Table 12.4 Different organic farming methods and their environmental benefits

Organic farming method	Environmental benefits
Rotation in crop	Improves soil quality; breaks the life cycles of weeds, insects, and diseases; sequesters nitrogen and carbon; and diversifies production (which may have market advantages)
Cover cropping	Improves quality of soil; lowers erosion, sequestering nitrogen and carbon and preventing dust (protecting air quality); increases nutrients in soil; and boosts productivity of crop
Insectaries and planting habitat corridors	Supports management of biological pest, provides habitat for wildlife, and enhances biodiversity
Synthetic pesticides avoidance	Quality of water, pest management, pollinator protection, and decreased costs associated with chemical inputs
Buffer areas	Enhances biodiversity, prevents wind erosion, and improves water quality
Use of green manure, compost	Sequesters carbon and nitrogen, contributes to boost productivity, and also enhances soil quality
Avoidance of synthetic fertilizers	Reduces salinity, improves soil quality, sequesters carbon, and prevents contamination of ground and surface waters

12.3.1.1 Biodiversity

Plants and animals have a place to live thanks to natural ecosystems. The present diversity of species is the result of such systems' million-year evolution. But when there is reduction in the use of natural ecosystems to artificial ones, there seem to be fewer places for unwanted plants and animals to live. Clear cutting is a practice used in modern agriculture to obliterate complex ecosystems like forests and rangelands. Additionally, using chemically based products to get rid from weeds and pests makes it a worse problem. Hedge bottom vegetation is less affected by organic farming, and hedge on farms of organic exhibits a great species diversity than that of which are grown conventionally. There is generally a trend toward greater abundance of earthworm under organic management, according to evidence from studies on comparative under arable regimes. There have been many studies that report the main cause of significantly increased butterfly, spider, and beetle abundance in organic rotations because of the presence of grass-clover leys (Nejadkoorki 2012).

12.3.1.2 Climate Change and Air Pollution

Dealing with close nutrient cycle to reduce losses to air and water reserve is a key component of organic practices. Air pollution is decreased due to both lower carbon footprint and the lack of chemical sprays that enter the atmosphere. In order to direct land to produce products that are desirable, and not weed and pest, a tremendous amount of chemicals have been used. Through the use of planes and material tanks, this can be specifically tracked after the revolution of agriculture. The Intergovernmental Panel on Climate Change (IPCC) estimate that agricultural sector released about six gigatons of CO_2 worth of greenhouse gases annually in 2005. This accounts for roughly 10–12% of all gases in greenhouse. As a result, using organic agriculture is the only way for producing healthier goods without having negative local (air pollution) or global effects (climate change) (Worthington 2021).

12.3.1.3 Reduce Soil and Water Pollution

A significant amount of nutrients and toxins may be left in water bodies by intensive aquaculture. The utilization and disposal of water in both plant and animal farming are largely responsible for water pollution. For example, when water is interchange in fish pond, water which is waste is released into the nearby surface water. The chosen indicators show that the wastewater contains a variety of pollutants. These pollutants eventually result from the fertilizers, chemicals, and feed that are added to ponds. As a result, water pollution is lesser in organic farming because there is significantly less chemical input eutrophication (Conacher and Conacher 1998). The soil in organic farm is better which makes nitrate pollution less and better for the crops that are planted, and the organic matter is better than other types and does not contain chemicals.

12.3.1.4 Benefits on Human Health

Currently there are about 200 different foodborne diseases in the world. Humans continue to suffer from eating disorders throughout their lives. Foodborne diseases can have many causes, including substances which are toxic, microorganisms which are pathogenic and parasites present in food (Haering et al. 2001). These include toxic substances produced in nature and other harmful substances resulting from human activity (such as industrial effluent). An irreplaceable contribution to organic farming for improving the quality of crops involves refraining from the use of pesticides to produce food crop. According to health and human service cancer expert panel's 2010 annual report, chemical used in nonorganic agriculture is associated with cancer in many body parts like the colon, lung, kidney, ovaries, pancreas, stomach, etc. It is suggested by US cancer experts that chemicals and pesticides in food increase the risk of cancer, and they advised consumers to take chemical-free and pesticide-free food. Modern farm workers and their families suffer from the negative effects of synthetic pesticides, which include elevated cancer rates among some farm workers and their blood relations, increase pesticide exposure for children residing in rural areas, and increase leukemia incidences. There will be no any health issues if you use organically grown food in which they do not use harmful synthetic pesticides, fungicides, and herbicides.

12.3.2 Nutritional Benefits and Food Safety

Due to disapproval of conventional fertilizer in production of organic crops, organic farming is preferred because it is typically based on fertilizers that are natural (such as green manure). This results that organic farming produces fewer plant nutrients overall than conventional farming, particularly less nitrogen (Bourn and Prescott 2002). Thus, there are fewer nutrients average to be had for plant growth; however, natural vegetation has a barely better dietary price than traditional ones, specifically on the subject of vitamins, minerals, and phenolic compounds. Organic foods had a significantly higher level of vitamin C as compared to food produced by other conventional methods. While some researchers discovered higher or comparable levels of carotenoids in tomatoes, carrots, yellow plums, and sweet peppers grown organically, others discovered lower or comparable levels in organically grown blanched tomatoes and carrots. Since the quality of amino acids in organic products is higher than that of conventional ones, hence in organic grains, there is the presence of large amount of essential amino acids. As organic fruits have less weight and size but more soluble solids and less acidity, organic cropping represented a stressful situation. Although the vitamin C and antioxidant enzyme activities of organically grown fruits are higher as compared to conventionally grown ones, they also had less total phenolic content.

Numerous studies on organic cropping show that they have less β-carotene, pesticide, residues, and nitrates but have more vitamin C, sugars, dry matter, certain amino acids, and minerals. In terms of antioxidant, phyto-micronutrient contents, nitrate build-up in vegetables, and toxic phytochemical residue levels, organic agriculture has capability to produce high-quality products with some pertinent improvements. The majority of crop compositional factors, including minerals and vitamins, are not impacted by management of agricultural system (Hansen et al. 2002). Phenolic compound concentration is 20% higher in crops which are organic than in traditional crops. Polyphenols, additionally called phenolic compounds, are secondary metabolites which are products of pentose phosphate pathway in plant. These substances, which belong to one of most prevalent phytochemical group, have significant morphological and physiological significance for plants. They also hold great potential for human diet because they have antioxidant properties.

Food quality refers to the suitability of food for its intentional use and is characterized by qualitative and quantitative traits which may vary between market, for example, consumers and geographic areas have an impact on the prices that producers receive and that consumers pay for fresh and processed goods. On other hand, food safety assured that food will not produce any reaction or effect on consumer when consumed in accordance with its intentional use. At level of the agroecosystem, agriculture through organic whether used for distribution, farming, processing, or consumption, it plays a great role in sustaining and advancing process of food safety and health. It is governed by both domestic and foreign laws. Food production practices most likely show significant impact on human health in terms of how they affect food quality. Organic foods are more safe than conventional ones, largely as a result of precautionary principle being applied in development of organic regulation and in evaluation of safety of food. Low nitrogen applications, which lower nitrate concentration, and a ban on pesticides lead to nearly undetectable pesticide residues, both of which promote high standards of product safety in organic foods. By reducing these factors, cancer risk and the spread of genes which are resistant from animal production systems to human pathogens can be reduced (Smith et al. 2015).

The stages of farming system have great impact on the nutrition value and food safety, and mounting evidences that show food via organic farming offer consumers a number of benefits over foods from conventional systems.

12.3.3 Economic Benefits and Profitability

Due to increasing global population, food supply must be increased without harming environment, and by selecting best agriculture method, economic benefit can be enhanced (Nemes 2009). Studies showed that organic farming is much more beneficial and environment friendly than traditional farming and has the capability to address legitimate development by applying technologies, leading to ecological and economical benefits (Cidon et al. 2021). The aim of profitability analysis can serve either to educate farmers about the outcomes of selecting organic management (farm profitability) or to enlighten policymakers about the benefits of investing more in organic research (system profitability). Profitability, which denotes the degree of farm earnings, should be differentiated from organic farm's relative profitability, which denotes the variations in the correlation between organic farm's agricultural income and nonorganic farm's agricultural income. Variations in yields, prices, expenses in production, direct payments, and outputs that are nonagricultural are recognized as primary factor which determine the relative profit of organic farming (Macrae et al. 2007) (Table 12.5).

12.4 Environmental Effects of Organic and Natural Farming

Organic agriculture methods can provide advantages in terms of both agronomy and the environment, achieved by implementing system-wide alterations and strategic farming management techniques. These advantages extend to developed countries, with benefits including preservation of the environment, increased biodiversity, and reduced energy consumption and greenhouse gas emissions. Developing nations, such as India, can also benefit from organic agriculture, including sustainable resource utilization, improved crop yield without excessive dependence on expensive external input, environmental protection, and biodiversity. The application of organic agriculture (OA) has ultimately reached consequences for environment, spanning from the local to global levels (Conacher and Conacher 1998). It is commonly understood that OA is more ecologically sound than traditional farming methods, as it avoids or minimizes the use of insecticides, herbicides, and synthetic fertilizers. Nevertheless, the impact of any agricultural practice on soil, air, water quality, and biodiversity cannot be overlooked. These are the following effects of natural and organic farming:

- **Effect on the soil quality**: Organic agriculture (OA) has demonstrated superior performance in maintaining or enhancing soil quality in terms of both physical (such as nutrient retention) and biological (such as biodiversity) aspects. In the United States, seven long-term organic agriculture comparison sites have reported higher soil quality in organic agriculture system, with increased carbon and nitrogen storage being notable examples. Organic farming results in higher water-holding capacity as compared to conventional method of agriculture. Organic agriculture also helps in reducing soil loss due to soil erosion. Organic agricultural farms show higher biomass of soil microorganisms especially of bacteria and nematodes and also enhances soil ecosystem functioning and nutrient cycling (Lee et al. 2015).
- **Effect on air quality**: Organic farming, like traditional farming, may lead to air pollution by discharging nutrients, heavy metals like copper, harmful microorganisms, dust particles, and harmful gases into the environment. Pollutants like particulate matter, carbon, nitrogen, and sulfur oxides, as well as the organic compounds which are volatile, methane, hydrogen sulfide, and the pathogens, are typically associated with animal manure handling and surface

Table 12.5 Economic key aspects and organic farming benefits

Economic key aspect	Organic farming benefits	References
Efficient use of biomass	Biochar utilization, developing biodigesters generation of bioenergy during anaerobic digestion.	Zhang et al. (2021) and Chen et al. (2011)
Greenhouse gas emissions	Reduced overall energy usage can be achieved via implementation of agroecological practices that utilize no-till roller crimper techniques. The integration of biogas technology within these practices can also aid in the promotion of energy which is renewable and reduces greenhouse gas emission. Conservation tillage methods and the use of biogas-powered machinery can further enhance these efforts.	Maier et al. (2017)
Crop improvement management	Agroecological benefit crops are inclined to improve effect on environment by boosting recycle energy output, reduced tillage based on cover crops, crop rotation diversification through the addition of legumes in rotation of cereal, and organic manure and catch crops based on legumes helped to enhance productivity.	Lori et al. (2017)
Use of water, land, and energy resources	If executed within a properly designed food system that includes decreased animal product consumption and fewer animals, as well as measures to reduce food waste, organic agriculture can supply ample food while also minimizing environmental harm. The use of techniques of the organic soil management also plays a great role in enhancing water and land quality.	Muller et al. (2017)
Assemblage education and training clusters	The widely accepted concept of open agriculture (OA) has been demonstrated to enhance the unity of a network and its gradual organization. This collaboration serves as a fundamental basis for generating novel ideas and implementing local modifications. Involving the integrating participatory, farmers and the modeling techniques can reveal the pros and cons of various options, enabling more informed decision-making.	Favilli et al. (2015)
Life cycle analysis	In organic farming, life cycle analyses are often used. It is used in estimation of efficient adaptation to organic agriculture.	Favilli et al. (2015)
Biotechnologies	Practice of organic farming uses inventive methods to implement conservation agriculture while avoiding the use of herbicides. It incorporates novel techniques such as vermicomposting, control in disease, rewilding, and utilization of the rhizosphere soil microbial technology.	Horrillo et al. (2016) and Andersen et al. (2015)

(continued)

Table 12.5 (continued)

Economic key aspect	Organic farming benefits	References
Well-being	Those who practice organic farming are deeply committed to their approach due to their beliefs and principles. Numerous farmers value the preservation of the environment. Organic agriculture, with importance on local and native knowledge, empowers small-scale farmer to enhance their standard of living. Organic agriculture guarantees the safety of food for the consuming population.	Kamau et al. (2019)

application, as well as emissions from feedlots. Additionally, the quality of air may deteriorate due to organic agriculture because of loss of nitrogen from compost or organic manures, which leads to volatilization of excess nitrogen that fails to meet the crop's requirements.

- **Effect on water quality**: Organic agriculture (OA) has the potential to outperform traditional farming methods in terms of maintaining high standards of the surface and quality of groundwater by eliminating the utilization of chemicals which are dangerous, specifically synthetic or manmade pesticides. In landscapes with subsurface drainage, OA techniques, such as utilizing composted manure of animals and incorporating forage legumes and organic manure in longer rotations of crop, can significantly enhance the quality of water.

12.5 Organic Agriculture and Food Security

In the past decades, the vacuity of epicure food has significantly increased due to population excrescence and intensification in agriculture. Still, this has redounded insignificant negative effects on the environment, including loses in biodiversity, an overabundance of reactive nitrogen, land and water eutrophication, and emigrations of greenhouse gases (GHGs). It is generally accepted that in order to feed the projected blanket population of over 9.7 billion people by 2050, agricultural affair will need to rise by another 50% (Meemken and Qaim 2018), and as the lifestyle changes, the challenge also increases. Thus, it is essential to reduce agriculture's harmful effects on the terrain while securing that the same volume of food can be produced and distributed. The adoption of holistic approaches like agroecology and organic product and a drop in the consumption of beast products and food waste are just a few suggestions for achieving this goal. Other ideologies include further perfecting product efficiency and duly resource use (Muller et al. 2017).

As we know, organic husbandry is constantly allowed of as being more sustainable than conventional farming. The demand for crops grown organically is rising as consumers seek out healthier food that is polluted by fungicides and genetically modified organisms. Also, organic husbandry is getting more extensively honored as a potential important factor to fight against poverty and attaining food and nutrition ammunition. It has been widely noted that organic farming performs well when compared to a variety of environmental pointers. Organic systems require larger land zones to provide the same amount of output as conventional production networks due to their lower yield (Seufert et al. 2012).

In light of this, organic agriculture's ability to sustainably feed the glove has been put to the test. It is typically necessary to take a holistic perspective to food systems rather than only concentrating on the production, yields, and environmental effects of individual commodities (Connor 2013). So we start by looking into the effects of switching to organic farming on a number of environmental and productivity variables. We also add to additional changes to the food system to this fictitious organic conversion scenario, videlicet

(a) Reductions in the stock of beasts and products (and accordingly mortal consumptions), as well as in the use of natural coffers and environmental goods associated with feeding livestock from pastoralist land (i.e., feed that competes with mortal food) (Bajzelj 2014).
(b) Reductions of food destruction, with similarly reduced product situations. The main question we are trying to answer is whether producing the same quality of food with organic farming has greater or lesser impact on the environment and then producing the same amount of food with conventional farming in terms of protein and calories (Reganold and Wachter 2016).

As a consequence, organic farming boosts nutritional diversity and healthy eating habits, prevents unstable diets that may cause malnutrition and degenerative diseases, and promotes customs and culinary traditions. This is because organic farming prioritizes more nutrients into food, especially those rich in micronutrients. Additionally, organic farming is a system of products that is ecologically sound and sustainable and gives developing countries a variety of social, cultural, economic, and esthetic advantages.

12.5.1 Adoption of Organic Agriculture

Through 4 million ha in 2011 to 9 million ha in 2022, certified organic farming has grown significantly over the previous 15 generations (Fig. 12.1). Despite this, hardly 1% of all agricultural land in the world is used for organic farming in 2015. According to a split of crop spices, organic products have a more significant impact for perennial crops like berries, coffee, and olives than their effect on transient crop like cereals and vegetables.

Sales of organic goods have grown together with the growth in the amount of organic land worldwide. In 2022, the organic food market in India had an approximate value of $1278 million.

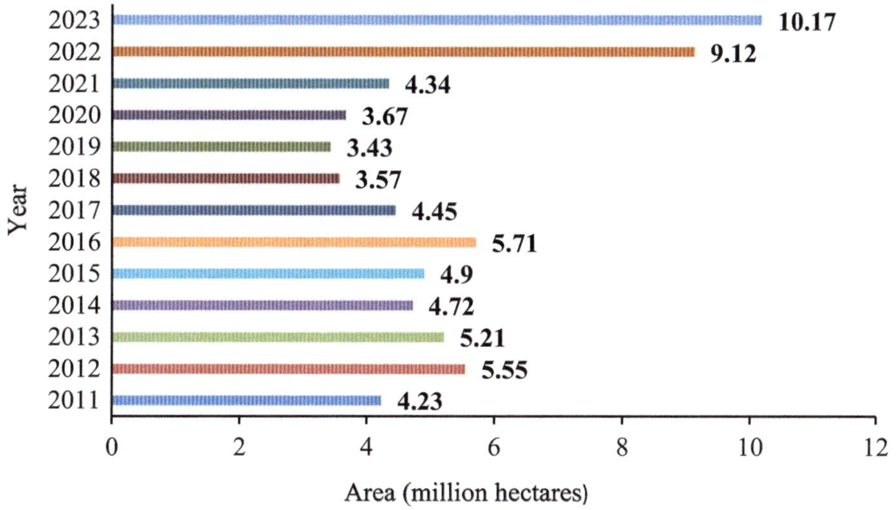

Fig. 12.1 Total organic area (million hectare) in India from years 2011 to 2023

12.5.2 Organic Farming in India: Production, Acceptance, and Economic Development

Global demand for organic food and farming is on the rise. The overall area of farmland used to produce organic goods has continuously expanded since 1985 during the past three decades. By 2017, there were a total of 69.8 million ha of organically modified land that have been extensively registered. This indicates a 20% increase in area compared to the year before 2016 according to Willer and Lernoud (2019).

In 2017, it was also noted that the availability of organic products is rising dramatically on daily bases all around the world. India has highest percentage of organic consumers (835,000), and Asia has the highest probability (40%) of organic products worldwide (Fig. 12.2).

The biggest obstacle to the growth of organic farming in India was the inadequacy of governance procedures for formulating an effective freedom to increase organic farming. Additionally, there were a number of significant hindrances to the development of agriculture in India, including a lack of consciousness, a lack of effective marketing strategies, a lack of biomass, a lack of agriculture organization, high expenses for inputs for farming, improper marketing of organic input, hampered agricultural programs, a lack of funding, an inability to meet acquire demand, and a lack of grade manures. So through a number of programs, the Indian government has expressed interest in encouraging organic farming. The National Programme for Organic Production (NPOP), which establishes standards for organic certification and maintains the quality of certified organic goods on the market, is one such project. Another program that promotes sustainable farming by

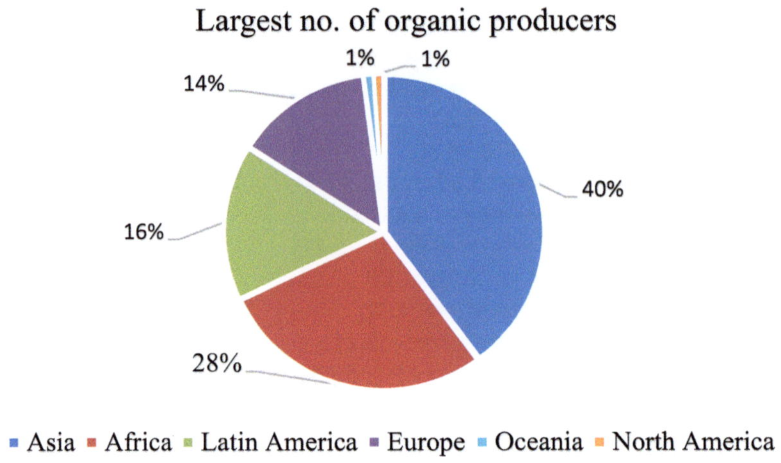

Fig. 12.2 Organic producers by region in 2017 (Willer and Lernoud 2019)

giving farmers monetary support and training includes the Paramparagat Krishi Vikas Yojana (PKVY) (Yadav 2017).

12.5.3 Yield Effects

The extent to which products from organic farming outperform those from conventional agriculture is a key factor in determining how well they can support environmental sustainability. For instance, when organic producers produce lesser yields than conventional farmers, this may be due to the organic practices, but it is also feasible that the growers are less skilled or operate in challenging conditions. In the most severe scenario, organic producers still achieve lower yields even when using the same technology as conventional farming. Hence, the yield difference of organic farming methods would be increased.

Even then, it is also possible that organic producers have more skill levels than conventional rivals, which would result in underestimated production gaps when just comparing observable organic and conventional outputs. Unexpectedly, little research has attempted to account for comparable selection prejudice when evaluating organically agricultural yield products based on experimental data.

12.5.4 Organic Foods' Impacts on Nutrition and Health

According to Seufert et al. (2017), consumers consistently think that organic products are healthy and higher in nutrients than conventional meals. In hypothesis, this may be due to the lower levels of toxic impurities in organic foods or the presence of more vital nutrients. Additionally, there has been a substantial amount of research done to determine whether there are recognized chemical compositional variations between conventional and organic food products. According to multiple systematic reviews, the amount of synthetic pesticide remains is lower in organic food (Huber et al. 2011).

1. Pesticide remains: One of the primary motives for choosing food that is organic is to protect oneself from being exposed to chemical-based pesticides, which are frequently employed in conventional agricultural practices.
2. GMOs (genetically modified organisms) and antibiotics: Both of these contaminants are absent among organic food products.
3. Quality and flavor: Some supporters of organic food claim that it tastes better or has more authentic flavors than those produced conventionally.

However, in organic cultivation, the chemistry of plants is influenced by a variety of factors, including the network of production, soil order, genotype (diversity), the stage of the crop at harvest, and post-harvest conditions specific to different crops (Brandt et al. 2011, Huber et al. 2011).

12.5.5 Organic Agriculture's Impacts on the Environment

According to Foley et al. (2011), agricultural practices have a role in a number of environmental problems, including climate change, biodiversity loss, deterioration of soil, and water contamination. Because it is widely believed that organic farming has less of a negative environmental impact than conventional farming, several governments provide subsidies to the organic sector, which further improves the following entities:

- Biodiversity: Diverse crop rotations and intercropping are frequently used in organic farming, which can boost the amount of biodiversity on farms. A broader variety of species, especially helpful insects and pollinators, can coexist, which helps keep the environment in balance.
- Water quality: Organic farming lowers the possibility of chemical contaminants entering water bodies by eliminating synthetic chemicals. This may result in better water quality, which will benefit aquatic life.
- Carbon sequestration: Several organic farming methodologies, such as growing cover crops and reduced tillage, help the soil store more carbon. By capturing carbon dioxide from the earth's atmosphere and storing it in the soil, this can help to mitigate climate change.

12.5.5.1 Land Use Effectiveness

Uninterrupted land use change, such as deforestation, is linked to a number of environmental issues, particularly the decline in biodiversity and the release of soil carbon into the atmosphere. To achieve a balance between agricultural output and environmental aspirations, it will thus become more important to use land and other natural resources more wisely (Foley et al. 2011).

12.5.5.2 Energy Use and Emigrations of Greenhouse Gases

About 25% of anthropogenic greenhouse gas (GHG) emigrations are caused by agricultural products (Edenhofer et al. 2014). In order to produce mineral fertilizer, for example, fossil fuels must be used. Additionally, soil chemical processes and animal digestion also produce carbon dioxide (CO_2), methane (CH_4), and nitrous oxide (N_2O) emissions (Gomiero et al. 2011). In organic agriculture, life cycle evaluations are often performed to evaluate energy consumption and greenhouse gas emigrations until the moment something goes away the farm (Meier et al. 2015; Smith et al. 2015).

12.5.6 Food Prices and Food Security

In organic agriculture, average yields are often lower which leads to rising expenses per unit of output, as a result increasing in prices. However, if all organic agricultural goods switched to organic farming practices, food costs would increase significantly across all over the world. Currently, only a tiny segment of the certifiable organic market is affected by the increased pricing. In developed nations, the majority of consumers are wealthy enough so that their food security is not jeopardized by high food costs. The situation is different in emerging nations, though. In developing countries, over 50% of the income of several underprivileged extended families is spent on food. According to Ecker and Qaim (2011), given these conditions, increased food costs are associated with an increase in food insecurity and undernourishment, particularly in densely populated regions. For rural residences, increased agricultural prices are welcomed in a similar way, but when higher prices are the consequence of reduced yields, they do not necessarily translate into greater revenues.

As a consequence, by increasing conventional agricultural yields without placing farmers at the whim of synthetic inputs, outputs per unit area of marginal land grow and stabilize as a result of enhanced resilience.

12.6 Organic Farming Methods

12.6.1 Natural Control

Modern agriculture involves the use of agrochemicals, compounds that utilized to control weeds, illnesses, or pests in crops. The use of agrochemicals in agriculture helps to amplify crop yields and fulfill the demands of the expanding human population. However, persistent and non-judicious application of agrochemicals can create chronic health problems in humans and can cause environment degradation and destroy biodiversity (Fianko et al. 2011). Organic farming has been pushed to restore soil health and fertility without producing pollution. Farmers need to return to natural as well as conventional practices employing agricultural residues and animal waste such as manure (Smith et al. 2019; Epule et al. 2015). Some of the natural control sources of organic manure are crop manure, animal manure, and compost, vermicompost, and biochar (Durán-Lara et al. 2020). Plants that are not photosynthetic and are left on farmed soils after crops have been harvested are called crop leftovers. They are regarded as a successful countermeasure to soil erosion due to their ability to enhance soil structure, boost soil organic matter content, lessen evaporation, and fix CO_2 in the soil (Laamrani et al. 2017). They can also add nitrogen to the soil through biological nitrogen fixation, which increases the amount of N available to succeeding crops and minimizes the need for chemical fertilizer treatments. They can also be employed in the creation of biofuel (Raheem et al. 2019). Animal dung is the primary source of nitrogen used to fertilize crops and grasslands, resulting in a decrease in the need of N fertilizer (Hadas and Rosenberg 1992). Several organic manures of animal origin include bird manure, bovine manure, sheep manure, and pig manure (Durán-Lara et al. 2020). Bird manure is main source of nitrogen and also contains components like nitrogen, phosphorous, and potassium (Ron and Rosenberg 2014). Composting is the process which involves the biooxidative decomposition of organic matter. Composting of bovine manure has a range of agricultural benefits such as decreasing the bulk and water content, suppressing infections, killing weed seeds, and producing stable and spreadable organic matter. Vermicomposting is another procedure that involves the use of specific species of earthworms to expedite the process of waste breakdown and create manure. Vermicomposting of biodegrade organic matter helps to restore essential nutrients back to the soil (Ferreyra et al. 2006). However, biochar can retain NH_3, NH_4^+, and NO_3^- found in animal dung and hence enhances nutrition in soil. Plant's essential oils (*Eucalyptus* oil) are utilized as biopesticides, an alternative for synthetic pesticides (Adekiya et al. 2019).

12.6.2 *Weed Management*

Unwanted plants or pests known as weeds can significantly lower crop output and production on any scale. The same resources that crops compete with include nutrients, water, air, light, and carbon dioxide. They also include insects and pathogens that could harm crops and ultimately have a significant impact on agricultural production and yield losses. In India, weeds are thought to reduce crop productivity by about 31.5% (22.7% in the winter and 36.5% in the summer and rainy season) (Gharde et al. 2018). Weed management is an organic, non-chemical approach of weed control that has many benefits for both people and the environment. The main goals of weed control in organic farming are to totally get rid of all weeds and to lessen weed competition by stopping the formation of weed seeds. In order to control weeds without using synthetic herbicides, organic farmers incorporate cultural, biological, mechanical, physical, and chemical strategies (Finney and Creamer 2008).

12.6.2.1 Culturable Methods

Crop Rotation

Since the Middle Ages and far into the twentieth century, crop rotation has been the cornerstone of organic weed management systems because of its proven impacts on weed populations. Crop rotation is the systematic alternation of several crops grown on the same plot of land. To create a wholesome, sustainable organic system and disrupt pest and weed life cycles, diverse crop rotations are crucial (Doucet et al. 1999). A crop rotation's primary goal is to create an unsteady and unbalanced environment that prevents weeds from taking over the area. Cropping and long rotational systems are combined in this type of operation to improve soil fertility and economic variety. Crop rotation generates a changing environment and avoids weed dominance of particular species; hence variation in crop species and planting periods is required (Schreiber 1992). Farmers should plan rotations for each field while creating a crop production plan, keeping weed species and potential issues in mind. Fast-growing plants that can outcompete weeds, such as tomatoes, winter squash, potatoes, and sweet corn, are a wise addition to your rotation (Clark 2008).

Cover Crops

Including cover crops in your crop and weed management has many advantages, including reducing soil erosion, enhancing soil fertility, and suppressing weeds. By planting cover crops such rice beans, sorghum Sudan grass, groundnuts, cereal rye, red clover, oats, and buckwheat, weeds are prevented from growing and from germinating. High-competitive crops can be developed as transient smother crops within the rotation. There are numerous strategies to utilize cover crops to control weeds, including preventing weed seed germination and lowering weed populations in the particular crop (Yeganehpoor et al. 2015). Including cover crops in your crop and weed management has many advantages, including reducing soil erosion, enhancing soil fertility, and suppressing weeds (Teasdale et al. 2007).

Mulch

Mulching is the practice of covering the soil's surface with weed-free straw, thick layers of leaf mulch, well-composted manure, or plastic mulch, all of which can inhibit the germination of weed seeds by obstructing light transmission (Coolong 2012). In addition to reducing light penetration, mulches also change the temperature and moisture cycles in the soil, which reduces weed competitiveness. According to Farooq et al. (2020), allelopathic compounds in the mulch can actually prevent seedling emergence. Mulches come in a variety of shapes and sizes. Here are a few prevalent examples:

(a) **Living mulch**: Clover is an example of a plant species that often grows densely low to the ground and is used as living mulch. Living mulches are used either prior to or following the establishment of a crop (Walters and Young 2008).To prevent the living mulch from competing with the actual crop, it must be eliminated, incorporated, or somehow managed. Before transplanting a broccoli crop, for instance, a living mulch made of *Portulaca oleracea* is utilized because it can control weed growth without reducing crop production. *Trifolium repens* (Latin clover) and *Festuca rubra* (red fescue) are two more types of live mulch that can prevent weeds from growing in a crop. Living mulch is frequently used to alleviate pest problems, increase soil fertility, improve soil structure and aeration, and suppress weed development (Fracchiolla et al. 2020).

(b) **Organic mulch**: The most popular type of mulch is organic mulch, which is made from living things like grasses, leaves, straws, pine needles, shredded bark, composite materials, and crop wastes (Monteiro and Santos 2022). Organic mulch can be utilized in a variety of ways to control weed growth. By obstructing light and lowering soil temperature, they prevent seed germination. Additionally, organic mulch can actually prevent seedling emergence. Utilizing organic matter (mulch) can improve soil quality, encourage soil biological activity, improve soil structure and fertility, increase water infiltration, and even encourage earthworms to aerate soil organically (Bàrberi 2002).

(c) **Inorganic mulch**: Inorganic mulch consists of materials such as plastics, landscape fabrics, and black polyethylene and are used for weed control in a range of crops in organic production systems. These materials are usually used as they do not decompose or break down after extended period of time. They are photodegradable and break down after 30–60 days of exposure to sunlight or are biodegradable and broken down by soil microorganisms (Sharma and Sharma 2019).

12.6.2.2 Mechanical Methods

Mechanical weed control is one of the most labor- and time-intensive procedures, but it is also one of the most successful. The following approaches for mechanical and physical weed control are commonly employed in organic farms:

Tillage

This refers to turning the soil to incorporate soil amendments and agricultural wastes between crops. This turning can get rid of weed growth that is already there and get a seedbed ready for planting. Cropping row crops is included in tilling the ground after seeding to eradicate weeds.

Mowing and Cutting

For many annual weeds, mowing is the most efficient technique of control. It also involves getting rid of weeds' top growth. As it lessens the number of flower stalks and stops the distribution of more seeds, it is successful when done before the weeds have a chance to set seed.

Flame Weeding and Thermal Weeding

Flame weeding and thermal weeding is the use of intense heat to kill weeds. It involves the use of flaming equipment to create direct contact between the flame and the plant. This technique works by rupturing plant cells when the sap rapidly expands in the cells. It is done prior to seeding the crop and can be used effectively on most organically produced crops. It is more effective on a smooth soil surface than a rough surface. However, flaming after crop emergence may damage the crop. Although the initial equipment cost may be high, flaming for weed control may prove cheaper than hand weeding (Cloutier et al. 2007; Wicks et al. 2017).

12.6.2.3 Biological Methods

Allelopathy

Allelopathy is an alternative and natural method of weed management. This involves using substances that a plant excretes to harm weeds either directly or indirectly by impairing their germination, growth, or development. One of nature's most powerful strategies for establishing plant dominance is this. Barley, rye, annual ryegrass, buckwheat, oats, sorghum, sudan-sorghum hybrids, alfalfa, wheat, red clover, and sunflower are some examples of allelopathic crops. In exceptionally weedy fields, choosing allelopathic crops can be helpful in reducing overall weed pressure. Allelopathic substances that reach the rhizosphere from the crop plant's roots or aerial portions might harm nearby weeds. Through such allelopathic interactions, crop residues from cover crops, such as fall rye, or other organic mulches can also be utilized to control weeds (Charudattan 2001; Sodaeizadeh and Hosseini 2012).

Mycoherbicides

In contrast to herbicides that need to be administered annually, mycoherbicide is a herbicide based on fungal diseases that may kill certain weeds and remain active on the weeds year after year. By creating phytotoxins, phytopathogens start infections in particular weeds and eventually destroy them within 3–5 weeks (Boyette and Abbas 1995). They are host-specific, environmentally benign alternatives to chemical herbicides (Templeton et al. 1979) (Table 12.6).

12.6.3 Soil Management

The use of procedures, methods, and treatments to preserve and improve the functionality and health of the soil is known as soil management. The biological activity of the soil is important in an organic system and is the primary contributor to the beneficial soil physical structure and soil fertility. Healthy and high-yielding organic crops can only be produced by a dynamic and diversified community of soil microorganisms. Organic soil material is a fantastic source of plant nutrients and has a high water-holding capacity. It is a usual practice to use methods like green manure and composting to replenish nutrients removed from the soil by previous harvests (Watson et al. 2002). Earthworms and soil microbes like mycorrhiza control these biological processes. As the growing season progresses, they produce nutrients that

Table 12.6 Some of mycoherbicides used to kill particular weeds

Sr. no.	Name	Weed
1.	*Colletotrichum gloeosporioides*	Aeschynomene
2.	*Colletotrichum furarioides*	*Asclepias sericea*
3.	*Cercospora rodmanii*	*Eichornea crassipes*
4.	*Puccinia chandrillus*	*Chondrilla juncea*
5.	*Puccinia abrupta*	*Parthenium hysterophorus*
6.	*Phytophthora palmivora*	*Morrenia odorata*
7.	*Alternaria cassia*	*Cassia angustifolia*
8.	*Alternaria crassa*	*Datura stramonium*
9.	*Alternaria helianthi*	*Xanthium strumarium*
10.	*Phomopsis convolvulus*	*Convolvulus arvensis*
11.	*Bipolaris halepense*	*Sorghum halepense*

are available to plants. Crop rotation, cover crops, decreased tillage, and compost application are just a few of the techniques farmers utilize to increase soil fertility. Reduced tillage enhances soil structure by minimizing soil erosion and lowers carbon emissions. Legumes, which can fix nitrogen from the atmosphere through symbiosis with rhizobial bacteria, can fix nitrogen from the atmosphere thanks to crop rotation and green manure. In addition to increasing soil nutrients, intercropping is utilized to manage insects and diseases. Crop leftovers can be plowed back into the soil, and the nitrogen levels in the various plant parts help in synchronization. In addition to using animal dung, organic farmers use certain industrial fertilizers like seed meal and a variety of mineral powders like rock phosphate and green sand, which is a naturally occurring form of potassium that adds minerals and nutrients to the soil, and also use animal manure. For enzymes and bacteria to work properly, the pH of the soil may occasionally need to be maintained. Lime and sulfur are two naturally occurring pH supplements, although in the United States, some substances including iron sulfate, aluminum sulfate, magnesium sulfate, and soluble boron products are permitted in organic farming. Mixed farms can be run as ley farms because they have both livestock and crops. This allows the land to meet its fertility needs by growing nitrogen-fixing fodder grasses like white clover or alfalfa, and then after fertility is established, it can grow cash crops like cereals. In order to maintain soil fertility, farms without livestock (referred to as "stockless") may need to rely more on external inputs, such as imported manure, grain legumes, and green manures, albeit grain legumes may only be able to fix a small amount of nitrogen because they are harvested. In protected environments, horticultural farms frequently depend even more on outside inputs (Shepherd et al. 2002). Manure is incredibly heavy and is frequently inefficient to transport more than a few miles from the source. In the event that a significant number of farms switch to organic management, manure for organic farms may become scarce.

12.6.4 Controlling Other Pathogenic Organisms

Nematodes, arthropods (insects, mites), fungus, and bacteria are other organisms that trouble farms and lower crop productivity. Beneficial insects that are predatory, such as ladybugs, big-eyed bugs, and little pirate bugs, consume a variety of pests. Both parasitoid wasps and predatory mites are adept in controlling their respective populations of prey. On organic farms, the use of naturally derived pesticides is permitted. Examples include *Bacillus thuringiensis* (a bacterial toxin), pyrethrum (a chrysanthemum extract), spinosad (a bacterial metabolite), neem (a tree extract), and rotenone (a legume root extract). These pesticides may cause harm; thus they are not always harmless or more environmentally friendly than synthetic pesticides. Although many organic pesticides come from natural sources, some naturally occurring compounds have generated debate. Copper, nicotine sulfate, pyrethrums, and rotenone are some of the contentious natural pesticides. The main reason rotenone and pyrethrum are notorious is that they affect the neurological system.

Rotenone can cause symptoms that are comparable to Parkinson's disease in mammals and is extremely harmful to fish (Crowder et al. 2010). The bacteria *Bacillus subtilis* and *Bacillus pumilus*, as well as the fungus *Trichoderma harzianum*, are naturally produced fungicides that can be used in organic farming (Ons et al. 2020). They work well against ailments that harm plant roots. Compost tea contains a variety of helpful bacteria that may fight off specific plant infections, but compost tea formulas and preparation methods may encourage the growth of hazardous microbes. On organic farms, it is forbidden to employ a number of naturally derived pesticides, including nicotine sulfate, arsenic, and strychnine. On organic farms, it is permitted to use synthetic pesticides such as insecticidal soaps, horticultural oils, Bordeaux mixture, copper hydroxide, and sodium bicarbonate to control fungus. Copper sulfate and Bordeaux mixture (copper sulfate with lime) are prohibited in organic farming because they can be more environmentally harmful than some synthetic fungicides. Similar to copper sulfate, copper hydroxide may eventually accumulate to dangerous amounts in soil if it is applied repeatedly as a fungicide. When such compounds are used for some crops at typical rates, environmental risks for many types of biotas occur (Meerburg et al. 2007).

12.7 Production of Inputs for Organic and Natural Farming

In organic and natural farming, biodegradable waste or the waste generated which can be further degraded to release utmost needed minerals are incorporated. Furthermore, degradation of organic matter by microbes resulting in the formation of a byproduct (such as compost) can be incorporated into the plant life to speed up its growth and overall development. The organic matter or the biodegradable waste when acted upon by microbes is a consortium of various PGPB (plant growth-promoting bacteria) which results in increasing the bioavailability of nutrients, uptaking of minerals with the help of various bacteria, symbionts of microbes helping in uplifting several essential micronutrients (such as Fe and Al).

12.7.1 Composting

With the need of higher crop production and population explosion, organic matter as a waste and by-product started increasing in size. In contemplation of decreasing organic matter load from environment and utilizing the end product for the greater good composting came up as a solution of solid waste management. Composting comprises of degradation of organic matter which can be agricultural waste, domestic waste, biodegradable waste, and waste from animal feces by incorporating microbial community to produce humus like substances which are active source of free ions and minerals. These ions and minerals are then utilized by plants and microbial community for their growth which affect directly and indirectly to food chain. During composting, several factors play an important role in degradation of this matter which can be biotic and abiotic.

12.7.1.1 Aerobic Composting

Aerobic composting is a process of degradation of organic matter by microorganisms in the presence of oxygen. Heat and solid substate is generated as a result of degradation of organic matter by consuming less carbon and nitrogen followed by humification of some of the organic matter resulting in the formation of compost (Roman et al. 2015; Azim et al. 2018). The organic matter consists of kitchen waste, leaves, tree bark and wood, and other biodegradable waste. Plant cell wall is made of cellulose and hemicellulose which is broken down by microbes using cellulase enzyme and using the simplified substrate for their energy requirements. During composting, cellulose-degrading bacteria plays an important role in the degradation of organic matter. In view of this, a study was conducted where effect of actinobacteria in cellulose degradation was noted at different inoculation time. The inoculants at different stages of composting helped the indigenous actinobacteria to degrade organic matter (Zhao et al. 2016).

Cow manure is an excellent source of phosphate, for example, phytate and phospholipids as it contains 70–80% of the originally consumed phosphate by the animal. Therefore, increasing P production for use in agriculture and ensuring a long-term sustainable P supply may be accomplished through the aerobic composting of manure. The release of phosphate from its bound state during composting can help in improving the quality of compost along with storage of N-based nutrients and reducing the toxicity of heavy metals. Thus, studying the changes in P segments during aerobic composting becomes a major concern (Zhang et al. 2021).

12.7.1.2 Anaerobic Composting

Composting process in the presence of negligible to no oxygen is termed as anaerobic composting. In other words, organic matter to be organically converted into other energy-rich complexes as end products, for example CO_2, NH_4^+, and stabilized organic matter (Su et al. 2016), in the absence/negligible amount of oxygen is said to be anaerobic composting. In places where manure is piled up with little aeration until the substrate can be used as a fertilizer or fully fledged compost, anaerobic composting is widely used as it takes littlest effort, cost, and maximum gain.

Although this composting process requires little oxygen and negligible number of efforts, the process of formation of compost through anaerobic respiration of microbes takes much longer time, that is, 40–60 days as compared to that of aerobic compost also, the compost has low stability (Yang et al. 2019). Table 12.7 depicts the difference in aerobic and anaerobic composting (Meena et al. 2021).

Table 12.7 Difference between aerobic and anaerobic composting

Characteristics	Aerobic composting	Anaerobic composting
Oxygen requirement	Necessary	Not necessary
Heat production	High	Low
Stages of decomposition	Three, namely, psychrophilic, mesophilic, and thermophilic phase of decomposition	Four, namely, hydrolysis, acidogenesis, acetogenesis, and methanogenesis
Temperature range	50–70 °C	>65 °C
Decomposition time	Ranges from 30 to 120 days	Ranges from 40 days up to 6–12 months
Pathogen suppression	High temperature facilitates killing of harmful pathogen and weed seeds	Low temperature makes it unable to kill pathogen and weed seeds
Thermophilic microbes	Present and important	Because of temperature requirement microbes are unable to grow
Nitrogen loss	High nitrogen loss	Low nitrogen loss
Decomposition of organic nitrogen and carbon compounds	$N_2 \rightarrow$ ammonia but in low concentration as it volatizes due to high temperature and $C \rightarrow$ carbon dioxide	Organic nitrogen converted to ammonia and organic acids whereas organic carbon changed to methane which is released into the air

12.7.1.3 Vermicomposting

Vermicompost is a compost which is obtained through the action of worms, that is, organic matter is degraded by worms to produce solid substrate which has high water-holding capacity, microbial load, and rich nutrient quantity. Vermicompost is a part of organic farming and contains various micro- and macronutrients which are absorbed in the crop roots easily. Vermicompost is a source of nitrogen-fixing and phosphate-solubilizing bacteria along with different plant growth-promoting substances such as gibberellins, auxins, cytokinins, vitamins, humic acid, and defensive enzymes (Yatoo et al. 2021). Joshi et al. (2020) did a comparative study of composting and vermicomposting with respect to bioconversion of poultry litters into degraded solid substrate and found that in comparison to compost, vermicompost had rich macronutrient concentration and reduced C:N ratio; thus vermicompost is said to be superior to compost.

The process of vermicompost formation includes ingestion and egestion of organic matter by earthworms. In the foregut of earthworm, calciferous glands increase the pH through which the population of fungi, bacteria, and trophozoite bacteria explodes leading to the high absorption of organic matter. C and N are assimilated in the mid gut which are then released as CO_2 and N_2O through the hind gut (Singh et al. 2020). Vermicompost results in the increase in dehydrogenase and phosphatase activity and microbial load and activity, has a positive effect on biological properties, stimulates the nitrogenase enzyme, and also suppresses plant diseases along with the increase in glucosidase and acid phosphatases (Singh et al. 2020). Vermicompost effect on different plant families, that is, Lamiaceae, Solanaceae, and Poaceae, for the commercial yield. Significant increase in the shoot and root biomass by 78% and 57% noted, respectively, is noted which added up to the total biomass increase of 13% along with the yield increase of 26%.

Vermiwash and vermicompost tea are a derivative of vermicompost. In vermiwash, vermicompost bed containing soil, organic matter, and worms are washed through water, and the end product is collected for the use of irrigation spray, whereas vermicompost elutriate or vermicompost tea is obtained by taking measured quantity of vermicompost in water and shaken for a definite period of time under aerobic and anaerobic conditions (Zarei et al. 2018).

12.7.2 Green Manure Production

Green manure is produced when undecomposed green plant tissues such as cover crops are ploughed in the soil to increase the soil's physical condition and fertility. Generally, nonessential crop or plant is weed out from the field and is ploughed or turned over in between crops for gaining maximum benefit. Velvet beans (*Mucuna pruriens*), lablab (*Lablab purpureus*), sunn hemp (*Crotalaria juncea*), *Crotalaria grahamiana*, and *Tephrosia vogelii* are some of the commonly used plants for green manuring in South Africa. Leguminous or nonleguminous crops are used for green manuring. Leguminous crops are often sown in between crops where nitrogen requirements are high (Thomas et al. 2019) and soil is either N-poor or is present in non-assimilatory form; thus N is provided for an extended period of time as the nitrogen fixers in the nodules of leguminous plants mineralize and fix nitrogen as long as the plant thrives.

With the gradual decrease in practicing green manure, there has been the increased use of chemical fertilizers which resulted in decreased crop productivity and soil toxicity and infertility over a period of time. Green manure production has recently been popularized as the understanding of the respective manure. In a study in green manure and cattle manure, combined treatment resulted in ten times increase in plant dry weight than that of unfertilized treatment, and the increase in antioxidant activity of thyme was noted. It has been concluded by Ma et al. (2021) that green manure can enhance soil biological activity, reduce soil bulk density, and increase the N, P, and K in soil. In their study, they also concluded that incorporation of leguminous crop as green manure increases wheat yield whereas both leguminous and nonleguminous green manure can be used for the higher yield of maize and the use of only nonleguminous green manure can increase the yield of potato in correspondence to Northern Chinese region.

12.7.3 Crop Rotation

Different sets of crops to be used to grow in different seasons of the year are referred to as crop rotation. Crop rotation is a simple change in the order of sowing crops without the use of any artificial product for the greater yield. It is vastly used to control weed infestations, manage insect, increase plant diversity (Zhao et al. 2020), and maintain the nutrition level in soil. Though crops are interchanged in different seasons, the yield of intercropped plant is highly dependent on the variability of previously introduced crop. On legume-based rotations, the yield can increase up to 14%. Zhao et al. (2020) studied the role of intercropping in the yield and found that it increased by 20.1 ± 3.5% as compared to monoculture practices. In a meta-analysis experiment by Weisberger et al. (2019), it was found that diversification of crop rotations as compared to that of simple crop rotation widely reduced the weed density.

12.7.4 Cow Dung-Based Bioformulation

12.7.4.1 *Ghanjeevamrit*

It is a thick paste prepared using the same components as to that of *Jeevamrit*, but the quantity of cow dung remains high as to that of cow urine which need to be low. *Ghanjeevamrit* is added into the soil before the sowing of respective crop to increase the nutrient supply, microbial community which has various plant growth-promoting traits. *Ghanjeevamrit* is prepared by mixing cow dung, urine, jaggery, gram flour, and fertile soil such that a paste is formed; this mixture has to be kept in shady conditions and incubated for 2–3 days with periodic sprinkling of water so as to not let the paste dry. After the respective incubation time, balls of *Ghanjeevamrit* are prepared which remains effective for 3–6 months. In a study by Jaya (2022) on the growth parameters of China aster to be affected by *Ghanjeevamrit*, it was found that *Ghanjeevamrit* along with chemical fertilizer and biofertilizer enhanced the growth of respective plant. A similar experiment was performed on *Oryza sativa* L. to check the effect of organic, inorganic, and integrated nutrient amendments on growth parameters of the respective crop, and it was concluded that *Ghanjeevamrit* + chemical fertilizer mixed treatments resulted in maximum growth of basmati rice crop in the span of 2 years (Rautela et al. 2022).

12.7.4.2 *Panchagavya*

Panchagavya is a mixture of five prime products of cow in which cow urine, dung, and milk are the direct products whereas curd and milk are derived products from the milk. Additionally, jaggery, tender coconut, and ripened banana are also mixed.

It is used as a fertilizer and pesticide in organic agriculture. It can either be used as a foliar spray or drenched in soil or as a seed treatment (Kumar et al. 2020a). This particular bioformulation is prepared in container made of plastic, mud, or concrete but not any kind of metal. The next step is to mix dung and ghee in the container which is to be stirred twice a day for 4–5 days. On the fifth day, add other three ingredients, left to be ferment for 7–8 days and stirred for 20–30 min twice daily. Now jaggery, tender coconut, and banana are added and mixed thoroughly. After 10 days of incubation, *Panchagavya* should be ready to use. On a field experiment on the growth attributes of cowpea using different concentrations of *Panchagavya* and *Jeevamrutha*, it was found that *Panchagavya* at 7.5% concentration improved the overall growth and yield of the plant. *Panchagavya* is also used in seed treatment, and such an experiment was designed to find the effect of *Panchagavya* on germination and length of plant, and it was noted that *Panchagavya* at 2% concentration for 24 h resulted in 86% growth percentage and root-shoot length explosion of 14 and 4 cm, respectively (Kumar et al. 2020b).

12.7.4.3 *Whapasa*

Whapasa is a part of zero-budget natural farming and focuses on the conservation of water by maintaining moisture in the soil with periodic irrigation of soil (Ranjan and Sow 2021). *Whapasa* maintains a microclimate in the soil with the mixture of water molecules and air in between the cavities/pores of soil (Mandla and Sharma 2022). *Whapasa* is achieved through decreasing the water quantity followed by irrigation in near places to where the crop is sown (Ranjan and Sow 2021), so the soil is well endowed with air and water molecules which can reduce the water requirement up to 90% (Khan et al. 2022). The soil microbes and plant root system are provided with enough water and air supply to maintain the soil structure and help in plant growth directly.

12.8 Implications of Organic and Natural Farming

12.8.1 Soil Quality

Sustainable crop production is the major goal of organic farming in order to preserve soil fertility over the long term. The control of soil health should be the top priority for all parties involved in agriculture growth. The continuous ability of soil to perform as an essential living system is known as soil health. The constant use of excessive chemical fertilizers is one of the main causes of degrading soil quality. Organic farming has brought a positive outcome in the soil quality as compared to conventional farming.

12.8.1.1 Impact on Physical Properties of Soil

Color, texture, structure, porosity, density, consistence, aggregate stability, and temperature are just a few physical characteristics of soil (Hillel 1982). Organic soils have greater biological activity and improved hydraulic processes because of greater pore roughness in comparison to normally managed soils (Giménez et al. 1997). Crop rotation being one of the important elements of organic farming affects the physical properties of soil either directly or indirectly. Crop rotation involves the beneficial use of various root systems from various crops, which helps to alter the soil's structure (Chan and Heenan 1991). Mulching of soil surface with organic matter is such a practice that makes the soil soft, pulverized, humid, and porous. Organic and natural practices of agriculture provide macro- and micronutrients, improve texture of soil, improve structure of soil, and enhance soil water-holding capacity, soil moisture conservation, etc.

12.8.1.2 Impact on Chemical Properties of Soil

Phosphorus, nitrogen, major cations, trace metals, cation exchange capacity, electrical conductivity, enzymes, organic matter and carbon, base saturation, salinity, sodium adsorption ration, and pH are among the chemical characteristics of soil. Organic farming increases organic carbon in soil which maintains soil fertility. The natural cycles are used to add essential nutrients to the soil for quality improvement. The use of good quality farm yard manure (FYM) enhances the nitrogen ration and organic matter in the soil. Organic plots have well-balanced amount of essential nutrients to the soil like sulfur, phosphorus, nitrogen, and carbon. Organic matter also helps in maintaining C:N ratio, nitrogen mineralization, total and available nitrogen, soil texture, and soil moisture. This is mainly due to slow and constant mineralization which continuously provides nutrients and prevents leaching.

12.8.1.3 Impact of Organic Inputs on Biological Properties of Soil

The biological characteristics of a given soil reflect both the direct and indirect effects of the living things that inhabit it. These biological characteristics are very vital while studying soil quality as soil is directly affected by the organisms existing in it. Microbial activities and biomass are crucial to sustain the productivity of soil. Microbial activity and biomass are intensified by 30–100% and 20–30%, respectively, due to organic farming. So the biological properties of soil can be enhanced by addition of more organic matter because organic matter ensures greater microbial activity. Soil organic matter increases the carbon and nitrogen contents in soil which intensifies the microbial biomass. Organically managed soils are rich in different types of microflora such as arbuscular mycorrhizal fungi which provide better crop nutrition and decreasing soilborne diseases. Nodule dry weight, photosynthetic rates, nitrogen fixation, nitrogen accumulation, and N concentration in many crops were all increased by the application of organic fertilizer. Many composts, including vermin-compost and farmyard manure, are used in organic farming to control nutrients. This increases the good macro- and microflora in the soil. Increased soil microbial biomass can also be achieved by applying organic inputs such as human urine, sewage sludge, municipal trash, deep litter, bovine slurry, and cattle manure, among others (Fig. 12.3).

12.8.2 Water Pollution

Water is a natural source which is vital for the sustenance of life. It is impossible for life to function without water. The quality of water is declining at a very fast rate, and one of the causes for this deterioration is agrochemicals. Organic and natural farming is one of the ways in which water deterioration could be reduced. A considerable number of insecticides and pesticides are identified in water systems (Esfandian et al. 2016). In organic agriculture, there is less or no usage of chemicals which makes the water less contaminated. Nitrate and phosphorus are two agricultural chemicals which are a major concern to water quality. They are subjected to leaching thus contaminating groundwater. Compost which is a substitute of chemical fertilizers decreases the phosphorus and nitrogen runoff, thereby making water less polluted (Trewavas 2004; Wolkowski 2003). Organic farming includes crop rotation as one of the practices which enhances water infiltration. Trap crops which are vital for integrated pest management in organic farming can boost soil moisture

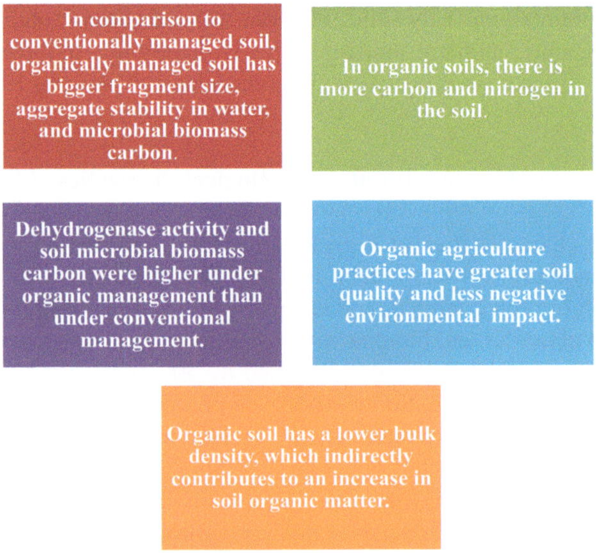

Fig. 12.3 Impact of organic and natural farming on soil quality

retention and reduce irrigation water requirements (Parizad and Bera 2021). Sediments transported by soil erosion make water cloudy affecting fish habitats, fluctuating water temperature, and becoming rooted in stream bank areas used for feeding and breeding. Increasing soil organic matter can help in preventing soil erosion. Rotation crops, cover crops, and intercrops help in recycling the nutrients that would otherwise cause water pollution due to leaching. Soil cover is provided by these crops which uplifts water penetration and reduces the capability for nutrient runoff and erosion. Transfer of nutrients to groundwater or surface waters can be reduced by the accumulation of organic matter and different communities of organisms found in soil as they serve as nutrient reserves. Compost is a better source of organic matter than fresh manure. Storing of compost and manure piles on solid slabs or soils with less leaching potential can prevent nutrient runoff (Sivaranjani and Rakshit 2019).

12.8.3 Air Pollution

The production of gaseous emissions from organic fertilizers and pesticides, such as ammonia, is the main cause of environmental risks in terms of air pollution. Organic agriculture causes air pollution due to the liberation of nutrients, heavy metals, pathogens, particulate matter, and poisonous gases into the atmosphere. The processing and surface application of manures releases particulate matters, oxides of nitrogen, carbon and sulfur, volatile organic compounds, and pathogens which are the main causes of air pollution (Udeigwe et al. 2015). Another cause of air pollution is the release of nitrogen from organic compost through evaporation of excess nitrogen. As compared to conventionally managed farms, organic farms release less airborne particulate concentrations because of reduced soil erosion (Gomiero 2013).

12.9 Present Status of Organic Farming

India was formerly one of the richest countries in the world until British took over. The main business of population of India is farming as 60% humans depend on it. In India, organic farming has been practiced for a long time. In conventional India, all farming was done using organic methods, and pesticides and fertilizers were made from items derived from plants and animals (wwoofindia.org 2007). India has diverse agro-climatic conditions which make India suitable to produce a variety of organic products. As per accessible data, India occupies ninth position in terms of world's organic agriculture land. In terms of cultivable land under organic certification, India is ranked tenth among the top ten nations by the "Agricultural and Processed Food Products Export Development Authority" (APEDA). 4.72 million ha are covered with organic certification overall (Khadda 2021). Different organic products, namely, sugarcane, cotton, oil seeds, basmati rice, pulses, spices, tea, fruits, dry fruits, vegetables, coffee, etc., are produced in India. Due to expeditious industrialization, agricultural land is becoming less in rural India. Also, the rapid increase in population has made food sufficiency a major concerning issue than ever before. Farmers who do not have enough financial resources rely on organic farming. Research suggests that the increase in the use of chemicals and intensive irrigation can be harmful for agriculture making organic farming increasingly popular. Recently, organic farming culture is being supported by the government on a large extent. There are main six organizations which provide accreditation to the organic products, namely, Agricultural and Processed Food Products Export Development Authority, Tea Board, Spices Board, Coconut Development Board, Directorate of Cashew and Cocoa, and Coffee Board. Certification from these authorization boards is very important for export of organic products. Madhya Pradesh has the largest area certified as organic, followed by Himachal Pradesh and Rajasthan (Singh et al. 2019). Different programs have been launched by Indian government to spread awareness regarding organic farming. The National Program for Organic Production (NPOP) has been implemented by the Ministries of Industries and Commerce, Government of India, during 2001. The main objectives of this program are development of organic farming and organic processing and certification of organic agriculture, livestock, and aquaculture. The National Institute of Organic Farming (NIOF) was established by the Government of India in 2003 with the main aim to develop rules, regulations, and certification of organic products (Wani et al. 2017). In 2015, the Paramparagat Krishi Vikas Yojana was launched with the main aim of promoting organic farming and providing the latest information on organic farm technologies (Layaraja and Layaraja 2020). The North Eastern Himalayan region is rain fed and poor in resources, so farming is mainly organic in these regions. Northeastern region covers 8% of the total land area of the country and contributes 4% to the total population. Main occupation of about 70% of the population is agriculture (Patel et al. 2020). Different agro-climatic conditions prevail in India which make it a suitable country for the production of different organic products.

12.10 Conclusion and Future Perspectives

India is an agricultural-based nation, with 55% of the labor force engaged in farming and related sectors, and 67% of the country's population as well practiced farming. The practice of organic farming was formerly thought to be native to India and was practiced in several farming and pastoral communities throughout the Renaissance. However, today's advancements in this field are comparatively laudable. With 1.78 million ha of organic farming area in the world in 2017 (Sharma and Goyal 2000), India is now the greatest producer of organic food on the planet (Willer and Lernoud 2017, 2019).

Compared with conventional farming, the produce from organic cultivation is safer and more nutrient-rich. As consumers look for organic meals that are believed to be healthier and reliable for our whole environment, the demand for organic food is rising drastically. Consequently, organic food may guarantee food safety and security from farm to plate. By preserving the soil's health and the purity of the climate, organic farming improves consumer health and immunity. In addition, the demand for organic products is now expanding at the quickest rate in every country, including India, which can help people to meet their nutritional needs. We may infer from the fact that India currently has the largest organic sector in the world (Willer and Lernoud 2019) that supporting organic farming there will soon result in a country that is nutrient-dense, sustainable, and prosperous. Because of this, organic farming is essential for long-term food security and conservation of resources.

Future polychromatic newer technologies have been developed in the field of organic agriculture, which include the incorporation of mycorrhizal fungi and nano-bio incitements (to increase productivity in agriculture through an environmentally conscious manner), by visualizing cultivated areas more intentionally through a system of sensors and spatial geodata and introduction of three-dimensional printers to assist the country's smallholder. A further improvement in organic farming is the Bee Scanning App, which helps beekeepers combat the parasitic mite Varroa destructor and serves as a foundation for population modeling and breeding initiatives. A comprehensive organic method for ensuring environmentally and economically sustainable agricultural products, Inhana Rational Farming Technology, was established on the premise of "Element Energy Activation." It is based on both ancient Indian scripture and cutting-edge scientific research to modernize organic farming for future generation.

References

Adekiya AO, Agbede TM, Aboyeji CM, Dunsin O, Simeon VT (2019) Effects of biochar and poultry manure on soil characteristics and the yield of radish. Sci Hortic 243:457–463

Andersen MM, Landes X, Xiang W, Anyshchenko A, Falhof J, Osterberg JT, Olsen LI, Edenbrandt AK, Vedel SE, Thorsen BJ (2015) Feasibility of new breeding techniques for organic farming. Trends Plant Sci 20:426–434

Araujo ASF, Santos VB, Monteiro RTR (2008) Responses of soil microbial biomass and activity for practices of organic and conventional farming systems in Piauistate, Brazil. Eur J Soil Biol 44:225–230

Azim K, Soudi B, Boukhari S (2018) Composting parameters and compost quality: a literature review. Org Agric 8:141–158. https://doi.org/10.1007/s13165-017-0180-z

Bajzelj B (2014) Importance of food-demand management for climate mitigation. Nat Clim Chang 4:924–929

Bàrberi P (2002) Weed management in organic agriculture: are we addressing the right issues? Weed Res 42(3):177–193

Bolwig S, Gibbon P, Jones S (2009) The economics of smallholder organic contract farming in tropical Africa. World Dev 37:1094–1104

Bourn D, Prescott J (2002) A comparison of the nutritional value, sensory qualities, and food safety of organically and conventionally produced foods. Crit Rev Food Sci Nutr 42(1):1–34

Boyette, C.D. & Abbas, H.K. 1995. Weed control with mycoherbicides and phytotoxins: a nontraditional application of allelopathy

Brandt K, Leifert C, Sanderson R, Seal CJ (2011) Agroecosystem management and nutritional quality of plant foods: the case of organic fruits and vegetables. Plant Sci J 30:177–197
Brenes-Munoz T, Lakner S, Brummer B (2016) What influences the growth of organic farms? Evidence from a panel of organic farms in Germany. Ger J Agric Econ 65:1–15
Chan KY, Heenan DP (1991) Differences in surface soil aggregation under six different crops. Aust J Exp Agric 31:683–686
Charudattan R (2001) Biological control of weeds by means of plant pathogens: significance for integrated weed management in modern agro-ecology. BioControl 46:229–260
Chen B, Chen Z, Lv S (2011) A novel magnetic biochar efficiently sorbs organic pollutants and phosphate. Bioresour Technol 102:716–723
Cidon CF, Figueiró PS, Schreiber D (2021) Benefits of organic agriculture under the perspective of the bioeconomy: a systematic review. Sustainability 13(12):6852
Clark A (ed) (2008) Managing cover crops profitably. Diane Publishing
Cloutier DC, Van der Weide RY, Peruzzi A, Leblanc ML (2007) Mechanical weed management. In: Nonchemical weed management: principles, concepts and technology. CAB International, Wallingford, pp 111–134
Conacher J, Conacher A (1998) Organic farming and the environment, with particular reference to Australia: a review. Biol Agric Hortic 16(2):145–171
Connor D (2013) Organically grown crops do not a cropping system make and nor can organic agriculture nearly feed the world. Field Crops Res J 144:145–147
Coolong T (2012) Mulches for weed management in vegetable production. In: Weed control, vol 5774. InTech
Crowder DW, Northfield TD, Strand MR, Snyder WE (2010) Organic agriculture promotes evenness and natural pest control. Nature 466(73):109–112
Deshmukh SM (2010) Organic farming: principles, prospects and problems. www.charoka.com/Organic-Farming-Principles-Prospects-and-Problems/b/60009. Accessed 25 Apr 2013
Doran JW, Zeiss MR, Zeiss MR (2000) Soil health and sustainability: managing the biotic component of soil quality. Appl Soil Ecol 15:3–11
Doucet C, Weaver SE, Hamill AS, Zhang J (1999) Separating the effects of crop rotation from weed management on weed density and diversity. Weed Sci 47(6):729–735
Duddigan S, Collins CD, Hussain Z, Osbahr H, Shaw LJ, Sinclair F, Winowiecki L (2022) Impact of zero budget natural farming on crop yields in Andhra Pradesh, SE, India. Sustainability 14(3):1689
Durán-Lara EF, Valderrama A, Marican A (2020) Natural organic compounds for application in organic farming. Agriculture 10(2):41
Ecker O, Qaim M (2011) Analyzing nutritional impacts of policies: an empirical study for Malawi. World Dev 39:412–428
Edenhofer O, Pichs-Madruga R, Sokona Y, Kadner S, Minx JC et al (2014) Technical summary. In: Climate change 2014: mitigation of climate change. Contribution of working group III to the fifth assessment report of the Intergovernmental Panel on Climate Change
Epule ET, Bryant CR, Akkari C, Daouda O (2015) Can organic fertilizers set the pace for a greener arable agricultural revolution in Africa? Analysis, synthesis and way forward. Land Use Policy 47:179–187
Esfandian H, Samadi-Maybodi A, Parvini M, Khoshandam B (2016) Development of a novel method for the removal of diazinon pesticide from aqueous solution and modeling by artificial neural networks (ANN). J Ind Eng Chem 35:295–308. https://doi.org/10.1016/j.jiec.2016.01.011
Farooq N, Abbas T, Tanveer A, Jabran K (2020) Allelopathy for weed management. In: Co-evolution of secondary metabolites. Springer, pp 505–519
Favilli E, Rossi A, Brunori G (2015) Food networks: collective action and local development. The role of organic farming as boundary object. Org Agric 5:235–243
Ferreyra, R., Ahumada, R., Santelices, M., García-Huidobro, J. 2006. Lombrices de tierra como agentes mejoradores de las propiedades del suelo en huertos frutales

Fianko JR, Donkor A, Lowor ST, Yeboah PO (2011) Agrochemicals and the Ghanaian environment, a review. J Environ Prot 2(3):221
Finney DM, Creamer N (2008) Weed management on organic farms. North Carolina Cooperative Extension Service, Raleigh, NC, p 34
Foley JA, Ramankutty N, Brauman KA, Cassidy ES, Gerber JS (2011) Solutions for a cultivated planet. Natures 478:337–342
Fracchiolla M, Renna M, D'Imperio M, Lasorella C, Santamaria P, Cazzato E (2020) Living mulch and organic fertilization to improve weed management, yield and quality of broccoli raab in organic farming. Plants 9(2):177
Gharde Y, Singh PK, Dubey RP, Gupta PK (2018) Assessment of yield and economic losses in agriculture due to weeds in India. Crop Prot 107:12–18
Giménez D, Allmaras RR, Nater EA, Huggins DR (1997) Fractal dimensions for volume and surface of interaggregate pores—scale effects. Geoderma 77(1):19–38
Gomiero T (2013) Alternative land management strategies and their impact on soil conservation. Agriculture 3(3):464–483
Gomiero T, Pimentel D, Paoletti MG (2011) Environmental impact of different agricultural management practices: conventional versus organic agriculture. Crit Rev Plant Sci 30:95–12
Hadas A, Rosenberg R (1992) Guano as a nitrogen source for fertigation in organic farming. Fertil Res 31:209–214
Haering A, Dabbert S, Offermann F, Nieberg H (2001) Benefits of organic farming for society. In: European conference-organic food and farming, Copenhagen
Hansen B, Alrøe HF, Kristensen ES, Wier M (2002) Assessment of food safety in organic farming. Health Saf 1:50–64
Hillel D (1982) Introduction to soil physics. Academic Press, San Diego, CA. Available online: http://soilhealth.cals.cornell.edu/
Horrillo A, Escribano M, Mesias F, Elghannam A, Gaspar P (2016) Is there a future for organic production in high ecological value ecosystems? Agric Syst 143:114–125. https://doi.org/10.1016/B978-0-12-814383-4.00007-4
Huber M, Rembiałkowska E, Srednicka D, Ugel S, Van de Vijver L (2011) Organic food and impact on human health: assessing the status quo and prospects of research. Wagening J Life Sci 58(3–4):103–109
Imran SM, Saxena D, Kazimi R, Pratap S (2022) Effects of various growing media, as well as jeevamrit, on the growth and production of strawberry: a review. Pharma Innov J 11:405
Jaya C (2022) Effect of *Ghanjeevamrit* on growth and flower yield attributes of China aster (*Callistephus chinensis* L. Nees) under organic production system. Pharma Innov J 11:294
Joshi M (2012) New vistas of organic farming. Scientific Publishers, New Delhi, p 140
Joshi TN, Nepali DB, Sah R, Bhattarai T, Midmore DJ (2020) A comparison of composting and vermicomposting for the disposal of poultry waste. Anim Prod Sci 60(7):986–992
Kamau JW, Biber-Freudenberger L, Lamers JP, Stellmacher T, Borgemeister C (2019) Soil fertility and biodiversity on organic and conventional smallholder farms in Kenya. Appl Soil Ecol 134:85–97
Kaur P, Saini JP, Meenakshi, Avnee (2021) Optimization of Jeevamrit doses and application time for enhancing productivity of wheat under natural farming system. J Pharmacogn Phytochem 10(1):405–408
Khadda BS (2021) Prospects of organic farming in India. Voice Agric 27:27–34
Khadse A, Rosset PM (2019) Zero budget natural farming in India-from inception to institutionalization. Agroecol Sustain Food Syst 43:848–871
Khan MR, Raja W, Bhat TA, Mir MS, Naikoo NB, Amin Z, Nazir A, Mir SA, Mohammad I, Wani AA, Patyal D (2022) Zero budget natural farming: a way forward towards sustainable agriculture. Curr J Appl Sci Technol 41(13):31–43
Korav S, Dhaka AK, Chaudhary A, Y.S.M. (2020) Zero budget natural farming a key to sustainable agriculture: challenges, opportunities and policy intervention. Ind J Pure Appl Biosci 8(3):285–295

Kumar K, Verma G, Veer R, Kumar S, Kumar P (2020a) Exploitation of *Panchagavya*, benefits and eco-friendly management of plant diseases: a review. J Entomol Zool Stud 8:2360–2364
Kumar S, Hariprabha S, Kamalakannan S, Sudhagar R, Sanjeevkumar K (2020b) Effect of *Panchagavya* on germination and seedling growth of balsam (*Impatiens balsamina*). Plant Arch 20(1):3735–3737
Kumari S, Meena L, Raghavendra KJ, Karwal M (2022) Jeevamrit/Jeevamrutha: organic concoctions for natural farming. Agric Food E-Newsl 4(5):40–42
Kumhar KC, Beniwal BS, Jat RD, Patel B, Kumar A, Raj H, Kumar M, Kumar N, Kumar S (2022) Plant disease management approaches for organic crop production in Indian scenario: a critical review. J Plant Bioinform Biotechnol 2(1):1–12. https://doi.org/10.22034/jpbb.2022.327017.1018
Laamrani A, Joosse P, Feisthauer N (2017) Determining the number of measurements required to estimate crop residue cover by different methods. J Soil Water Conserv 72(5):471–479
Lapple D (2010) Adoption and abandonment of organic farming: an empirical investigation of the Irish dry stock sector. J Agric Econ 61:697–714
Layaraja M, Layaraja C (2020) Organic farming in India: benefits and challenges. Eur J Mol Clin Med 7(11):3021–3029
Lee KS, Choe YC, Park SH (2015) Measuring the environmental effects of organic farming: a meta-analysis of structural variables in empirical research. J Environ Manag 162:263–274
Lichtfouse E, Navarrete M, Debaeke P, Véronique S, Alberola C (eds) (2009) Sustainable agriculture, vol 1. Springer, Dordrecht
Lori M, Symnaczik S, Mäder P, DeDeyn G, Gattinger A (2017) Organic farming enhances soil microbial abundance and activity—a meta-analysis and meta-regression. PLoS One 12:0180442
Ma D, Yin L, Ju W, Li X, Liu X, Deng X, Wang S (2021) Meta-analysis of green manure effects on soil properties and crop yield in northern China. Field Crop Res 266:108146
MacRae RJ, Frick B, Martin RC (2007) Economic and social impacts of organic production systems. Can J Plant Sci 87(5):1037–1044
Maier S, Szerencsits M, Narodoslawsky M, Ismail IMI, Shahzad K (2017) Current potential of more sustainable biomass production using eco-efficient farming practices in Austria. J Clean Prod 155:23–27
Mandla I, Sharma S (2022) Subhash Palekar natural farming: introduction and it's four pillars. Just Agric 3:39
Meemken EM, Qaim M (2018) Annual review of resource economics. Organic agriculture, food security and the environment. Annu Rev 10:39–63
Meena AL, Karwal M, Kj R, Narwal E (2021) Aerobic composting versus anaerobic composting: comparison and differences. Food Sci Rep 2:23–26
Meerburg BG, Vermeer HM, Kijlstra A (2007) Controlling risks of pathogen transmission by flies on organic pig farms: a review. Outlook Agric 36(3):193–197
Megali L, Glauser G, Rasmann S (2013) Fertilization with beneficial microorganisms decreases tomato defenses against insect pests. Agron Sustain Dev 34:649. https://doi.org/10.1007/s13593-013-0187-0
Meier MS, Stoessel F, Jungbluth N, Juraske R, Schader C, Stolze M (2015) Environmental impacts of organic and conventional agricultural products—are the differences captured by life cycle assessment. J Environ Manag 149:193–208
Monteiro A, Santos S (2022) Sustainable approach to weed management: the role of precision weed management. Agronomy 12(1):118
Mukherjee S, Sain S, Ali MN (2022) Microbiological properties of *Beejamrit*, an ancient Indian traditional knowledge, uncover a dynamic plant beneficial microbial network. World J Microbiol Biotechnol 38:111. https://doi.org/10.1007/s11274-022-03296-3
Muller A, Schader C, Scialabba NH, Niggli UR (2017) Strategies for feeding the world more sustainably with organic agriculture. Nat Commun 18:1290. https://doi.org/10.1038/s41467-017-01410-w. www.nature.com/naturecommunication
Nejadkoorki F (2012) Environmental benefits of organic farming. IntechOpen

Nemes N (2009) Comparative analysis of organic and non-organic farming systems: a critical assessment of farm profitability. Food and Agriculture Organization of the United Nations, Rome, p 33

Offermann F, Nieberg H (2000) Economic performance of organic farms in Europe, vol 5. University of Hohenheim, Department of Farm Economics

Ons L, Bylemans D, Thevissen K, Cammue BP (2020) Combining biocontrol agents with chemical fungicides for integrated plant fungal disease control. Microorganisms 8(12):1930

Palekar S (2009) How to practice natural farming? All India Pingalwara Charitable Society, pp 22–27

Parizad S, Bera S (2021) The effect of organic farming on water reusability, sustainable ecosystem, and food toxicity. Environ Sci Pollut Res 30:1–12

Patel GT, Kharpude SN, Dabral PP, Kumar V (2020) Impact of organic farming on sustainable agriculture system and marketing potential: a review. Int J Environ Clim Chang 10(11):100–120

Prasad K (2015) Biofertilizers: a new dimension for agriculture and environmental development to improve production in sustainable manner. J Basic Appl Mycol 11(1& II):5–13

Prasad K, Warke RV, Khadke K (2019) Management of soilborne pathogens to improve productivity of pulses using organic technologies for sustainable agriculture. Int J Res Anal Rev 6(2):82–101

Raheem A, Zhang J, Huang J, Jiang Y, Siddik MA, Deng A, Zhang W (2019) Greenhouse gas emissions from a rice-rice-green manure cropping system in South China. Geoderma 353:331–339

Ramesh P, Singh M, Rao AS (2005) Organic farming: its relevance to the Indian context. Curr Sci 88(4):561–568

Ranjan S, Sow S (2021) A way towards sustainable agriculture through zero budget natural farming. Food Sci Rep 2(3):30–32

Rautela S, Bains G, Singh DK, Jahan S, Thi T (2022) Effect of organic, inorganic and integrated nutrient amendments on growth parameters of basmati rice (*Oryza sativa* L.). Agric Ecosyst Environ 140:339–353

Reddy BS (2010) Organic farming: status, issues and prospects—a review. Agric Econ Res Rev 23(2):343–358

Reganold JP, Wachter JM (2016) Organic agriculture in the twenty-first century. Nat Plants 2:1–8

Roman P, Martinez MM, Pantoja A (2015) Farmer's compost handbook: experiences in Latin America. FAO, Rome. ISBN: 978-92-5-107845-7. http://www.fao.org/3/a-i3388e.pdf

Ron EZ, Rosenberg E (2014) Enhanced bioremediation of oil spills in the sea. Curr Opin Biotechnol 27:191–194

Sahoo RK, Ansari MW, Pradhan M, Dangar TK, Mohanty S, Tuteja N (2014) Phenotypic and molecular characterization of efficient native Azospirillum strains from rice fields for crop improvement. Protoplasma 251:943–953

Schreiber MM (1992) Influence of tillage, crop rotation, and weed management on giant foxtail (*Setaria faberi*) population dynamics and corn yield. Weed Sci 40(4):645–653

Seufert V, Ramankutty N, Foley JA (2012) Comparing the yields of organic and conventional agriculture. Natures 485:229–232

Seufert V, Ramankutty N, Mayerhofer T (2017) What is this thing called organic? – how organic farming is codified in regulations. Food Policy 68:10–20. https://doi.org/10.1016/j.foodpol.2016.12.009

Shaikh NF, Gachande BD (2015) Effect of organic bio-booster and inorganic inputs on rhizosphere mycoflora population and species diversity of wheat. Int J Sci Res 4:295–302

Sharma AK, Goyal RK (2000) Addition in tradition on agroforestry in arid zone. LEISA-INDIA 2:19–20

Sharma S, Sharma DP (2019) Weed management in stone fruit nectarine orchard with inorganic mulches and herbicides. Karnataka Agric Sci J 17:72–77

Shepherd MA, Harrison R, Webb J (2002) Managing soil organic matter–implications for soil structure on organic farms. Soil Use Manag 18:284–292

Singh JS, Pandey VC, Singh DP (2011) Efficient soil microorganisms: a new dimension for sustainable agriculture and environmental development. Sustain Agric 3:399–424

Singh R, Jat NK, Ravisankar N, Kumar S, Ram T, Yadav RS (2019) Present status and future prospects of organic farming in India. In: Sustainable agriculture. Scientific Publishers, pp 275–299

Singh A, Karmegam N, Singh GS, Bhadauria T, Chang SW, Awasthi MK, Sudhakar S, Arunachalam KD, Biruntha M, Ravindran B (2020) Earthworms and vermicompost: an eco-friendly approach for repaying nature's debt. Environ Geochem Health 42:1617–1642

Sinha RK, Valani D, Chauhan K, Agarwal S (2010) Embarking on a second green revolution for sustainable agriculture by vermiculture biotechnology using earthworms: reviving the dreams of Sir Charles Darwin. J Agric Biotechnol Sustain Dev 2(7):113

Sivaranjani S, Rakshit A (2019) Organic farming in protecting water quality. In: Organic farming: new advances towards sustainable agricultural systems. Springer, pp 1–9

Smith LG, Williams AG, Pearce BD (2015) The energy efficiency of organic agriculture: a review. Renew Agric Food Syst 30:280–301

Smith OM, Cohen AL, Rieser CJ, Davis AG, Taylor JM, Adesanya AW, Crowder DW (2019) Organic farming provides reliable environmental benefits but increases variability in crop yields: a global meta-analysis. Front Sustain Food Syst 3:82

Sodaeizadeh H, Hosseini Z (2012) Allelopathy an environmentally friendly method for weed control. InTech Open, London, p 387

Sreenivasa MN, Naik NM, Bhat SN (2010) Beejamruth: a source for beneficial bacteria. Karnataka J Agric Sci 22:1038–1040

Su LL, Lee LH, Wu TY (2016) Sustainability of using composting and vermicomposting technologies for organic solid waste biotransformation: recent overview, greenhouse gases emissions and economic analysis. J Clean Prod 111:262–278. https://doi.org/10.1016/j.jclepro.2015.08.083

Teasdale JR, Brandsaeter LO, Calegari A, Neto FS, Upadhyaya MK, Blackshaw RE (2007) Cover crops and weed management. In: Non-chemical weed management: principles, concepts and technology. CABI, pp 49–64

Templeton GE, TeBeest DO, Smith RJ Jr (1979) Biological weed control with mycoherbicides. Annu Rev Phytopathol 17(1):301–310

Thomas CL, Acquah GE, Whitmore AP, McGrath SP, Haefele SM (2019) The effect of different organic fertilizers on yield and soil and crop nutrient concentrations. Agronomy 9(12):776

Trewavas A (2004) A critical assessment of organic farming-and-food assertions with particular respect to the UK and the potential environmental benefits of no-till agriculture. Crop Prot 23:757–781. https://doi.org/10.1016/j.cropro.2004.01.009

Udeigwe TK, Teboh JM, Eze PN, Stietiya MH, Kumar V, Hendrix J, Kandakji T (2015) Implications of leading crop production practices on environmental quality and human health. J Environ Manag 151:267–279

Walters SA, Young BG (2008) Utility of winter rye living mulch for weed management in zucchini squash production. Weed Technol 22(4):724–728

Wani SA, Wani MA, Mehraj S (2017) Organic farming: present status, scope and prospects in northern India. J Appl Nat Sci 9(4):2272–2279

Watson CA, Atkinson D, Gosling P, Jackson LR, Rayns FW (2002) Managing soil fertility in organic farming systems. Soil Use Manag 18:239–247

Weisberger D, Nichols V, Liebman M (2019) Does diversifying crop rotations suppress weeds? A meta-analysis. PLoS One 14(7):e0219847

Wicks GA, Burnside OC, Felton WL (2017) Mechanical weed management. In: Handbook of weed management systems. Routledge, pp 51–99

Willer H, Lernoud J (2017) The world of organic agriculture. Statistics and emerging trends. FIBL & IFOAM-Organics International, Bonn

Willer H, Lernoud J (2019) The world of organic agriculture. Statistics and emerging trends. In: Research Institute of Organic Agriculture (FiBL), Frick and IFOAM—Organics International, Bonn. https://www.organicworld.net/yearbook/yearbook-2019.html

Wolkowski R (2003) Nitrogen management considerations for land spreading municipal solid waste compost. J Environ Qual 32:1844–1850. https://doi.org/10.2134/jeq2003.1844
Wollni M, Andersson C (2014) Spatial patterns of organic agriculture adoption: evidence from Honduras. Ecol Econ 97:120–128
Worthington V (2021) Nutritional quality of organic versus conventional fruits, vegetables, and grains. J Altern Complement Med 7(2):161–173. https://doi.org/10.1089/107555301750164244
wwoofindia.org (2007) History of organic farming. Retrieved from wwoofindia.org, http://www.wwoofindia.org/history.htm
Yadav M (2017) Towards a healthier nation: organic farming and government policies in India. Int J Adv Res Dev 2:153–159
Yang B, Ma Y, Xiong Z (2019) Effects of different composting strategies on methane, nitrous oxide, and carbon dioxide emissions and nutrient loss during small-scale anaerobic composting. Environ Sci Pollut Res 26:446–455. https://doi.org/10.1007/s11356-018-3646-y
Yatoo AM, Ali MN, Baba ZA (2021) Sustainable management of diseases and pests in crops by vermicompost and vermicompost tea. A review. Agron Sustain Dev 41:7. https://doi.org/10.1007/s13593-020-00657-w
Yeganehpoor F, Salmasi SZ, Abedi G, Samadiyan F, Beyginiya V (2015) Effects of cover crops and weed management on corn yield. J Saudi Soc Agric Sci 14(2):178–181
Zarei M, Jahandideh Mahjen Abadi VA, Moridi A (2018) Comparison of vermiwash and vermicompost tea properties produced from different organic beds under greenhouse conditions. Int J Recycl Org Waste Agric 7:25–32. https://doi.org/10.1007/s40093-017-0186-2
Zhang T, Wu X, Shaheen SM, Rinklebe J, Bolan NS, Ali EF, Li G, Tsang DC (2021) Effects of microorganism-mediated inoculants on humification processes and phosphorus dynamics during the aerobic composting of swine manure. J Hazard Mater 416:125–738
Zhao Y, Lu Q, Wei Y, Cui H, Zhang X, Wang X, Shan S, Wei Z (2016) Effect of actinobacteria agent inoculation methods on cellulose degradation during composting based on redundancy analysis. Bioresour Technol 219:196–203
Zhao J, Yang Y, Zhang K, Jeong J, Zeng Z, Zang H (2020) Does crop rotation yield more in China. A meta-analysis. Field Crops Res 245:107–659

Chapter 13
Detoxification of Contaminated Soil to Restore Its Health for Sustainable Agriculture

Pushpak Dahiya, Manya Behl, Deeksha Kumari, Ekta Arya, Ranju Kumari Rathour, Vikash Kumar, and Ravi Kant Bhatia

13.1 Introduction

Environmental contamination is the planet's most pressing problem for living things right now. Pollutants in soil pose a global threat to human health and food security (Qin et al. 2021). Toxic elements exist in soil through both natural- and human-induced means. The origins of these pollutants from natural sources encompass processes such as rock erosion, volcanic activities, dust movement, and smelting, as noted by Chen et al. (2020). Conversely, human activities play a significant role in introducing these harmful substances. Common anthropogenic sources entail the industrial application of toxic elements, electronic usage, combustion of fossil fuels, and military exercises, as outlined by Wang et al. (2021). Additionally, the dissemination of these toxic agents into the environment can also be attributed to the utilisation of biosolids, including materials like manures, compost, and sewage sludge. These activities cause ecological imbalance and pollution of soil that causes land degradation every year (Vaverková et al. 2019). It primarily entails declining soil fertility, soil contamination, salinisation, soil erosion, and land desertification. These biosolids release toxicants in the soil and eventually pollute it. Some pollutants like heavy metals (Pb, Cd, Hg) are released into the soil because of the use of fungicides and pesticides in agriculture (Havugimana et al. 2017). As soil is so intimately linked to human health, it is crucial to protect the soil from the deleterious effect of pollutants (Tsvetnov et al. 2019). Different physical, chemical, and

P. Dahiya · M. Behl · D. Kumari · E. Arya · R. K. Rathour · R. K. Bhatia (✉)
Department of Biotechnology, Himachal Pradesh University, Summer Hill, Shimla, India

V. Kumar
Division of Life Sciences, Department of Bio & Medical Big Data, Research Institute of Natural Science (RINS), Gyeongsang National University (GNU), Jinju, Republic of Korea

R. K. Bhatia, A. Walia (eds.), *Advancements in Microbial Biotechnology for Soil Health*, Microorganisms for Sustainability 50,
https://doi.org/10.1007/978-981-99-9482-3_13

biological remediation techniques are being employed for the decontamination of soil and to improve the quality of soil as well (Fig. 13.1).

13.2 Different Kinds of Soil Contamination

13.2.1 Chemical Contamination

This situation arises when substances like pesticides, herbicides, industrial chemicals, or heavy metals gather in the soil, reaching levels that could potentially endanger both human well-being and the surrounding ecosystem (Virkutyte et al. 2002). These types of pollution fall into two categories: point pollution and non-point pollution. Point pollution occurs when identifiable sources of chemical contaminants are responsible, such as processing plants and factories. On the other hand, non-point pollution represents the second type, where pinpointing the exact sources of chemical contaminants becomes challenging.

13.2.1.1 Petroleum Hydrocarbons

Crude oil and natural gas are typically the causes of soil contamination from petroleum hydrocarbons. Lands close to oil wells or pipelines contain crude oil. Natural gas contamination frequently results from leaks in metropolitan pipeline systems (Das and Chandran 2011). Petroleum hydrocarbons induce the organic matter formation and nitrogen fixation, which have a significant impact on the microbiological, physical, and chemical characteristics of the soil (Stepanova et al. 2022). Higher organic matter in soil boosts its ability to hold water to the level of

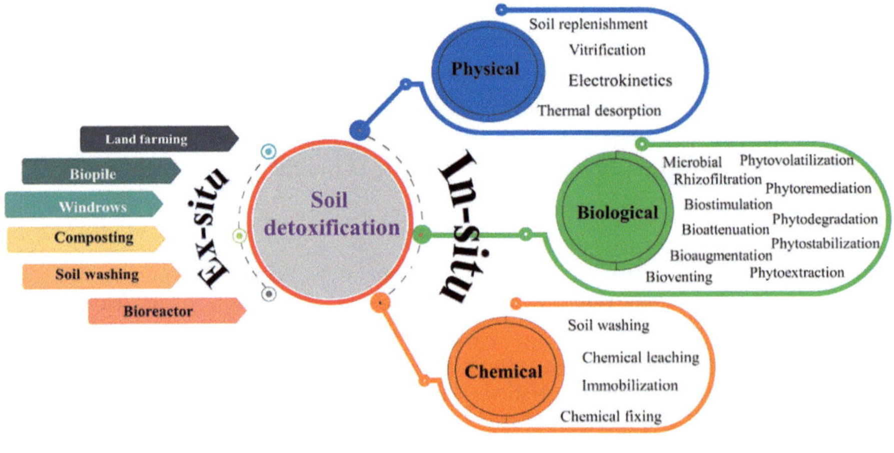

Fig. 13.1 Flowchart showing different forms of detoxification methods

deflocculation, causing the soil to lose structure and become more prone to erosion (Steliga and Kluk 2020).

13.2.1.2 Agrochemicals (Pesticides, Herbicides, and Fertilizers)

Nematicides, herbicides, pesticides, insecticides, and fungicides are the main categories of agrochemicals, also known as agricultural chemicals. It might also be referring to artificial fertilizers and hormones. Improper bulk chemical storage puts the environment at risk from chemical spills, leaks, and leaching. Due to different wind patterns, pesticides applied by tractors or planes frequently end up on other plants and animals instead of intended targets (Speight 2017). Overuse of pesticides can harm earthworms, bacteria, fungi, and biomass in the soil—organisms vital to maintaining the cycle of nutrients and soil fertility. As pests frequently acquire immunity to or resistance to pesticides, they are often just a short-term solution. It is frequently determined that using integrated pest management techniques seems to be preferable to the use of agrochemicals, which can contaminate soil and even groundwater (Issaka et al. 2022).

13.2.1.3 Polycyclic or Polynuclear Aromatic Hydrocarbons (PAHs)

The incomplete combustion of organic materials, waste, or fuels results in the production of polycyclic aromatic hydrocarbons, also known as polynuclear aromatic hydrocarbons (PAHs), which are cancer-causing micropollutants. In addition, it could result from industrial-scale fossil fuel combustion, volcanic eruptions, and forest fires. Due to their hydrophobic characteristics, PAHs pose a formidable challenge when it comes to treatment and exhibit resilience against degradation (Barathi et al. 2023). When PAH contamination infiltrates the food chain, it escalates the peril to both the environment and human well-being. These compounds are not only toxic but also possess carcinogenic and mutagenic properties. The presence of PAHs in soil adversely affects its permeability, water retention capacity, and particle size. Furthermore, PAHs exert a notable influence on the composition, permeability, plasticity, diversity, and population of microorganisms inhabiting the soil, as highlighted by Singh and Haritash (2019).

13.2.1.4 Solvents

At clean-up sites, solvents, encompassing halogenated non-polar aromatics, aliphatics, heterocyclics, and other polar organic compounds, are commonly encountered. This prevalence can be attributed to their widespread application as cleaning agents or as constituents in the production of various goods across diverse commercial and industrial processes. Notably, halogenated solvents find prominent use in activities such as dry cleaning, adhesive formulation, and the cleansing and degreasing of metal surfaces (Barathi et al. 2023). Among the array of solvents, four prominent ones belong to the chlorinated category, including trichloroethylene, trichloroethane, methyl chloride, and tetrachloroethylene. Solvents have contaminated soil and groundwater as a result of poor handling, use, and disposal. Soil fertility and nitrogen fixation are decreased by solvent pollution. Options for removing chlorinated solvents from the environment include air stripping, carbon adsorption, soil ventilation, aeration, and detoxification (Trellu et al. 2021).

13.2.1.5 Heavy Metals

Lead

Although lead naturally exists in soil, industrial processes and the use of products containing lead cause lead concentrations to exceed normal ranges. Lead does not deteriorate over time, and high exposure rates to the substance have been linked to significant health problems in people, particularly in children, whose physical and mental developments are negatively impacted by lead exposure. Due to its anti-knock qualities, lead was frequently added to petroleum fuel in the past. Additionally, it is applied when decorating homes. Heavy traffic regions tend to have a lot of lead traces (Ren et al. 2023).

Asbestos

The building substance asbestos is frequently used in wall cladding and roofing tiles. Poor demolition techniques and waste handling of asbestos cement sheets, fragments, fibres, thermal pipes, and other detritus frequently result in asbestos soil pollution. Lung scarring, lung cancer, and mesothelioma are caused by prolonged exposure to elevated asbestos concentrations. In addition to metal, paint, and petrochemical plants, industrial regions, mining sites, and milling plants are frequently found to have heavy metal pollution in their soils (Yao et al. 2012). To increase the soil's metal content for plant development, they can also come from fertilizers. Metals do not deteriorate chemically or microbiologically, in contrast to organic substances. Through ingestion, direct contact, or introduction to the food chain, they accumulate over time and pose health dangers to people and ecological life. The productivity and fertility of the contaminated soil are also impacted by heavy metals, which causes issues with agricultural production and land use (Doyle et al. 2023).

13.2.2 Biological Contaminants

This happens when pathogens that can harm humans, such as bacteria, viruses, or fungi, are prevalent in the soil. Antibiotic resistance genes (ARGs) and antibiotic-resistant bacteria (ARB) have grown and spread more quickly as a result of excessive use of antibiotics. ARB and ARGs are acknowledged as emerging environmental contaminants that pollute soil and pose significant health risks to the general populace. One of the main methods for spreading ARB and ARGs in the soil is the application of manure in cultivation (Ondon et al. 2021) (Fig. 13.2).

13.3 Ex Situ Detoxification

13.3.1 Land Farming

Detoxification might not need excavation if a contaminant is less than 1 m below the soil's surface. In contrast, if the contaminant is deeper than 1.7 m, it is raised to the surface for successful soil restoration, which is accomplished by native microbes. This process is widely utilized and has been used to dispose of many different waste kinds, including oily sludge and other waste products from petroleum refineries. The purpose of land treatment is to maximise degradation in an aerobic biological process employing soil as the inoculum and support medium for biological growth (Cerqueira et al. 2014). The activity of microorganisms is increased in land farming

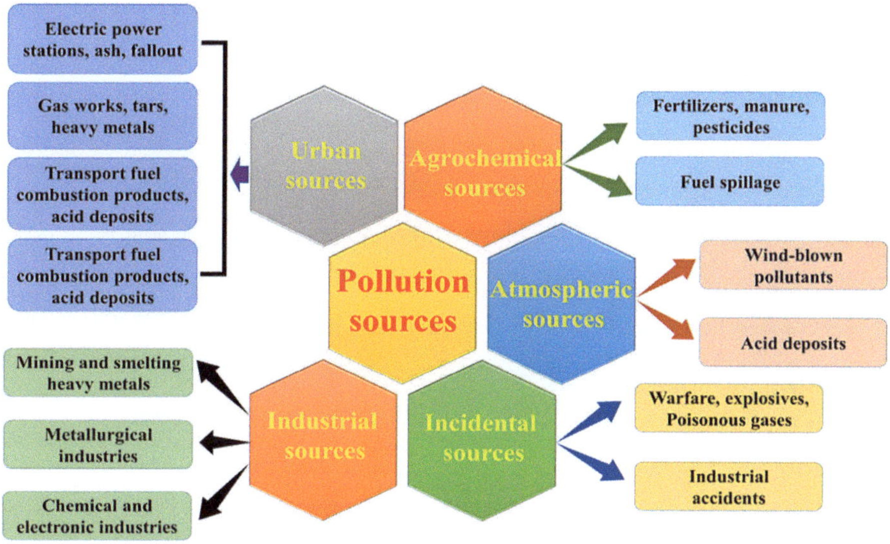

Fig. 13.2 Different types of soil contaminants

by using optimal irrigation, fertilizers, and aeration, which also improves the detoxification process. It is typically utilized for PAH detoxification. Because of its simplicity, it may be used to remediate huge amounts of polluted soils with little energy expenditure. This method has some drawbacks, such as the need for a large space, environmental circumstances that cannot be avoided that inhibit microbial activity, the expense of excavation, and ineffective removal of toxic volatiles and inorganic contaminants (Mastnak et al. 2023).

13.3.2 Biopile

It is an ex situ therapy approach that makes use of biological processes to transform pollutants into by-products with low toxicity. It is a combination of both landfarming and composting. In this technique, contaminated soils are piled or stacked to encourage microbial activity using efficient and controlled nutrient addition, aeration, and irrigation as shown in Fig. 13.3. The usual operation of a pile is to raise the temperature of the pile material due to heat created by the process of biodegradation, and the greater temperature enhances the rate of degradation. Non-halogenated volatile organic compounds (VOCs) and semivolatile organic compounds (SVOCs) are examples of aerobically biodegradable organic chemicals that are treated with piles. Biopiles are widely utilized to treat soils polluted with petroleum hydrocarbons. A pile arrangement has the benefit of treating a lot of polluted soil in a small space whereas radionuclides and inorganic pollutants cannot be eliminated by this technique (Barathi et al. 2023).

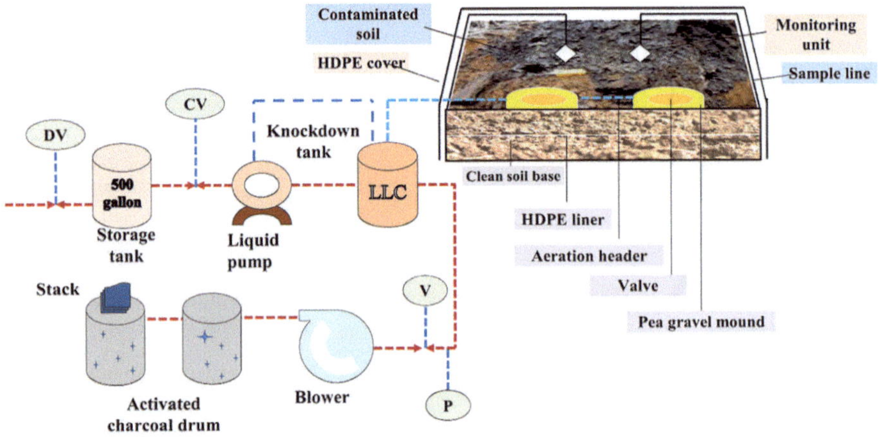

Fig. 13.3 Biopile set-up for treatment of contaminated soil

13.3.3 Windrows

Windrow treatment represents an evolution of pile treatment methodology. This approach amplifies the microbial breakdown of both native and transient hydrocarbon components present in contaminated soil. This is achieved through the periodic turning of the stacked polluted soil, thereby augmenting the detoxification process. In comparison with the pile technique, windrow treatment exhibits a heightened capacity for hydrocarbon removal, the pace of which is contingent upon the soil type, as noted by Coulon et al. (2010). However, it is important to note that the efficacy of this method is compromised if the contaminated soil harbours perilous volatile substances or releases methane (a greenhouse gas) during the treatment, as indicated by Saeed et al. (2023).

13.3.4 Composting

Composting is a regulated biological process involving the utilisation of microbes to convert organic pollutants like PAHs into secure and stabilized end products. To foster the growth of microbial communities capable of breaking down contaminants within the soil through metabolic pathways, the contaminated soil is blended with innocuous organic additives (Ambaye et al. 2022). These additives often take the form of additional solid wastes such as manure and agricultural remnants, chosen for their compatibility with the composting process. By combining factors such as a robust microbial population, minimal levels of hydrocarbons and salts, ample nutrients, and appropriate moisture levels, the rate of biodegradation for organic pollutants can be effectively elevated, as outlined by Lukić et al. (2017).

13.3.5 Soil Washing

Soil washing involves an ex situ water-based procedure designed to effectively cleanse soils of contaminants. This process entails freeing contaminants adhered to soil particles by immersing the polluted soil in a cleansing solution, as depicted in Fig. 13.4. To dissolve or suspend pollutants present in the soil, a range of additives such as surfactants, acids, and chelating agents are introduced. These additives facilitate the chemical or physical bonding of these pollutants, spanning both organic and inorganic varieties, with components like silt, clay, and organic matter within the soil. Subsequently, a substantial portion of the clay and silt components become affixed to larger particles, such as gravel and sand. The washing processes separate the particles by breaking the adhesive connections between larger and smaller particles. To make their disposal more practical, pollutants are concentrated into small volumes using gravity separation, particle size separation, and attrition

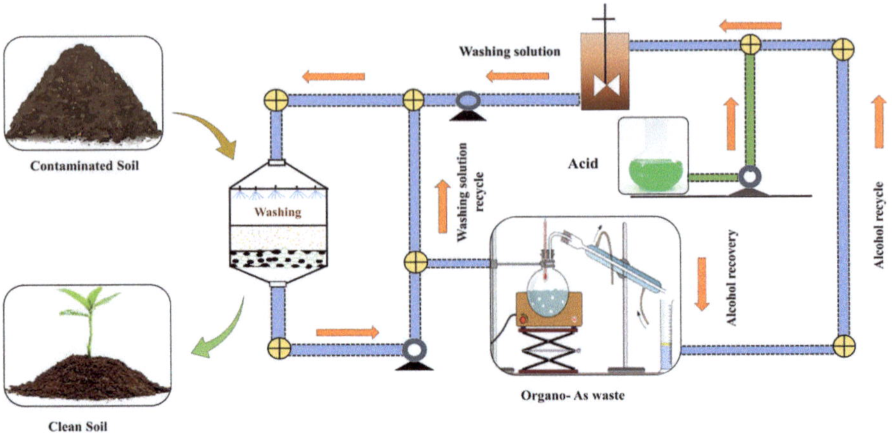

Fig. 13.4 Soil washing set-up for treatment of contaminated soils

scrubbing. However, various limitations could limit the process' applicability and efficacy (Caetano et al. 2023).

13.3.6 *Bioreactor*

In the realm of remediating soil contaminated with persistent pollutants, bioreactors have emerged as a leading ex situ methodology. These advanced systems operate within controlled and automated environmental conditions, as highlighted by Ambaye et al. (2022). Through a bioreactor-driven detoxification approach, meticulous regulation of factors such as temperature, agitation, aeration, pH levels, substrate concentrations, and inoculum concentrations significantly reduces the required detoxification duration. Bioreactors are frequently employed for treating polluted soils in either slurry or solid phases, depicted in Fig. 13.5.

The key advantage of bioreactor-based detoxification lies in the comprehensive management of essential parameters within the bioreactor environment, encompassing pH levels, substrate and inoculum concentrations, temperature, agitation intensity, and aeration rates. Successful instances of remediating soil contaminated with volatile organic compounds (VOCs), including toluene, xylenes, benzene, and ethylbenzene, have been well documented through the application of bioreactors, as evidenced by Uddin et al. (2017).

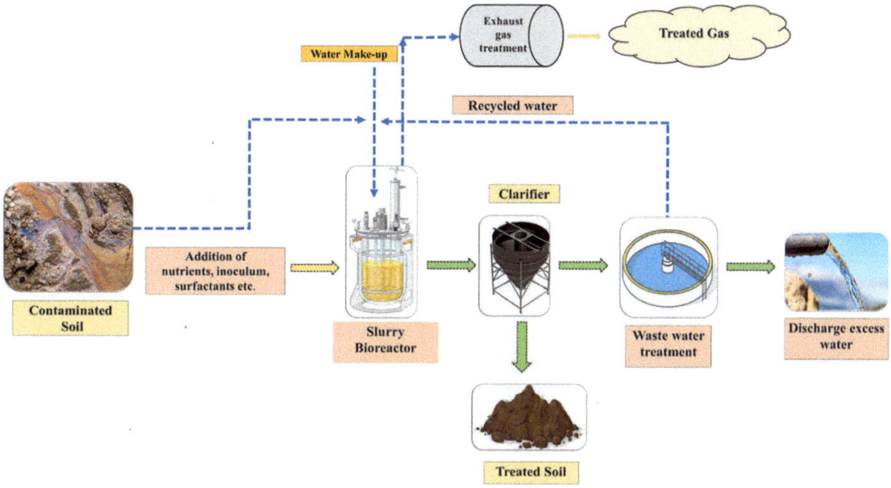

Fig. 13.5 Slurry phase bioreactor

13.3.7 Pyrolysis

Pyrolysis is a viable technique for remediation of contaminated soil in which the chemical breakdown of toxic compounds occurs with the utilisation of high temperature (>430 °C), high pressure, and anoxic conditions (Venderbosch and Prins 2010). Through this process, the pollutants or harmful substances are converted into organic liquids, water, char, and syngas (Gao et al. 2022). According to the rate at which the biomass is changed, pyrolysis operations can be categorised into two basic types: fast and slow. Oil spillage from industrial activity can contaminate the soil with petroleum hydrocarbons (Abbasian et al. 2016). The petroleum chemicals that are spilled raise the level of organic carbon and lower the accessibility of phosphorus. Petroleum hydrocarbons that have been leaked on the ground also coat the soil's surface, making it impervious, which results in poor water circulation in the soil as well as impeded gas exchange with the air, which turns the soil anaerobic (Streche et al. 2018).

The decrease in bacterial population and metabolic activity in anaerobic soil results in ecological imbalances. Most organic and inorganic contaminants, for example, PCBs (polychlorinated biphenyls), SVOCs, PAHs, cyanides, refinery wastes, paint wastes, pesticides, wastes from processing of synthetic rubber, dioxins, mercury, soils contaminated with creosote and hydrocarbons, wastes from coal tar and wood treatment, and radioactive wastes are most commonly remediated using pyrolysis treatment. However, this method encounters several limitations. Elevated soil moisture content contributes to higher treatment expenses, while the release of specific gases like CO, H_2, and CH_4 during the process presents another challenge. Additionally, when contrasted with inorganic pollutants, this approach demonstrates greater suitability for remediating organic pollutants present in soils and oily sludges (Ambaye et al. 2022).

13.4 In Situ Remediation

13.4.1 Physical Remediation

13.4.1.1 Soil Replenishment

Clean soil is used to restore or partially restore the contaminated soil in order to lower contaminant concentration, increase the soil's environmental capacity, and then remediate the soil (Kos and Leštan 2003). There are three types of soil replenishment: soil importation, soil spading, and soil replacement. In order to reduce the contaminant quantity, increase the soil's environmental capacity, and subsequently remediate the soil, soil replacement refers to the use of clean soil to replace or partly substitute the contaminated soil (Qian and Liu 2000). The soil replenishment method is appropriate for soil that is cost-prohibitive, severely polluted, and of a small area. The environment will be less affected by soil replacement since it can effectively isolate the ecosystem and soil (Raza et al. 2023).

Soil Spading

To avoid a second environmental damage, the replaced soil must be managed affordably. Deep excavation of contaminated soil, known as soil spading, is done to spread the contamination throughout the soil's depths and accomplish the desired dilution and natural degradation. A contaminated soil must be completely replaced with a clean soil, from the base to the surface, or it must be mixed to reduce the concentration of contaminants while importing new soil. This technique is most suited for soil that is both small in area and severely polluted, has a high working volume, and is expensive (Priya et al. 2023).

Soil Import

The process of melding clean soil into severely metalloid-polluted soil is known as importation. To reduce the metal concentration, the additional soil is either mixed in or covered on the surface. The environment's impact can be reduced by using the soil replacement approach to adequately isolate the contaminated ecosystem. However, this approach is costly considering of its high labour costs and the fact that it is only effective in small areas of very contaminated soil. The price of transporting excavated soil across large distances could be significantly higher. Also, because there is a chance that soil fertility will be lost, this procedure might not be appropriate for agricultural locations (Deepika and Haritash 2023).

Soil Isolation

However, extra auxiliary engineering techniques are still needed for a thorough clean-up. It refers to the separation of soil polluted with heavy metalloids from uncontaminated soil. Isolation technologies are primarily used to prohibit the off-site transport of heavy metals and other contaminants by containing them to a specific location (Zheng and Wang 2002). Soil isolation technology is used to stop further heavy metal groundwater pollution when standard clean-up methods are neither technically or financially feasible. Hazardous locations may occasionally be temporarily isolated to stop transfer while site assessment and clean-up are being carried out. Subsurface barriers are widely used to segregate polluted water and soil by restricting the flow of ground- and/or surface water at a contaminated site. Subsurface obstructions obstruct both the flow of uncontaminated groundwater away from the contaminated site and the flow of dirty groundwater into the uncontaminated site (Zhu et al. 2012). Groundwater cannot travel laterally due to vertical subsurface impediments. These vertical barriers might be built upstream, downstream, or all the way around the construction site. Vertical barriers are commonly built in conjunction with a capping system to prevent the intrusion of clean surface water. These barriers are typically only able to be buried about 30 ft deep. For the contaminated area of the soil to be successfully isolated, the barrier must continue into a layer of poor permeability. The primary strategy for doing this is to cover the polluted region of soil with a layer of low-permeable material, such as clay. Some of the materials used for subsurface barriers include sheet piles, grout curbs, and slurry walls (Mulligan et al. 2001)

13.4.1.2 Thermal Desorption

Based on the temperatures, thermal desorption can be classified as high temperature (320–560 °C) and low temperature (90–320 °C). This technique has a number of advantages, including a simple methodology, portability, and the ability to reuse repaired soil. Thermal desorption heats the contaminated soil using steam, microwaves, and infrared radiation to make the pollutant (for instance, Hg and As) volatile based on the pollutant's volatility. The heavy metals are then eliminated by gathering the flammable heavy metals with the use of a carrier gas, hoover, or both. This technology was used for in situ clean-up, but its use is limited by a number of factors, such as expensive equipment and a protracted desorption phase (Li et al. 2010).

13.4.1.3 Vitrification

Vitrify technology speeds up the process by heating the soil to a temperature between 1400 and 2000 °C at which organic matter volatilises or decomposes. The off-gas treatment system collects the steam and the pyrolysis product. Following cooling, the melted substance solidifies into a rock-like vitreous substance, which prevents the movement of heavy metals. According to some estimates, vitreous materials have ten times the strength of concrete (Yao et al. 2012). Both in situ and ex situ methods of vitrification are feasible, but the in situ method is recommended because of how inexpensive and energy-efficient it is. For in situ vitrification, only moist soil with a low alkali concentration is suitable. Ex situ vitrification involves a number of steps, including excavation, mixing, pre-treatment, feeding the melting product, and casting the molten result (Dellisanti 2016). The cost of ex situ vitrification is high since melting requires a lot of energy. Ex situ clean-up requires a lot of energy, which is produced by burning fossil fuels or by directly heating electrodes and then transferred through arc, plasma, and microwave. Electrodes can be inserted into the contaminated soil to generate heat for in situ clean-up. In a nutshell, this technique removed heavy metals quite effectively. The vitrified material may occasionally be mixed with additives like clay, native soil, or sand to produce a product with specific qualities. The vitrified material can be recycled and used as aggregate, clean fill, and reusable materials (Madhav et al. 2023) (Fig. 13.6).

13.4.1.4 Electrokinetics

By providing a voltage to the soil's two sides, the electrokinetic remediation method of detoxification creates an electric field gradient. Using electromigration, electro-osmotic flow, or electrophoresis, the pollutant is moved to a space with two poles and then given additional treatment (Virkutyte et al. 2002). This procedure makes it possible to protect the original ecotope and restore its habitat. According to Barathi et al. (2023), the main improved processes include adding buffer solution to the cathode and anode to adjust pH, using an ion exchange membrane to control pH, adding complexant to improve migration, etc. (Fig. 13.7).

13.4.2 Chemical Remediation

13.4.2.1 Immobilisation

In order to lessen the mobility, bioavailability, and bioaccessibility of heavy metals or metalloids in the soil, immobilising compounds are added to polluted soils. Heavy metals and metalloids can get immobilised in soil by reactions such as complexation, precipitation, and adsorption (Ashraf et al. 2016). Among the most often utilised additives are cement, clay, zeolites, phosphates, minerals, microbes, and

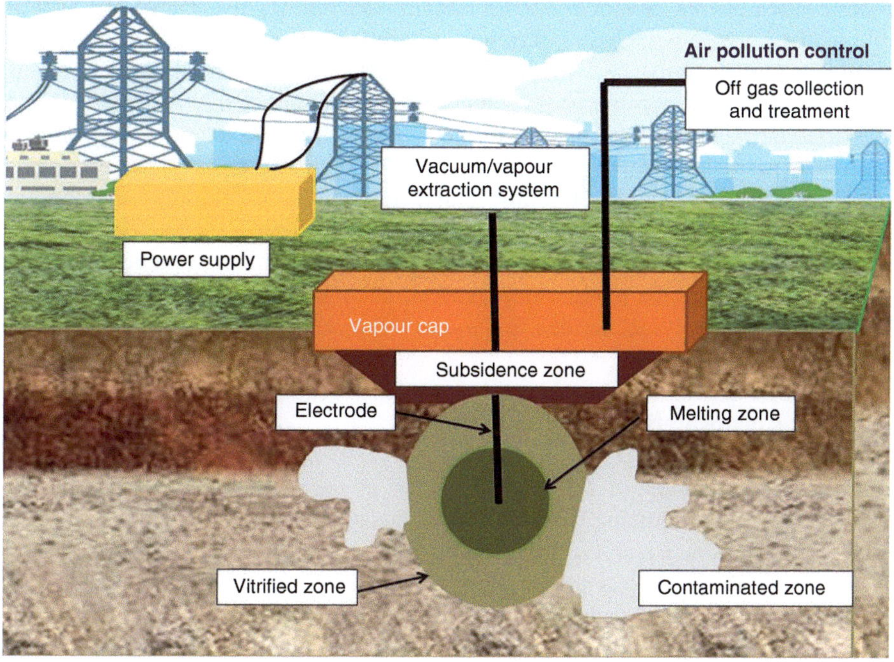

Fig. 13.6 Vitrification

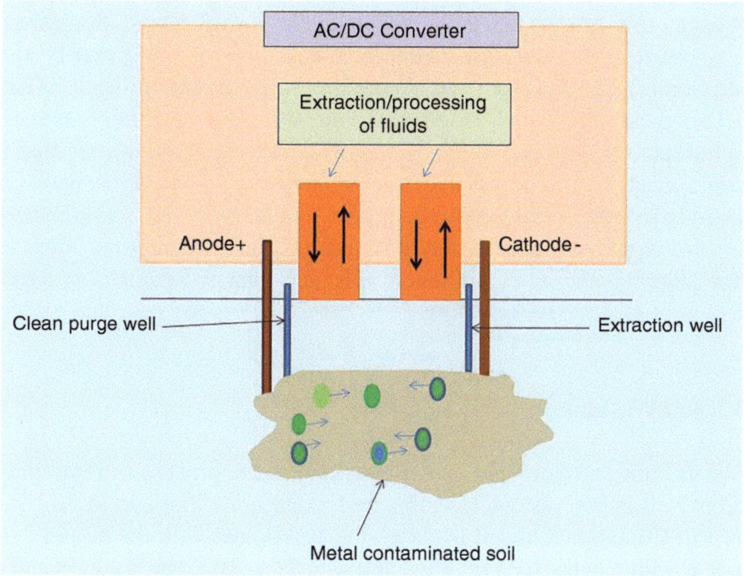

Fig. 13.7 Electrokinetic remediation of soil

organic amendments (Sun et al. 2016). Recent research has also shown that affordable industrial wastes including termitaria, industrial eggshells, and red mud have the ability to immobilise heavy metals in contaminated soils. The two main organic additions used to immobilise heavy metals are animal dung and biosolids (Cele and Maboeta 2016). These days, manure wastes with low metal concentrations are used to immobilise metal in soils (Venegas et al. 2015). Typically, the majority of organic amendments are composed of cellulose and lignin. The list of additional components also includes extractives, hemicellulose, proteins, lipids, starches, simple sugars, hydrocarbons, and a wide range of compounds that contain functional groups with a high affinity for metal complexation, such as carbonyl, phenolic, amido, amino, structural polysaccharides, and esters (Niazi et al. 2016).

13.4.2.2 Encapsulation

Immobilisation of toxic metal solutions is one of the remedial approaches that is most efficient in minimising hazardous materials and assuring their safe disposal in landfills by encasing hazardous materials in controllable solid blocks (Ucaroglu and Talinli 2012). Polluted soils are encapsulated by being mixed with other substances like concrete, lime, or asphalt. The contaminated soil keeps the neighbouring materials from becoming contaminated by becoming immobile. Cement is used in the creation of solid blocks instead of other binding materials because it is widely available, versatile, and inexpensive (Pandey et al. 2012). Calcium aluminate cement (CAC) has been found to be particularly efficient and effective at encasing dangerous substances (Navarro et al. 2013). When acidic phosphate solutions are used as the base reactant in an acid-base reaction, a new type of CAC matrix known as phosphate-bonded CAC is produced. According to study and utilisation (Kinoshita et al. 2013), these phosphate-added CAC are effective cementing materials for encasing radioactive wastes and avoiding corrosion in nuclear wastes that contain aluminium. Encapsulation makes use of a number of immobilisation agents, including polyvinyl alcohol, chitosan, alginate, agar, polyacrylamide, and polyurethanes. The main drawback of these methods is that the finished product must have an immediate market; otherwise, the result will be haphazard patches of asphalt and concrete (Liu et al. 2023) (Fig. 13.8).

13.4.2.3 Soil Washing

According to Park and Son (2017), soil washing is the process of removing heavy metals from soil using various reagents and extractants that can leach the metals from the soil. Contaminated soil is excavated and mixed with the proper extractant solution depending on the type of metal and soil. The extractant solution and dirt are thoroughly mixed over a set period of time. Through precipitation, ion exchange, chelation, or adsorption, heavy metals in soil are transferred from the solid phase to the liquid phase and subsequently removed from the leachate (Ferraro et al. 2015).

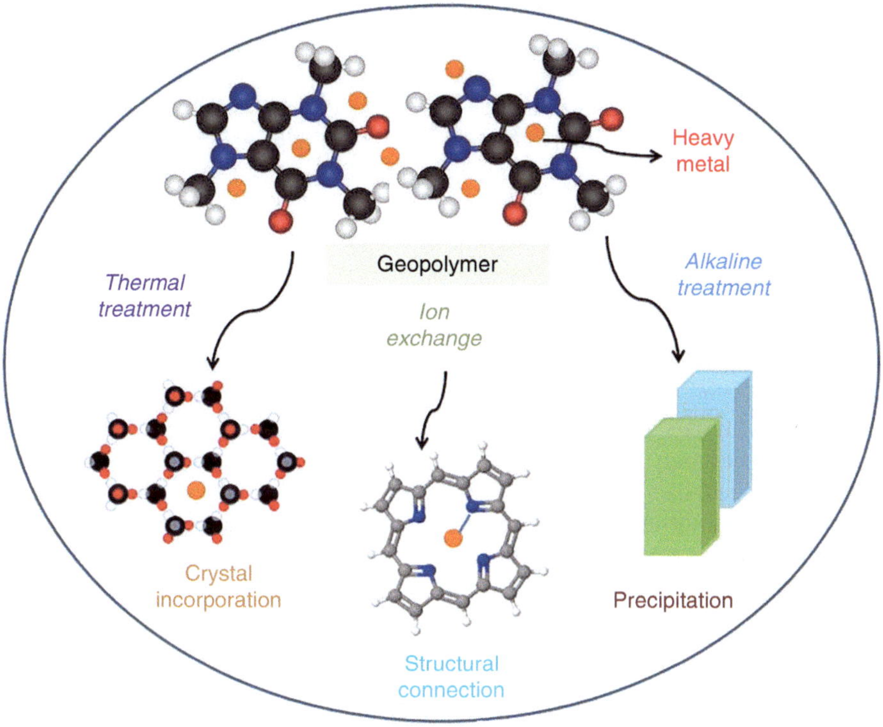

Fig. 13.8 Encapsulation

Because soil washing removes metals from the soil, it is widely used to detoxify areas that have been contaminated with heavy metals. Additionally, soil washing is an efficient process that may be tailored to specific needs without creating long-term liabilities (Park and Son 2017). It is widely known that the type of soil and metal will affect the exchange, extraction, and solubilisation of heavy metals during soil washing (Liao et al. 2015). Because they can form stable complexes with the majority of heavy metals over a wide pH range, synthetic chelates like EDTA (ethylenediaminetetraacetic acid) and EDDS (ethylenediamine-*N*,*N*′-disuccinic acid) are regarded to be the most effective extractants for soil washing (Ullah et al. 2015). Repeated washing and rinsing methods are employed to boost the efficiency of heavy metal removal, which can reduce the consumption of washing agents and washing costs (Torres et al. 2012).

13.4.2.4 Chemical Leaching

According to Tokunaga and Hakuta (2002), chemical leaching is the process of eliminating pollutants from contaminated soil using chemicals, freshwater, and other fluids or gases. Through ion exchange, precipitation, adsorption, and chelation, heavy metals in soil are transported from the soil to the liquid phase and then eliminated from the leachate. Inorganic eluent, chelation agents, and surfactants are a few of the main components employed in leachate. The most effective extractant was found to be phosphoric acid (Alam et al. 2001). The results show that Na_2EDTA solutions are far more efficient at removing heavy metals from soil samples than $Na_2S_2O_5$. Chelation compounds like EDTA have a poor biological degradability and are expensive (Ehsan et al. 2007).

13.4.2.5 Chemical Fixing

Chemical fixation, which reduces the migration of heavy metals to water, plants, and other environmental media while achieving soil remediation, involves adding reagents or materials to contaminated soil and using them with heavy metals to create insoluble or barely movable, low-toxic matter (Kos and Leštan 2003). The addition of bone meal decreased the availability of the metals in the soils, according to batch testing and subsequent metal extraction from control and bone meal-modified soils using 0.01 M $CaCl_2$ and DTPA (diethylenetriaminepentaacetic acid) (Hodson et al. 2000).

13.4.3 Biological Detoxification

13.4.3.1 Microbial Detoxification

The removal or immobilisation of pollutants, including all potential poisons such as hydrocarbons, agrochemicals, and other organic toxicants, is accomplished by a microbe-mediated process (Ahmad 2012). In detoxification procedures, microorganisms convert organic pollutants into either metabolic intermediates or final by-product like carbon dioxide and water that serve as the main building blocks for cellular growth. Detoxification provides a high degree of practicality and economic application when compared to physical and chemical clean-up processes. Microorganisms like bacteria and fungi are utilized more frequently to remove heavy metals from contaminated soils; however, yeast and algae are also regularly used (He and Chen 2014). In comparison to single-strain cultures, it was discovered that the bacterial combinations had higher levels of resistance and efficacy for the removal of heavy metals. Microorganisms use precipitation processes, biosorption, desolation by internal metal-binding proteins, and enzymes for the metal conversion to harmless forms (Ojuederie and Babalola 2017). Microorganisms like *Streptomyces clavuligerus*, *Alcaligenes faecalis*, *Enterobacter cloacae*, and *Micrococcus* sp. are used for chromium removal, *Pseudomonas aeruginosa* and *Brevibacterium iodinum* for the removal of lead, and *Streptomyces rimosus* for nickel and zinc removal (De et al. 2008). Remediation by microbes is comparatively cheap, and it can totally transform complicated organic contaminants into other simpler organics or entirely mineralise organic pollutants into water, inorganic chemicals, carbon dioxide, and cell proteins. Studies proved that species like *Scedosporium boydii*, *Rhodococcus* sp., and *Pseudomonas* sp. are capable to degrade petroleum contaminants that exist in the environment (Emenike et al. 2018).

13.4.3.2 Bioengineered Microbial Strains for Soil Pollution Detoxification

Engineered microorganisms have the potential to effectively remediate a wide range of contaminants. The incorporation of newly identified genes possessing robust degradation capabilities, when introduced and activated, can substantially expedite the recovery process. The strategies for engineering microorganisms involve the following steps: (a) modifying enzyme affinity and specificity; (b) inducing changes in genes and regulatory pathways; (c) creating, overseeing, and supervising bioremediation procedures; and (d) implementing sensor-based bio-affinity indicators to identify pollutants, diminish toxicity, and forecast endpoints. Engineered organisms like *Caulobacter crescentus* JS4022/p723-6 H, *Escherichia coli* DH5α, recombinant *Rhodococcus erythropolis*, and recombinant *Deinococcus radiodurans* are used for the bioremediation of soil pollution (Meagher et al. 2020).

13.4.3.3 Detoxification Using Fungi

Fungi hold significant prominence in various industrial applications due to their extensive presence in the natural environment. Their capacity for comprehensive mineralisation of contaminants offers a holistic remediation solution. Notably, studies indicate that *Saccharomyces cerevisiae* could potentially eliminate around 65–79% of Pb and Cd from polluted soil. This process of biosorption is driven by mechanisms such as ion exchange, complexation, and adsorption, facilitated by fungal cell walls containing constituents like glucans, lipids, polysaccharides, pigments, and proteins. Functional groups such as carboxyl, sulphate, hydroxyl, phosphate, or amino also play a pivotal role in this process (Kapahi and Sachdeva 2019). The abundant presence of cell wall materials equipped with robust binding capabilities for heavy metals establishes fungi as a superior choice for the detoxification of heavy metals. In particular, the chitin and chitosan found in fungal cell walls aid for the sequestration of heavy metal ions. Fungal species like *Mucor rouxii*, *Saccharomyces cerevisiae*, *Penicillium chrysogenum*, and *Aspergillus niger* are used for elimination of heavy metals like cadmium, copper, zinc, and lead. Similar to bacteria, fungi also create metal-specific proteins and polypeptides in response to HM exposure. These proteins and polypeptides help fungi survive in harsh conditions by binding a significant amount of internal metals like Cd or Cu (Tobin 2001). The fundamental mechanism of metal tolerance in ectomycorrhizal fungi is bioaccumulation, which is accomplished through subcellular compartmentalisation and intracellular chelation by metallothionein and glutathione (GSH). The ectomycorrhizal fungi's mode of action is identical to that of other eukaryotes like algae (Bellion et al. 2006).

13.4.3.4 Bioventing

Bioventing involves a deliberate approach of enhancing air circulation in the unsaturated (vadose) zone by introducing oxygen. This serves to amplify the activity of nearby bacteria, fostering the process of detoxification. The ultimate objective of bioventing is to facilitate the microbial conversion of contaminants into harmless forms. Enhancements are achieved by introducing nutrients and moisture to enhance the detoxification process (Atlas and Philp 2005). With this approach, a contaminated environment is infused with a large amount of air that has undergone environmental degradation. Many different types of soil contaminants, mostly petroleum hydrocarbons and BTEX (benzene, toluene, ethylbenzene, xylene) chemicals, have been successfully treated using this method. Bioventing emerged as a brand-new and practical technology that cleans up the dangerous wastes in the soil of vadose area (Mihopoulos et al. 2002). Bioventing is a cost-effective technique that improves microbial activity.

13.4.3.5 Biostimulation

Every contamination in the soil can be remedied depending on several factors, including temperature, moisture content, pH, soil properties, nutrients, accessible oxygen, and pollutant concentration (Atagana 2008). Biostimulation has the potential to increase the native microorganisms' detoxification abilities in contaminated soil. It entails boosting the growth of the local bacteria by introducing limiting nutrients and electron acceptors like carbon, phosphorus, or nitrogen (Payne et al. 2013). By using the process of biostimulation, the environment of microorganisms can be modified to stimulate their activity. We can carry out this process by adding nutrients such as carbon, nitrogen, oxygen, or phosphorus. The addition of these nutrients brings an increase in the activity of microorganisms that can be used for the detoxification of soil. As a result of their flexibility to the subsurface ecosystem and ideal spatial distribution within the subsurface, the native bacteria existing in the soil cause detoxification, which is a benefit of the biostimulation method (Al-Sulaimani et al. 2010). A study showed that the bacterial consortia that were biostimulated were effective in removing large number of contaminants like cadmium, iron, copper, etc. The carbon source that is used most commonly for biostimulation is molasses. Cheese whey and waste glycerol can also be used as an alternative to molasses. Biostimulation is a cost-effective technique that improves the nutrient content of the soil (Fulekar et al. 2012).

13.4.3.6 Bioaugmentation

Bioaugmentation is described as a method for enhancing the ability of contaminated regions to be removed by introducing particular suitable strains or consortia of microorganisms to the polluted site, hence promoting the process of biodegradation (Derz et al. 2006). The justification for this technology is that the local bacteria might not be able to break down the explosives there or that they might be stressed from exposure to high explosive concentrations. To shorten the organisms' lag phase and, consequently, the explosive concentration, bioaugmentation may be used to accelerate clean-up. A study showed that the bioaugmentation of bacteria *Rhizobium trifolii* resulted in 60% degradation of an explosive named trinitrotoluene (Adams et al. 2015).

13.4.3.7 Bioattenuation

Bioattenuation is the in situ technique used for heavy metal detoxification. This technique makes use of organic processes to lower the concentration of contaminants in contaminated areas and control the spread of contamination from chemical spills. As a result, the environmental toxins are left alone, giving them a chance to naturally degrade, diminish, or convert. The rate of bioattenuation is influenced by a multitude of factors including various contaminants as well as the associated chemical, biological, and physical attributes of the soil and groundwater, as emphasized by Emenike et al. (2018). This process involves reducing the concentration of contaminants through mechanisms like straightforward dilution or dispersion or by binding contaminants to soil particles to restrict their movement and dispersion. As a result, the overall mass of contaminants is diminished, encompassing both biodegradation and chemical transformations. Bioattenuation, being a cost-effective detoxification technique, predominantly relies on natural processes. The process of bioattenuation employs a spectrum of chemical, physical, and biological mechanisms with the objective of reducing the mass, toxicity, volume, or overall presence of pollutants. These procedures involve the transformation of pollutants as well as aerobic and anaerobic decomposition, volatilisation, sorption, and chemical or biological stabilisation (Jan-Roblero et al. 2016).

13.4.4 Phytoremediation

Using specific plants and trees to eliminate, confine, or deliver innocuous environmental toxins is known as phytoremediation (Anju 2017). Usually, plant species with a heightened capacity for accumulating heavy metals exhibit slower growth rates. Nevertheless, their expansive and deeply penetrating root systems offer a unique mechanism for extracting heavy metals from the soil. As suggested by Chen et al. (2019), plants can enhance their resistance to heavy metals by expediting the breakdown of associated phenolic compounds. A study conducted by Manquián-Cerda et al. (2018) revealed that the presence of heavy metals led to an augmentation of phenolic content and antioxidant properties in blueberry plants. These phenolic compounds play a crucial role in mitigating oxidative damage within plants by neutralising reactive oxygen species (ROS) through carboxyl and hydroxyl groups and by forming chelates with metal ions. Employing phytoremediation provides a means to rehabilitate contaminated areas without jeopardising the topsoil. Furthermore, it bolsters the organic content of the soil, thereby fostering plant growth, preventing soil erosion, and curbing the leaching of pollutants (Raskin and Ensley 2000). Plant species utilized for phytoremediation produce a lot of biomasses that can be used to make bioenergy and other products that can be sold after being harvested and processed, which helps to control the costs of implementing the technology (Steliga and Kluk 2020). Plants function akin to solar-driven pumping and filtration systems, drawing in pollutants, particularly those soluble in water, through their root systems. These contaminants are then transported through diverse plant tissues, where they undergo processes such as digestion, sequestration, or volatilisation (Turnau et al. 2006). Hyperaccumulator plants possess a unique capability to not only detoxify but also store heavy metals, enabling them to accumulate substantial levels in aboveground plant parts without triggering phytotoxic effects (Fig. 13.9).

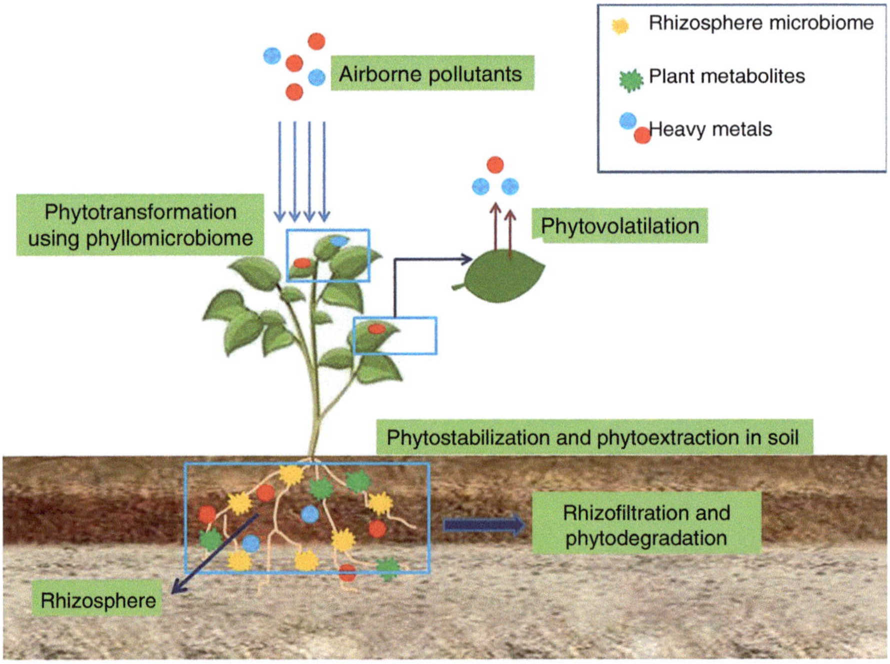

Fig. 13.9 Phytoremediation

Hyperaccumulators possess the ability to sequester harmful heavy metals within the vacuoles of leaf cells in above-ground plant parts, while non-hyperaccumulators store them within the vacuoles of root cells in below-ground plant sections, as highlighted by Sharma et al. (2016). These regions of heavy metal detoxification and sequestration are often strategically located to minimise potential damage to photosynthetic processes. This includes areas such as the cuticle, trichomes, and even the epidermis, as discussed by Shahid et al. (2017). There are many plants that can collect Cd, Zn, Ni, Cr, Pb, and Cu, including *Brassica juncea*, *Arabidopsis thaliana*, *Cynodon dactylon*, *Bidens pilosa*, *Helianthus annuus*, and *Calendula officinalis* (Wei et al. 2018). Several categories have been established within the topic of phytoremediation, including phytoextraction, phytofiltration (rhizofiltration, plant-assisted detoxification, phytostabilisation, phytovolatilisation, phytodegradation (also known as phytotransformation), etc. a number of potential plant species (i.e., *Ipomea carnea*, *Azolla caroliniana*, *Ricinus communis*, *Typha latifolia*, *Saccharum munja*, and *Thelypteris dentata*) have been identified to remediate metalloids from fly ash polluted sites or fly ash dumpsites (Pandey et al. 2014). Phytoremediation can be classified into different categories:

13.4.4.1 Phytoextraction

Plants absorb metals from the soil and transmit them to the branches that can be harvested, where they accumulate. The soil impurities are then removed by harvesting the roots and shoots (Moosavi and Seghatoleslami 2013). It is the process through which roots absorb pollutants from contaminated areas (soil, water, and sediments), move those contaminants, and then deposit them in the portions of the plant that grow above ground. Plants with the genetic capacity to withstand greater quantities of unfavourable pollutants are employed for it (Hasanuzzaman and Fujita 2013). These plants are gathered after being grown on polluted land for a while and then burned to reprocess the toxins. Translocating metal pollutants from contaminated locations using plants is a time-tested and effective method (EPA 2000). Phytoextraction, a biological process carried out by plants, removes contaminants from contaminated areas without adversely changing the properties of the soil, water, or sediments (Yanai et al. 2006).

13.4.4.2 Phytostabilisation

Phytostabilisation is the use of plants to change the mobility and bioavailability of pollutants in the soil, water, and/or sediments in a physical, chemical, and biological manner. Using specific plants is an easy and affordable in situ technique with little environmental incursion (Jadia and Fulekar 2009). The other objectives of phytostabilisation include preventing harmful toxins from entering the food chain as well as their leaching (which can contaminate groundwater) and erosion (which can transport toxic elements by water and/or wind to other areas) (Raskin and Ensley 2000). Phytostabilisation is used for both heavy metals and radioactive compounds.

13.4.4.3 Phytodegradation

"Phytodegradation" refers to the enzymatic alteration of pollutants by plants, leading to their conversion into simpler or less harmful compounds. This transformation can occur either in the rhizosphere before absorption or, in certain cases, within the root after uptake, followed by further synthesis. This detoxification method, also known as phytotransformation, holds significant importance. During phytodegradation in the rhizosphere, plant roots release essential enzymes (Rugh 2001). However, when degradation happens within the plants themselves, the plants must have the ability to absorb pollutants and transport them in their original form to the transformation site, all without causing harm to their cells (Hasanuzzaman and Fujita 2013). Various plant enzymes are involved in this process, including nitrilase, phosphatase, dehalogenase, nitroreductase, and oxidoreductase, to name a few common examples (EPA 2000).

13.4.4.4 Phytovolatisation

Plants collect pollutants from contaminated places through their roots, transform the more harmful elements inside the plant into less harmful, and then discharge these less toxic substances into the atmosphere through their leaves. This process is known as phytovolatilisation (Bizily et al. 1999). The environmental factors at remediation sites and the genetic power of the utilised plant species both affect how effective phytovolatilisation is. In addition, the level of these pollutants at the contaminated locations has a significant impact on their absorption, translocation, and chemical destiny (Meagher 2000).

13.4.4.5 Rhizofiltration

Rhizofiltration is the mechanism through which plant roots absorb contaminants from the rhizosphere. This remediation method decontaminates contaminated aquatic areas by using aquatic or terrestrial plants (Ensley et al. 1995). During the course of this method, plants are cultivated on the contaminated site (in situ) or in an ex situ environment. When plant roots are saturated with contaminants, the plants are finally harvested together with their roots. The likelihood of atmospheric contamination is quite minimal because contaminants only build up in the roots and do not go to the shoots. Some significant plants employed in this method are Indian mustard, rye, sunflower, tobacco, and spinach (Henry 2000). Terrestrial plants have been used extensively in this method due to their quick development and fibrous root systems. These plants have been grown in ex situ environments using designed tank or pipe systems, as well as in situ environments where they are used as floating rafts. Moreover, it has been shown that the introduction of microorganisms to the rhizosphere can promote the uptake of pollutants (Turnau et al. 2006).

13.5 Conclusion and Future Prospective

Combining bioremediation with traditional methods would be considerably more beneficial to get biotechnology for soil decontamination with an efficiency of >95% toxicant removal with incidental advantages. Although biological remediation appears to be a promising method, ensuring its sustainability remains a difficult task. Physical and chemical characteristics of the environment that microbes are exposed to have a big impact on the metabolic paths they take. In order to comprehend microbial degradation mechanisms and their interactions with contaminated soils in various environmental contexts, more research is required. The field of molecular biology can assist us in identifying the crucial genes involved in biodegradation and advancing research. To get over the current obstacles, more cooperation between brilliant minds from other professions is needed, including genetic engineers, biochemists, environmental engineers, and microbiologists. This would enhance current research and give the necessary field new development directions. Many methods such as physical, chemical, and biological have been developed for the decontamination of soil. These methods are beneficial according to their effectiveness in different areas. Treatment of soils depends on cost-effectiveness, restoration performance, time, site of pollution, etc. In the above-mentioned methods, decontamination using bacteria and plants proved to be both economical and environmentally responsible. The strategies described in this chapter are being employed for tackling the polluted soil issue in innovative ways maintaining the sustainable detoxification of polluted soils to reduce the ecological and health risks.

References

Abbasian S, Lockington R, Megharaj M, Naidu R (2016) The biodiversity changes in the microbial population of soils contaminated with crude oil. Curr Microbiol 72:663–670

Adams GO, Fufeyin PT, Okoro SE, Ehinomen I (2015) Bioremediation, biostimulation, and bioaugmentation: a review. Int J Environ Bioremediat Biodegrad 3(1):28–39

Ahmad M (2012) Implications of bacterial resistance against heavy metals in bioremediation: a review. J Inst Integr Omics Appl Biotechnol 3(3):39

Alam GM, Tokunaga S, Maekawa T (2001) Extraction of arsenic in a synthetic arsenic contaminated soil using phosphate. Chemosphere 43(8):1035–1041

Al-Sulaimani H, Al-Wahaibi Y, Al-Bahry S, Elshafie A, Al-Bemani A, Joshi S, Zargari S (2010) Experimental investigation of biosurfactants produced by Bacillus species and their potential for MEOR in Omani oil field. In: SPE EOR conference at oil and gas West Asia

Ambaye TG, Chebbi A, Formicola F, Rosatelli A, Prasad S, Gomez FH, Vaccari M (2022) Ex-situ bioremediation of petroleum hydrocarbon contaminated soil using mixed stimulants: response and dynamics of bacterial community and phytotoxicity. J Environ Chem Eng 10(6):108814

Anju M (2017) Biotechnological strategies for remediation of toxic metal (loid) s from environment. In: Plant biotechnology: recent advancements and developments. Springer, pp 315–359

Ashraf A, Bibi I, Niazi NK, Ok YS, Murtaza G, Shahid M, Kunhikrishnan A, Mahmood T (2016) Chromium (VI) immobilisation efficiency of acid-1 activated banana peel over organo-montmorillonite in aquatic environments. Int J Phytoremed 19:605

Atlas RM, Philp J (2005) Bioremediation: applied microbial solutions for real-world environmental cleanup. ASM Press

Atagana HI (2008) Compost bioremediation of hydrocarboncontaminated soil inoculated with organic manure. Afr J Biotechnol 7(10):1516–1525

Barathi S, Gitanjali J, Rathinasamy G, Sabapathi N, Aruljothi KN, Lee J, Kandasamy S (2023) Recent trends in polycyclic aromatic hydrocarbons pollution distribution and counteracting bio-remediation strategies. Chemosphere 337:139396

Bellion M, Courbot M, Jacob C, Blaudez D, Chalot M (2006) Extracellular and cellular mechanisms sustaining metal tolerance in ectomycorrhizal fungi. FEMS Microbiol Lett 254(2):173–181

Bizily SP, Rugh CL, Summers AO, Meagher RB (1999) Phytoremediation of methylmercury pollution: merB expression in Arabidopsis thaliana confers resistance to organomercurials. Proc Natl Acad Sci 96(12):6808–6813

Caetano G, Machado RDM, Correia MJN, Marrucho IM (2023) Remediation of soils contaminated with total petroleum hydrocarbons through soil washing with surfactant solutions. Environ Technol:1–14. https://doi.org/10.1080/09593330.2023.2198733

Cele EN, Maboeta M (2016) A greenhouse trial to investigate the ameliorative properties of biosolids and plants on physicochemical conditions of iron ore tailings: implications for an iron ore mine site remediation. J Environ Manag 165:167–174

Cerqueira VS, Peralba MR, Camargo FAO, Bento FM (2014) Comparison of bioremediation strategies for soil impacted with petrochemical oily sludge. Int Biodeterior Biodegrad 95:338–345

Chen S, Wang Q, Lu H, Li J, Yang D, Liu J, Yan C (2019) Phenolic metabolism and related heavy metal tolerance mechanism in Kandelia obovata under Cd and Zn stress. Ecotoxicol Environ Saf 169:134–143

Chen X, Kumari D, Cao CJ, Plaza G, Achal V (2020) A review on remediation technologies for nickel-contaminated soil. Hum Ecol Risk Assess Int J 26:571–585

Coulon F, Al Awadi M, Cowie W, Mardlin D, Pollard S, Cunningham C, Risdon G, Arthur P, Semple KT, Paton GI (2010) When is a soil remediated? Comparison of biopile and windrowed soils contaminated with bunker-fuel in a full-scale trial. Environ Pollut 158:3032–3040

Das N, Chandran P (2011) Microbial degradation of petroleum hydrocarbon contaminants: an overview. Biotechnol Res Int 2011:1–30

De J, Ramaiah N, Vardanyan L (2008) Detoxification of toxic heavy metals by marine bacteria highly resistant to mercury. Mar Biotechnol 10:471–477

Deepika, Haritash AK (2023) Phytoremediation potential of ornamental plants for heavy metal removal from contaminated soil: a critical review. Hortic Environ Biotechnol 64:1–26

Dellisanti F (2016) In-field remediation of tons of heavy metal-rich waste by Joule heating vitrification. Int J Miner Process 93:239–245

Derz K, Schmidt B, Schwiening S, Schuphan I (2006) Comparison of microbial pyrene and benzo [a] pyrene mineralization in liquid medium, soil slurry, and soil. J Environ Sci Health B 41(5):471–484

Doyle E, Blanchon D, Wells S, de Lange P, Lockhart P, Waipara N, Berry TA (2023) Internal transcribed spacer and 16s amplicon sequencing identifies microbial species associated with asbestos in New Zealand. Genes 14(3):729

Ehsan S, Prasher SO, Marshall WD (2007) Simultaneous mobilization of heavy metals and polychlorinated biphenyl (PCB) compounds from soil with cyclodextrin and EDTA in admixture. Chemosphere 68:150–158

Emenike CU, Jayanthi B, Agamuthu P, Fauziah SH (2018) Biotransformation and removal of heavy metals: a review of phytoremediation and microbial remediation assessment on contaminated soil. Environ Rev 26(2):156–168

Ensley B, Dushenkov V, Raskin I, Salt DE (1995) Rhizofiltration: a new technology to remove heavy metals from aqueous streams. In: New remediation technology in the changing environmental arena, p 4

Ent AV, Baker AJM, Reeves RD, Pollard AJ, Schat H (2013) Hyperaccumulators of metal and metalloid trace elements: facts and fiction. Plant Soil 362(1–2):319–334

Ferraro A, van Hullebusch ED, Huguenot D, Fabbricino M, Esposito G (2015) Application of an electrochemical treatment for EDDS soil washing solution regeneration and reuse in a multistep soil washing process: case of a Cu contaminated soil. J Environ Manag 163:62–69

Fulekar MH, Sharma J, Tendulkar A (2012) Bioremediation of heavy metals using biostimulation in laboratory bioreactor. Environ Monit Assess 184:7299–7307

Gao N, Wang F, Quan C, Santamaria L, Lopez G, Williams P (2022) Tire pyrolysis char: processes, properties, upgrading and applications. Prog Energy Combust Sci 93:1–37

Hasanuzzaman, M., & Fujita, M. 2013. Heavy metals in the environment: current status, toxic effects on plants and phytoremediation. Phytotechnologies—remediation of environmental contaminants. Edited by NA Anjum, ME Pereira, I. Ahmad, AC Duarte, S. Umar, and NA Khan. CRC Press, Boca Raton, 7–73.

Havugimana E, Bhople BS, Kumar A, Byiringiro E, Mugabo JP, Kumar A (2017) Soil pollution–major sources and types of soil pollutants. Environ Sci Eng 11:53–86

He J, Chen JP (2014) A comprehensive review on biosorption of heavy metals by algal biomass: materials, performances, chemistry, and modeling simulation tools. Bioresour Technol 160:67–78

Henry JR (2000) An overview of the phytoremediation of lead and mercury. US Environmental Protection Agency, Office of Solid Waste and Emergency Response, Technology Innovation Office, Washington, DC, pp 1–31

Hodson ME, Valsami-Jones É, Cotter-Howells JD (2000) Bonemeal additions as a remediation treatment for metal contaminated soil. Environ Sci Technol 34:3501–3507

Issaka E, Fapohunda FO, Amu-Darko JNO, Yeboah L, Yakubu S, Varjani S, Bilal M (2022) Biochar-based composites for remediation of polluted wastewater and soil environments: challenges and prospects. Chemosphere 297:134163

Jadia CD, Fulekar MH (2009) Phytoremediation of heavy metals: recent techniques. Afr J Biotechnol 8(6):921

Jan-Roblero J, Vasquez-Murrieta M, Hernandez-Hernandez O, Cruz-Maya J, Cancino-Diaz JC (2016) Approaches for removal of PAHs in soils: bioaugmentation, biostimulation and bioattenuation. InTech

Kapahi M, Sachdeva S (2019) Bioremediation options for heavy metal pollution. J Health Pollut 9(24):191203

Kinoshita H, Swift P, Utton C, Carro-Mateo B, Marchand G, Collier N, Milestone N (2013) Corrosion of aluminium metal in OPC- and CAC-based. Cement Concrete Res 50:11–18

Kos B, Leštan D (2003) Induced phytoextraction/soil washing of lead using biodegradable chelate and permeable barriers. Environ Sci Technol 37:624–629

Li J, Zhang GN, Li Y (2010) Review on the remediation technologies of POPs. Hebei Environ Sci 65:1295–1299

Liao X, Li Y, Yan X (2015) Removal of heavy metals and arsenic from a co-contaminated soil by sieving combined with washing process. J Environ Sci 41:1–9

Liu L, Song Z, Tang J, Li Q, Sarkar B, Ellam RM, Wang H (2023) New insight into the mechanisms of preferential encapsulation of metal (loid) s by wheat phytoliths under silicon nanoparticle amendment. Sci Total Environ 875:162680

Lukić B, Panico A, Huguenot D, Fabbricino M, Van Hullebusch ED, Esposito G (2017) A review on the efficiency of landfarming integrated with composting as a soil remediation treatment. Environ Technol Rev 6(1):94–116

Madhav S, Mishra R, Kumari A, Srivastav AL, Ahamad A, Singh P, Sillanpää M (2023) A review on sources identification of heavy metals in soil and remediation measures by phytoremediation-induced methods. Int J Environ Sci Technol. https://doi.org/10.1007/s13762-023-04950-5

Manquián-Cerda K, Cruces E, Escudey M, Zúñiga G, Calderón R (2018) Interactive effects of aluminium and cadmium on phenolic compounds, antioxidant enzyme activity and oxidative stress in blueberry (Vaccinium corymbosum L.) plantlets cultivated in vitro. Ecotoxicol Environ Saf 150:320–326

Mastnak T, Cenčič Predikaka T, Svoljšak Jerman M, Finšgar M (2023) Ex situ bioremediation of diesel fuel-contaminated soil in two different climates. Int J Phytoremediation 25:1–9

Meagher RBP (2000) Phytoremediation of toxic elemental and organic pollutants. Curr Opin Plant Biotechnol 3(2):153–162

Meagher RB, Rugh CL, Kandasamy MK, Gragson G, Wang NJ (2020) Engineered phytoremediation of mercury pollution in soil and water using bacterial genes. In: Phytoremediation of contaminated soil and water. CRC Press, pp 201–219

Mihopoulos PG, Suidan MT, Sayles GD, Kaskassian S (2002) Numerical modeling of oxygen exclusion experiments of anaerobic bioventing. J Contam Hydrol 58(3–4):209–220

Moosavi SG, Seghatoleslami MJ (2013) Phytoremediation: a review. Adv Agric Biol 1(1):5–11

Mulligan CN, Yong RN, Gibbs BF (2001) Remediation technologies for metal-contaminated soils and groundwater: an evaluation. Eng Geol 60(1–4):193–207

National Risk Management Research Laboratory (US) (2000) Introduction to phytoremediation. National Risk Management Research Laboratory, Office of Research and Development, US Environmental Protection Agency

Navarro A, Cardellach E, Cañadas I, Rodríguez J (2013) Solar thermal vitrification of mining contaminated soils. Int J Miner Process 119:65–74

Niazi NK, Murtaza B, Bibi I, Shahid M, White JC, Nawaz MF, Bashir S, Murtaza G (2016) Removal and recovery of metals by biosorbents and biochars derived from biowastes. In: Environmental materials and waste: resource recovery and pollution prevention. Academic Press

Ojuederie OB, Babalola OO (2017) Microbial and plant-assisted bioremediation of heavy metal polluted environments: a review. Int J Environ Res Public Health 14(12):1504

Ondon BS, Li S, Zhou Q, Li F (2021) Sources of antibiotic resistant bacteria (ARB) and antibiotic resistance genes (ARGs) in the soil: a review of the spreading mechanism and human health risks. Rev Environ Contam Toxicol 256:121–153

Pandey B, Kinrade SD, Catalan LJJ (2012) Effects of carbonation on the leachability and compressive strength of cement-solidified and geopolymer-solidified synthetic metal wastes. J Environ Manag 101:9–67

Pandey VC, Singh N, Singh RP, Singh DP (2014) Rhizoremediation potential of spontaneously grown Typha latifolia on fly ash basins: study from the field. Ecol Eng 71:722–727

Park B, Son Y (2017) Ultrasonic and mechanical soil washing processes for the removal of heavy metals from soils. Ultrason Sonochem 35:640–645

Payne ZM, Lamichhane KM, Babcock RW, Turnbull SJ (2013) Pilot-scale in situ bioremediation of HMX and RDX in soil pore water in Hawaii. Environ Sci Processes Impacts 15(11):2023–2029

Priya AK, Muruganandam M, Ali SS, Kornaros M (2023) Clean-up of heavy metals from contaminated soil by phytoremediation: a multidisciplinary and eco-friendly approach. Toxics 11(5):422

Qian SQ, Liu Z (2000) An overview of development in the soil-remediation technologies. Chem Ind Eng Process 4:10–20

Qin G, Niu Z, Yu J, Li Z, Ma J, Xiang P (2021) Soil heavy metal pollution and food safety in China: effects, sources and removing technology. Chemosphere 267:129205

Raskin I, Ensley BD (2000) Phytoremediation of toxic metals. John Wiley and Sons

Raza T, Ali I, Khan N, Eash NS, Qadir MF, Jatav HS (2023) Detoxification of sewage sludge by natural attenuation and application as a fertilizer. In: Environmental pollution impact on plants. Apple Academic Press, pp 245–270

Ren Z, Wang L, Wang H, Liu S, Liu M (2023) Solidification/stabilization of lead-contaminated soils by phosphogypsum slag-based cementitious materials. Sci Total Environ 857:159552

Rugh CL (2001) Mercury detoxification with transgenic plants and other biotechnological breakthroughs for phytoremediation. In Vitro Cell Dev Biol Plant 37:321–325

Saeed MU, Hussain N, Javaid M, Zaman H (2023) Microbial remediation for environmental cleanup. In: Advanced microbial technology for sustainable agriculture and environment. Academic Press, pp 247–274

Shahid M, Dumat C, Khalid S, Schreck E, Xiong T, Niazi NK (2017) Foliar heavy metal uptake, toxicity and detoxification in plants: a comparison of foliar and root metal uptake. J Hazard Mater 325:36–58

Sharma SS, Dietz KJ, Mimura T (2016) Vacuolar compartmentalization as indispensable component of heavy metal detoxification in plants. Plant Cell Environ 39(5):1112–1126

Singh SK, Haritash AK (2019) Polycyclic aromatic hydrocarbons: soil pollution and remediation. Int J Environ Sci Technol 16:6489–6512

Speight JG (2017) Sources and types of organic pollutants. In: Environmental organic chemistry for engineers, pp 153–201

Steliga T, Kluk D (2020) Application of Festuca arundinacea in phytoremediation of soils contaminated with Pb, Ni, Cd and petroleum hydrocarbons. Ecotoxicol Environ Saf 194:110409

Stepanova AY, Gladkov EA, Osipova ES, Gladkova OV, Tereshonok DV (2022) Bioremediation of soil from petroleum contamination. Processes 10(6):12–24

Streche C, Istrate IA, Badea AA (2018) Decontamination of petroleum-contaminated soils using the electrochemical technique: remediation degree and energy consumption. Sci Rep 8:32–72

Sun L, Wu Q, Liao K, Yu P, Cui Q, Rui Q, Wang D (2016) Contribution of heavy metals to toxicity of coal combustion related fine particulate matter (PM2.5) in *Caenorhabditis elegans* with wild-type or susceptible genetic background. Chemosphere 144:2392–2400

Tobin JM (2001) Fungal metal biosorption. In: British mycological society symposium series, vol 23, pp 424–444

Tokunaga S, Hakuta T (2002) Acid washing and stabilization of an artificial arsenic-contaminated soil. Chemosphere 46(1):31–38

Torres LG, Lopez RB, Beltran M (2012) Removal of As, Cd, Cu, Ni, Pb, and Zn from a highly contaminated industrial soil using surfactant enhanced soil washing. Phys Chem Earth 37–39:30–36

Trellu C, Pechaud Y, Oturan N, Mousset E, van Hullebusch ED, Huguenot D, Oturan MA (2021) Remediation of soils contaminated by hydrophobic organic compounds: how to recover extracting agents from soil washing solutions? J Hazard Mater 404:124137

Tsvetnov EV, Marakhova NA, Makarov OA, Strokov AS, Abdulkhanova DR (2019) Experience in approbation of societal land value as a basis for ecological and economic assessment of damage from land degradation. Eur Soil Sci 52:1298

Turnau K, Orlowska E, Ryszka P, Zubek S, Anielska T, Gawronski S, Jurkiewicz A (2006) Role of mycorrhizal fungi in phytoremediation and toxicity monitoring of heavy metal rich industrial wastes in southern Poland. In: Soil and water pollution monitoring, protection and remediation. Springer, Dordrecht, pp 533–551

Ucaroglu S, Talinli İ (2012) Recovery and safer disposal of phosphate coating sludge by solidification/stabilization. J Environ Manag 105:131–137

Uddin MJ, Aditya Saga G, Jagdeeshwar J (2017) Soil pollution and soil remediation techniques. Int J Adv Res Ideas Innov Technol 3(1):582–593

Ullah S, Shahid M, Zia-Ur-Rehman M, Sabir M, Ahmad HR (2015) Phytoremediation of Pb-contaminated soils using synthetic chelates. In: Soil remediation and plants. Elsevier Inc.

Vaverková MD, Maxianová A, Winkler J, Adamcová D, Podlasek A (2019) Environmental consequences and the role of illegal waste dumps and their impact on land degradation. Land Use Policy 89:104234

Venderbosch RH, Prins W (2010) Fast pyrolysis technology development. Biofuels Bioprod Biorefin 4(2):178–208

Venegas A, Rigol A, Vidal M (2015) Viability of organic wastes and biochar as amendments for the bioremediation of heavy metal contaminated soils. Chemosphere 119:190–198

Virkutyte J, Sillanpaa M, Latostenmaa P (2002) Electrokinetic soil remediation—critical overview. Sci Total Environ 289:97–121

Wang Q, Shaheen SM, Jiang Y, Li R, Slaný M, Abdelrahman H, Kwon E, Bolan N, Rinklebe J, Zhang Z (2021) Fe/Mn- and P-modified drinking water treatment residuals reduced Cu and Pb phytoavailability and uptake in a mining soil. J Hazard Mater 403:123628

Wei H, Huang M, Quan G, Zhang J, Liu Z, Ma R (2018) Turn bane into a boon: application of invasive plant species to remedy soil cadmium contamination. Chemosphere 210:1013–1020

Yanai J, Zhao F-J, McGrath SP, Kosaki T (2006) Effect of soil characteristics on Cd uptake by the hyperaccumulator Thlaspi caerulescens. Environ Pollut 139(1):167–175

Yao Z, Li J, Xie H, Yu C (2012) Review on remediation technologies of soil contaminated by heavy metals. Proc Environ Sci 16:722–729

Zheng C, Wang PP (2002) A field demonstration of the simulation optimization approach for remediation system design. Ground Water 40:258–265

Zhu L, Ding W, Feng L, Kong J, Xu Y, Xu J, Yang X (2012) Isolation of aerobic denitrifiers and characterization for their potential application in the bioremediation of oligotrophic ecosystem. Bioresour Technol 108:1–7

Chapter 14
Nano-biotechnology and Its Applications in Maintaining Soil Health

Aparajita Sharma, Aditya Banyal, Naimi Sirjohn, Saurabh Kulshreshtha, and Pradeep Kumar

14.1 Introduction

According to the analysis of the United Nations Department of Economic and Social Affairs Population Division, there will be a rise in population of approximately 9–10 billion by the year 2050 (https://population.un.org/wpp/Graphs/DemographicProfiles/Line/900). Increased population means increased food requirements for which products of the crop (higher yields) should have to be doubled for achieving the Sustainable Development Agenda by 2030 (Cuevas et al. 2019), and by 2050, it is necessary to increase the production of grains by almost a billion tonnes annually (De Vries et al. 2023). Being the foundation of the economy and a source of food security, agriculture is a way of life, a tradition, and a source of general opportunities for human livelihood (Singh et al. 2019). Annually, 25% of anthropogenic greenhouse gas emissions are released by the global food system, causing terrestrial acidification (one-third) and global surface water eutrophication. If the use of synthetic pesticides, fossil fuels, food waste production, and artificial fertilizers continues, then the planet's capacity will soon be outpaced (Schreefel et al. 2020). Global sustainability is likely to be threatened by the upcoming load on natural resources because of increasing population, global climatic changes, fertile agricultural acreage loss, and soil deterioration. Crop production on agricultural acreages is sustainable based on the variables of soil quality (an intrinsic characteristic), which is further assessed by soil health (an intrinsic characteristic) (Tahat et al. 2020). Agriculture sustainability has to go hand in hand with sustainable use and care of soil for the development of agriculture to decrease soil degradation, loss

A. Sharma · A. Banyal · N. Sirjohn · S. Kulshreshtha · P. Kumar (✉)
Faculty of Applied Sciences and Biotechnology, Shoolini University of Biotechnology and Management Sciences, Solan, Himachal Pradesh, India
e-mail: pradeep.kumar@shooliniuniversity.com

R. K. Bhatia, A. Walia (eds.), *Advancements in Microbial Biotechnology for Soil Health*, Microorganisms for Sustainability 50,
https://doi.org/10.1007/978-981-99-9482-3_14

of ecosystem services, and productivity (Montanarella et al. 2016). Unexpected technological manufacturing issues, a lack of marketing expertise, and insufficient preclinical safety/risk management are some of the major hurdles encountered during research and development efforts (Seth et al. 2023).

Therefore, there is a need to address these pertinent issues by applying new, emerging eco-friendly strategies and technologies. Before now, nanotechnology has been used in a variety of industries, including chemistry, pharmaceutical research, diagnosis, and therapies. A cutting-edge technology of the century ahead is nanobiotechnology. Rising nanobiotechnology blends the salient characteristics of nanotechnology and biomolecules in live cells. Nanotechnology's concepts and traits could be applied to biological mechanisms to create new goods. Products and processes including membranes and filters, nanobiosensors, nanofluids, and nanoproteomics serve as examples of this (Sahu et al. 2021). Particles that are tiny, between 1 and 100 nm in size, and that are either naturally occurring or artificially created and display markedly distinct chemical and physical characteristics from the larger ones are known as nanoparticles (Chugh et al. 2021). Due to their enormous potential in comparison to their bulk counterparts, nanoparticles (NPs), at the cutting edge of nanotechnology, are used in a variety of consumer items. In light of this, the food and agriculture industries have paid close attention to the distinctive and emerging qualities of these new-age materials (Mittal et al. 2020). Detection of soil atmosphere parameters, namely, moisture content, pH, temperature, pathogenic interactions, and availability of nutrients by nano-sensors, will further help for their timely treatment, soil health update, and least deprivation of crop production. The use of chemical pesticides and fertilizers harms the soil quality and efficiency of fertilizers, thus generating serious issues correlated to environmental impact, sustainability, and health problems (Singh et al. 2018; Rajput et al. 2021). Different environmental toxins from the soil may be converted and detoxified through bioremediation with nanoparticles. The use of nanoparticles can increase the impacts of bioremediation by speeding up the rate at which pollutants are taken up and accumulated by plants as well as how quickly they are broken down by microorganisms. The detrimental effects of nanoparticles on organisms inevitably have an impact on the bioremediation process in soil. In order to increase the efficacy of soil restoration, nanoparticles must be carefully examined in conjunction with biotechnology to determine their potential for interacting with plants and bacteria in contaminated soils, their likely fate, and whether nanoparticles have an impact on pollution (Aliyari Rad et al. 2023). In the majority of domains, phytoremediation, which is well known in the USA and Europe, is used to remove pollutants from soil using highly engineered nanomaterials. Recently, a strategy for cleaning up contaminated land was described that uses the proper nanomaterials like the ability of nano-zerovalent iron (nZVI), bimetallic nanoparticles (Pd/Fe), and magnetite nanoparticles (nFe_3O_4) to degrade toxic organic from the polluted soil including chlorpyrifos, lindane, trichloroethylene, polychlorinated biphenyls, pyrene, 2,4-dinitrotoluene, atrazine, pentachlorophenol, and ibuprofen (Biswal 2023). The development of nanoparticle-mediated herbicides, fertilizers, and pesticides can enhance the soil's capability for the production of healthy and efficient crops. Hence, the main aim of this chapter is to highlight the major role of nano-biotechnology in the field of soil improvement, following the future opportunities to be dealt with.

14.2 Soil for Agriculture

The sophisticated system of soil, which is at the meeting point of the atmosphere, hydrosphere, biosphere, and lithosphere, is essential for sustainability and essential for food production since it supports crucial societal and ecosystem functions (Brevik et al. 2015; Baveye et al. 2016). When soil is used as humans would like, its extrinsic quality changes. It might have to do with agricultural productivity and the ability to support wildlife, safeguard watersheds, or produce recreational goods (Lichtfouse et al. 2009). Several additional similar terms emphasize the function or role of soil in the ecosystem and agriculture, such as soil fertility and soil quality (Bünemann et al. 2018). In terms of soil fertility, crop yield, and environmentally friendly agriculture practices, soil organic matter is crucial (Johnston et al. 2009). To supply the world's population with food, agriculture has always been a potent tool. The population is expanding quickly right now; thus agriculture must be able to keep up with the demand. Thus, it is critical to raise crop productivity and quality (Sansinenea 2021; Zambrano-Mendoza et al. 2021). Therefore, it is important to manipulate the.

For the plant to develop and grow properly, the soil where interactions between the plant and its surroundings take place must be of sufficient quality (Ortiz and Sansinenea 2021). The term soil quality which originally meant soil health refers to a soil's capacity to support agriculture and its immediate surroundings. Hence, the health of plants, animals, and entire ecosystems is all impacted by soil quality (Doran et al. 1994). Millions of fungi, billions of bacteria, and other macro-organisms can be found in soil, making it the most varied and intricate ecosystem on earth. The soil's microorganisms are crucial for the cycling of nutrients and for protecting plants from the negative impacts of biotic and abiotic stress (Bardgett and Van Der Putten 2014; Kumar et al. 2010). The degree of compaction, the capacity for water storage, and the ease of drainage are all important factors in the health of soil and plants. Salt content, pH, and nutrient availability are a few of the crucial chemical indicators of soil health (Usharani et al. 2019). High-yielding, superior-quality, and healthy crops can grow on healthy soil. Crop yields are higher when the soil is healthy because the roots can spread out easily, there is enough water entering and being stored in the soil, there is a sufficient supply of nutrients, there are no harmful chemicals in the soil, and beneficial organisms are very active and able to control potentially harmful ones while also promoting plant growth (Fu et al. 2021). Many ecosystem services that support human needs are provided by soils. These include provisioning services, which include the supply of food, wood, fibre, raw materials, and physical support for infrastructure; regulating services, which include flood mitigation, nutrient and contaminant filtering, waste detoxification and recycling, and population control of pests and diseases, carbon storage, and greenhouse gas regulation; and cultural services, which include recreation, aesthetics, heritage values, and cultural identity (Dominati et al. 2014).

14.3 Significant Soil Activities That Affect Crop Output, Human Health, and Environmental Quality

14.3.1 *Sustainable Plant Production*

Since it influences water use and quality, human and animal health, climate, and biodiversity, plant production, the primary objective of intensive agriculture, is a key area of soil health management (Jian et al. 2020). Soil health takes into account interactions between plants and the microbial populations surrounding roots in addition to controlling the physical, chemical, and biological aspects of soil for plant development (Berendsen et al. 2012). Rhizosphere-dwelling bacteria and fungi can act as bio-fertilizers that aid in the tolerance of biotic and abiotic stress and boost the nutrition of host plants. This promotes the growth and development of plants. As many microorganisms destroy insects and other pests that endanger crops, they can also serve as biopesticides. Moreover, dangerous organic and inorganic compounds that build up in the soil as contaminants as a result of many activities, including agriculture production, can be degraded and detoxified by microbes. They execute bioremediation, which is beneficial to the health of the soil and plants (Medfu Tarekegn et al. 2020). Figure 14.1 shows the significance of beneficial soil microbes and their interactions in the advancement of sustainable agriculture and the environment.

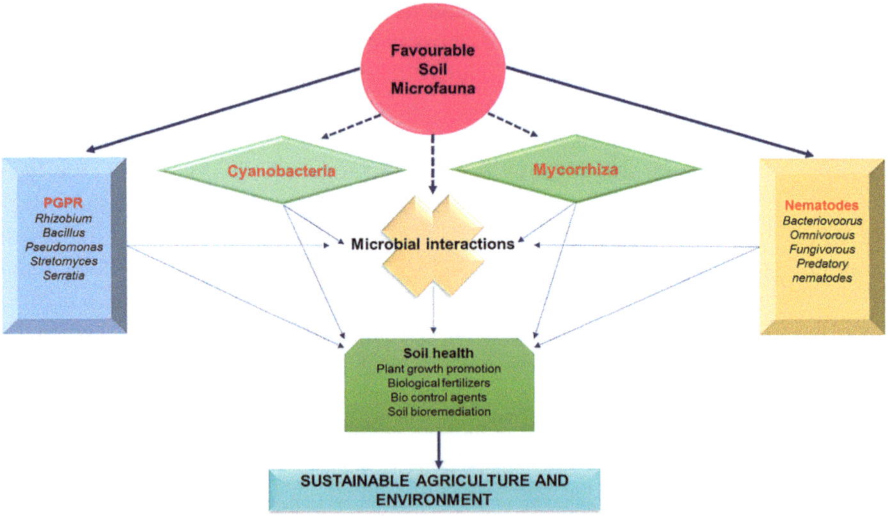

Fig. 14.1 Significance of beneficial soil microbes and their interactions in the advancement of sustainable agriculture and the environment

14.3.2 Human Health

In addition to and beyond the evident relationship between soil and human health through food production, human health is heavily dependent on soil health. Human health can suffer from the effects of soil. For instance, crops can become contaminated by soil pollutants by direct contact, dust, suspension, or rain splash. Certain substances, including arsenic and the majority of inorganic contaminants, can also be absorbed by the root system and build-up in grains or fruits (Oliver and Gregory 2015). Many ecosystem services, including water filtration, nutrient cycling, carbon storage, pest control, and soil stabilization, which are all crucial worldwide issues for human well-being, can be facilitated by soil biodiversity (Elramady et al. 2022). In general, determining the components of soil health is crucial for the effective use of national and international agricultural monitoring systems and consequently for the sustainable growth of our agricultural systems (Tahat et al. 2020).

14.4 Nano-delivery System: Current Scenario (in Agriculture)

The emergence of nanobiotechnology as an interdisciplinary field of study is the result of the integration of biological sciences, surface science, chemistry of organic compounds, the field of proteomics, pharmaceutical products, molecular science, and semiconductor physics with biotechnology in general and nanotechnology. Fundamentally, the newly created, efficient tools and procedures of nanotechnology have transformed the entire scientific sector, including agriculture. As a result, a new frontier in modern agriculture is now being presented, and due to its enormous potential, it has the potential to become a big push (Tripathi and Prakash 2023). Nanotechnology is a cutting-edge, developing discipline of research that examines changes to matter at a specific size scale of 1–100 nm. Nanoparticles are essentially tiny metal-based matter particles with the power to combine systems of different active principle-consisting particles to conduct functioning within an experimental organism as a means of improving the functioning of the chosen organism in all respects (Baker et al. 2015). Products made from nanotechnology that are used in agriculture are essential for improving plant development and crop productivity. They are a common substitute for conventional pesticides and approaches for farmers because of several features, including their small size, ease of handling, long-term storage, high effectiveness, and non-toxicity (Singh et al. 2021). The extensive use of agricultural pesticides for crop protection and maximum crop output has a negative influence on the environment and causes many health problems, some of which are even life-threatening for humans and other animals (Sangeetha et al. 2019). For the administration of fertilizers and pesticides as well as for plant development mediated by genetic engineering, delivery systems are crucial in agriculture. Agrochemical formulations based on nanomaterials or nano-sized preparations have the potential to deliver chemical and biological insecticides effectively. In comparison to commonly used pesticides and their formulations, the advantages of nanomaterial-based formulations include improved efficacy due to higher surface area, higher solubility, induction of systemic activity due to smaller particle size and higher mobility, and lower toxicity due to the elimination of organic solvents (Sasson et al. 2007). Sometimes referred to as "smart delivery systems," nanomaterials have a wide range of special uses. They have distinctive qualities such as a high surface-to-volume ratio, which increases holding capacity, and small size, which facilitates simple entry into cells (Solanki et al. 2015; Kah et al. 2018). The administration of nanoagrochemicals as nano-pesticides or nano-fertilizers for agriculture has the ability to fundamentally alter productivity in agriculture and safeguard biodiversity since they have beneficial effects on plant health and vitality (An et al. 2022). Recently, *Trichoderma harzianum* filtrate was used to create organic titanium nanoparticles that were crucial in promoting the growth of *T. harzianum* or maintaining structures vital for biological control, suggesting that this may be a key tactic in fostering the growth of biocontrol organisms to support more sustainable agriculture (Pasquoto-Stigliani et al. 2023). Table 14.1 gives insight into the approaches that nanomaterials are being employed to boost agricultural productivity and efficiency.

14.5 Role of Nanotechnology in Soil Health/Applications

Due to their improved reactivity, using nanoparticles and nanotechnology can reduce the application of large volumes of pesticides at a reduced cost while enhancing yields. These nanoparticles work well against fungi, insects, and nematodes such as fungicides, herbicides, and nematicides (Verger and Boobis 2013). With the use of suitable nano-pesticide formulations (Pimentel 1995) and the solubility rate qualities of nanoparticles, the poorly soluble active ingredient can be made more soluble. This aids the active ingredient's slow and effective release which lasts for a longer period (Hu et al. 2015). Targeting a particular gene of interest is made easier by the development of nano-encapsulated pesticide formulations (Chareonviriyaphap et al. 2013), which have indistinguishable stability and specificity. Moreover, it shields/encapsulates the active ingredient against premature deterioration, improving its ability to control pests. Karate ZEON and Ospray's chyella (Gouin 2004) are two significant micro-encapsulated nano-pesticides/nano-emulsions that are frequently employed in the manufacturing of organic nanoparticles containing active agrochemicals.

Hence, nanoparticles have a comparative advantage over regular pesticides because the latter are accountable for absorption in plants and animals, which leads to them entering the food chain and producing a variety of health problems (Tran et al. 2017) like Parkinson's disease, cancer, and neurological problems. The most

Table 14.1 Improved agricultural productivity and practices using nanoparticles

Nanomaterial	Mechanism	Applications	Advantages	Reference
Silver nanoparticles	Size on the nanometre scale determines several properties	Defence against plant diseases, nanobactericides, nano-fungicides, and nano-insecticides	Their effectiveness is enhanced by size, shape, and composition, bacterial phyla enrichment with silver resistance property, and alteration in the composition of soil microbiome	Rosarin and Mirunalini (2011) and Ahmed et al. (2023)
Carbon nanotubes	Put appropriate molecules within the seed	Defence against plant diseases by measuring the hydrogen peroxide levels and transmitting chemical signals	Combined mechanical, chemical, and bio-compatible qualities that are incomparable	Gandhi et al. (2010), Lobo et al. (2011), and Bala et al. (2023)
Nano-sensors	The key to creating robust and well-structured electrode materials is an effective biomolecule immobilization without affecting bioactivity	To identify and measure minute levels of pollutants, such as viruses, bacteria, poisons, and biohazardous materials, increase soil quality and soil fertility	Extremely sensitive, quick to react, nutrient runoff reduction, water conservation, provide proper optimized soil data including moisture, pH, and nutrient level	Tothill (2011) and Yadav et al. (2023)
Nonviral gene transfer through nanoparticles	DNA delivery using polymeric systems and therapeutic ultrasound	Delivering gene	Potentially safe	Shih et al. (2015)
Nanoemulsion	By encapsulating functional substances within their droplets, nanoemulsions can reduce the rate of chemical deterioration	To supply nutrients and treat plant infections	A broad spectrum of particle sizes at the nanoscale	Díaz-Blancas et al. (2016)
Mesoporous silica nanoparticles	Chemicals, genes, or herbicides as magic bullets	Introducing chemicals and DNA into isolated plant cells	Large surface area and pore volume; uniform pore size distribution	Douroumis et al. (2011) and Patil et al. (2011)

crucial area of nanomedicine, whose size can range from 1 to 10 nm, is presently nanotechnology (Mirza and Siddiqui 2014). They are preferred to conventional medicine because they overcome the drawback of conventional drugs' poor bioavailability (Nene et al. 2021). As a result, nanoparticles provide a good carrier that, when combined with ligands, increases therapeutic selectivity and minimizes drug toxicity (Large et al. 2021) and side effects at nearby sites. Moreover, magnetic nanoparticles are becoming more popular since they can be used for illness diagnosis and treatment (Attwood et al. 2021). For instance, Ferumoxytol is the iron oxide that has been the subject of most clinical trials and is being used to treat various cancers (Sato et al. 2008) and the diseases associated with them. Numerous more clinical trials are being investigated for diseases outside cancer. For instance, a different strategy is being used to treat hepatitis-B. Liposomal siRNA (Bobbin and Rossi 2016) targets three genes in a genome to enable the suppression of viral mRNA transcripts. The following list summarizes several further future techniques utilized in the soil health and clinical sector (Table 14.2).

14.6 Clinical or Lab Trials of Nano-deliveries for Improving Soil Health

Nanotechnology has been applied in many different sectors, including medicine, chemistry, and agriculture, among others. As a result, using a nanoparticle-based product has been demonstrated to be a revolutionary breakthrough in the field of modern agriculture (Wang et al. 2017). This is because of their high level of reactivity and big surface area to volume ratio (Kolenčík et al. 2020) and highly reactive nature. The increasing population worldwide results in a higher demand for calories, placing significant pressure (Liu and Lal 2015) on the agri-food industry. Many experts believe that nanotechnology has the potential to significantly enhance agri-food security, thereby boosting (Fatima et al. 2021) agricultural productivity. The main objective of utilizing nanomaterials in agriculture is to enhance crop productivity through effective pest control and fertilization using nano-sensors, nano-pesticides, and nano-fertilizers (Table 14.3). The encapsulation of nutraceuticals (Jakhar et al. 2022) and food preservation, as well as the encapsulation of fertilizers and pesticides, are a few important applications that will be covered in this section.

14.6.1 Interaction of Nanoparticles with Plants

The roots of the plant or the aerial section of the plant are the two entry points for nanoparticles into the plant, where the aerial portion is naturally exposed to several nanomaterials (Anjum et al. 2016). Unfortunately, these nanomaterials are unable to overcome the plant membrane barrier because of the presence of a natural barrier,

Table 14.2 Mechanistic action of nanoparticle's driven properties

Properties	Technologies enabled by nanoparticles	Mechanism of action	References
Nano-sized formulations of mineral micronutrients can enhance their bioavailability, reduce soil absorption, and increase the solubility and dispersion of insoluble minerals in the soil	Natural zeolite—the addition of amino acids, vitamin C, and natural aluminosilicate to the soil increased the yield of spinach plants and enhanced the bioavailability of iron not only in the soil but also in surrounding plants	The application of zeolite coated with vitamin C in spinach plants reduced the release of ammonium and potassium from the soil and enhanced the bioavailability of iron in the soil	Li et al. (2013)
Nanostructured formulations have the potential to control the release rate and pattern of nutrients in the soil, which can extend the duration of nutrient input from fertilizers	According to Yuvaraj and Subramanian, nano-zeolite showed higher adsorption of Zn compared to $ZnSO_4$, and the release of Zn from nano-zeolite lasted for up to 800 h, while that from $ZnSO_4$ was observed for up to 200 h		Li et al. (2013)
Controlled-release modes	Semipermeable membranes can be enclosed in envelope forms and coated with resin polymer, waxes, and sulphur to precisely control the rate and pattern of nutrient release for water-soluble fertilizers		Babadi et al. (2015)
Controlling plant pests	A combination of nanoparticles including CuO, ZnO, $Mg(OH)_2$, and MgO was found to be effective in increasing the mortality rate of green peach aphids	As green aphids are considered the key pest on peaches, the nanomaterial directly attaches to the enzyme carboxyl transferase thus inhibiting its growth. These nanomaterials are gaining recognition as aphid resistance has become a major concern	Madanat et al. (2016)
Decrease metal accumulation in plants	The use of a foliar spray containing nanoparticles of Mg-OH on green sweet pepper resulted in no observed metal residues in the plant	For healthier plants with greener leaves and higher fruit quality, a foliar spray was prepared. It included nanoparticles of Mg-OH which showed no metal precipitation in the plants even after multiple uses	Ghidan et al. (2017)

(continued)

Table 14.2 (continued)

Properties	Technologies enabled by nanoparticles	Mechanism of action	References
Nanomaterials exhibiting anti-microbial activity	Methicillin-resistant *Staphylococcus aureus* was successfully treated using silver nanoparticles	In a study, cellulose acetate phthalate-coated silver nanoparticles (Ag-NPs) demonstrated antibacterial efficacy against fungi including *Candida albicans* and *Aspergillus niger* as well as certain yeast	Necula et al. (2010)
Use of nanoparticles as nano-fungicides	To reduce spore development in *F. oxysporum* f. sp. *Lactucae* and *F. oxysporum* f. sp. *Lycopersici* in the soil, 100 ppm Ni nanoparticles were utilized	The use of Ni nanoparticles was assessed in a concentration-dependent manner, and the most effective concentration was found at 100 ppm which significantly reduced the mycelium growth of tested pathogens	Pulit et al. (2013)
Copper nanomaterials for soil health improvements	Cu-based nanomaterials (10 mg/L) showed increased root exudation in *Cucumis sativus* and released Cu-ions that are used by the plant	The Cu-based nanoparticles improved the soil pH in acidic soil and aided in increasing root exudation in *Cucumis sativa*. It also helps in mobilizing nutrients in saline or alkaline soil	Zhao et al. (2016)
Reducing soil volatility of N-compounds	Using zeolite crystal nanoparticles, an increase in nitrogen utilization was observed in the surrounding soil	By decreasing ammonia volatilization and enhancing the interchange of NH_4^+ and NO_3^-, clinoptilolite zeolite improves the efficiency of nitrogen utilization. The zeolite crystals' tiny pores make it easier for ammonia cations to be absorbed while preventing nitrifying microbes from having access to them	Gupta et al. (2013)

Table 14.3 Applications of nanotechnology in soil health

Activity	Employment of nanoparticles	Applications	References
Fertilizers for plant protection	Nanoparticles/nano-emulsion for protection against disease and pest control	Neem (*Azadirachta indica*) oil nano-emulsion is used as a larvicidal agent	Anjali et al. (2012)
Controlled release for delivery of nutrients	Controlled release/delivery of nutrients to a specific site of a plant	Use of micronutrient fertilizers coated with sulphur and chitosan for the delivery of micronutrients	Corradini et al. (2010)
Nano-sensors	Carbon nanotubes and nano-fibres are used to detect the environmental and chemical changes occurring in the soil	Monitoring of organophosphorus pesticides, that is, dichlorvos and paraoxon using liposome-based nano-biosensors	Vamvakaki and Chaniotakis (2007)

namely, the cuticles (Heredia-Guerrero et al. 2014) of the plant. Therefore, the stomata and hydathodes, among other places, are the most typical entry points into the plant (Kaiser 2014). Moreover, nanoparticle size is controlled to give them simple access to the plasma membrane. The process of entry translocation and adsorption depends majorly (Magniont et al. 2012) on the type of plant but also depends on the nature of the nanoparticles used. Any modification in the physical/chemical nature of the nanoparticle (Kaiser 2014) can affect the transport of such materials across the membrane. Nanoparticle uptake mechanisms have been categorized as polar (Gomes Martins et al. 2011) and non-polar. Polar pathways involve entry points on the leaf surface, like trichomes, hydathodes, stomata, or plant surface lesions (Avellan et al. 2021) that serve as polar apertures, and in contrast the non-polar route uses the pores and cuticles of leaves to enter the plant (Hubbard et al. 2020). The diameter of the vessels in the intracellular space ranges from 2 to 20 μm whereas the vessels surrounding the plasma membrane show a pore diameter of (Zhu et al. 2020) less than 1 μm. Nanoparticles and nano-aggregates smaller than 1 μm in diameter can easily overcome the initial barriers (Anselmo and Mitragotri 2019).

14.6.2 Antibiotic Properties Exhibited by Nanoparticles

The majority of the research that is now available mostly focuses on the antibacterial, antiviral, and fungicidal qualities that various nanoparticles demonstrate. One such type of nanomaterial especially copper (Xiong et al. 2017) and silver nanoparticles has demonstrated a considerable inhibitory effect against a wide variety of plant diseases. For instance, silver nanoparticles were tested for their ability to provide protection (Apodaca et al. 2017) against *Xanthomonas campestris* and *Xanthomonas axonopodis* pv. *malvaceae* in cowpea plants. Moreover, when grown on graphene oxide, DNA-detected silver nanoparticles demonstrated resistance to

Xanthomonas perforans in *Lycopersicon esculentum* (Raliya et al. 2015). Reactive oxygen species are known to be produced when copper interacts with organelles (Omara et al. 2019) and enters the intracellular space. The breakdown of lipids and proteins is one of the key metabolic pathways/processes that these species later activate.

The antibacterial activity of silver nanoparticles produced from extracts of Tulsi leaves was tested against different types of microorganisms such as *Staphylococcus epidermis* (Ako-Adounvo et al. 2017) (gram-positive) and *Pseudomonas aeruginosa* (gram-negative). These nanoparticles caused leakage from the cell thus further leading to cell death. For example, when *Staphylococcus aureus* (Ingle et al. 2014) was treated with cephalon-reduced gold nanoparticles, the peptidoglycan synthesis was inhibited thus making the membrane previous and leading to leakage of cellular contents (Ako-Adounvo et al. 2017). Metal oxide combinations have demonstrated the potential to combat a wide range of pathogenic microorganisms, including strains that are resistant to multiple drugs (Nuruzzaman et al. 2016). Chitosan nanoparticles with a size of 40 nm have been tested on *Staphylococcus choleraesuis* and were found to chelate trace elements (Ali et al. 2014) that inhibit growth. This results in the formation of a barrier layer around the cell, preventing the transport of essential fluids across the cell.

14.7 Plant Productivity and Stress

As the demand for food rises, more fertilizers are being used, which causes soil leaching and acidity. To combat this issue, fertilizer alternatives are being explored, such as acidulated organic fertilizers (Servin et al. 2015), animal manure, leftover food, and nitrogen-rich grass clippings. Additionally, nanoparticles are being researched for their potential to reduce the reliance (Peng et al. 2020) on fertilizers and their negative impact on the environment.

To easily enter the wheat plants, Abdel-Aziz et al. employed chitosan-based nano-fertilizers comprising nitrogen, phosphorous, and potassium as foliar sprays (Kumar et al. 2020). The use of such foliar spray increased plant output (Khodakovskaya et al. 2009) and exhibited accelerated growth. When TiO_2 nanobionics (nanoparticles) stimulate the enzyme RuBisCO, they improve photosynthetic C-assimilation (Parisi et al. 2015). The rate of RuBisCO carboxylation was increased (Reddy et al. 2007), which specifically targeted mRNA and increased protein synthesis. Several substances, including fullerene, fullarol, multi-walled carbon nanotubes, silver, and gold nanoparticles, are employed in different ways to promote plant development (Parisi et al. 2015). For instance, carbon nanotubes are frequently utilized for seed germination, with an increase in water retention serving as the likely mechanism (Verma et al. 2019). Moreover, *Brassica juncea*, which revealed a rise in the root length and root hair at 20 ppm, exhibited remarkable germination in response to these carbon nanotubes (Chaudhary et al. 2018). Thus, nanoparticles show a wide range of applications in various plants (Fig. 14.2).

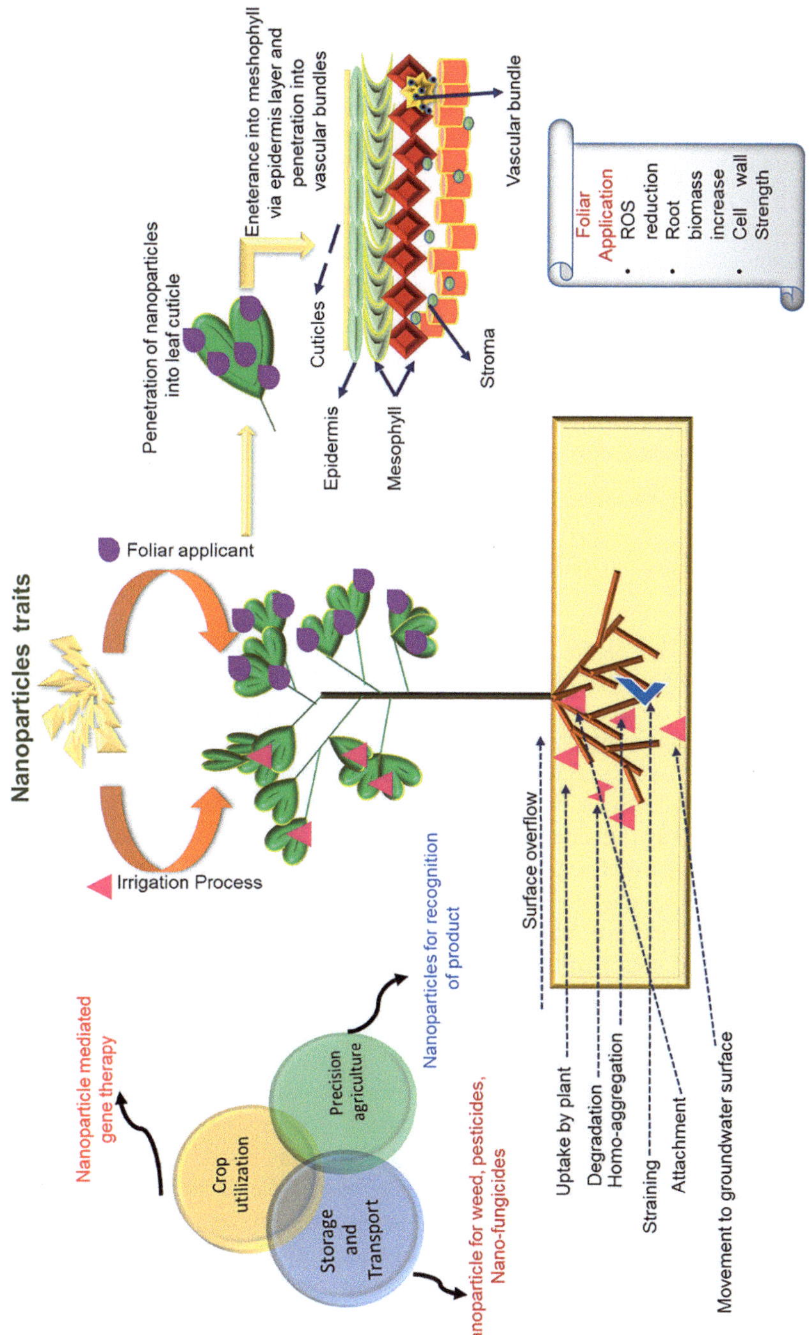

Fig. 14.2 Schematic diagram for nanoparticle traits

The high efficacy of nanoparticles could lead to a significant decrease in the usage of agrochemicals, which would provide environmental benefits. This advantage becomes more noticeable when NP-based substitutes are employed instead of traditional agrochemicals (Liu and Lal 2015). In recent years, nanoparticles have demonstrated their usefulness in a wide range of agricultural applications (Medina-Pérez et al. 2019), including reducing soil acidity, decreasing the leaching of heavy metals, and improving food production. Furthermore, both chemical and biological methods for synthesizing nanoparticles have been extensively researched (Lv et al. 2019). However, there are also certain drawbacks to their use which consist of the lack of standardization in the formulation process of nanoparticles (Khan et al. 2019) and the varying pedoclimatic conditions that can lead to varying effects even with the same nanomaterial and also the high cost of manufacturing nanoparticles.

Hence, in recent years, these nanoparticles have proved to be very useful in a broad range of areas in agriculture such as reducing soil acidity, reduction in leaching of heavy metals, and enhancing food productivity. The chemical and biological synthesis of these nanoparticles has also been studied extensively.

14.8 Limitations and Future Prospects

Nanoparticles are still being explored worldwide but are limited to laboratory practices rather than going for clinical trials. There is a lack of risk assessments and toxicological shreds of evidence. No direct evidence of nanoparticles in the food web undertaking the trophic transfer. Research is not well focused on mechanisms of phytotoxicity, ROS accumulation, and lysosomal damage by nano-formulations. People are showing less acceptance of this new technology because of fear of compromising crop production and lack of awareness, policies, and regulations. The use of nanoparticles in agriculture is a low-margin sector that is unable to produce large economic returns on the initial manufacturing investment because of the high production costs and complicated production procedures.

So in the future, nanoparticle performance comparison with the traditional formulations should be focused on in-depth research at biochemical, physiological, and molecular (using metabolomics and proteomic tools) levels. There should be awareness programmes by the government to upskill society about nanotechnology's role in the agriculture sector with their useful applications concluded with the communal and food industrialists. In the context of commercialization, nanotechnology should be made low cost for large-scale manufacturing. To develop a more workable strategy, a thorough analysis of upcoming studies is required to ascertain the danger connected with the use of nano-formulated products in the agricultural sector. The mechanistic interaction of plant-soil, plant-water-soil, and water-soil should be thoroughly studied to formulate new nano-products based on novel research findings so that the soil microbiota should not get affected in any way. To evaluate nano-formulations from the perspective of both the benefits and drawbacks throughout the entirety of the life span of produced or current models, proper theories and tools are expected. Environmental remediation is aided by numerous nano-phytoremediation techniques. The pace of restoration, nevertheless, differs from area to area because of variations in the soil type, pH, temperature, dampness, moisture, and minerals that are in the soil as a whole. Thus, research is needed to improve the efficacy and compatibility of nanoparticles using appropriate species of plants.

14.9 Conclusion

Through a variety of tightly interwoven methods, biotechnological technologies have redesigned unanticipated agricultural liabilities that affect crop productivity. One of the adaptable innovations with numerous uses in both the physical and biological worlds is nano-biotechnology. Nanotech is evolving rapidly in plenty of other domains, notably energy and medicine; ultimately, the data gleaned in these kinds of sectors might find its way towards expanding into the agricultural industry, given the numerous advantages of employing nano-biotechnology, including increased vegetation tolerance, increased nutrient efficacy, decreased pesticide use, and increased growth and yield of plants. Nano-biotechnology has made it very simple to transport DNA for crop studies. As an organic scaffold, soil is abundant in core nanoparticles as well as clusters and aggregations of these particles. Nanoparticles introduced artificially into the earth's soil have a substantial impact since they are very resilient to deterioration and possess the capacity to build up in the soil. Protracted in situ experimental studies are necessary to advance nano-biotechnology studies and development, and universal healthcare also needs to be ensured to influence the green economy.

Acknowledgement The authors would like to thank the Prof. Prem Kumar Khosla, Chancellor, Shoolini University of Biotechnology and Management Sciences, Solan, Himachal Pradesh, India, and the School of Biotechnology here for providing the laboratory and technical facilities in the field of nano-biotechnology.

References

Ahmed A, He P, He P, Wu Y, He Y, Munir S (2023) Environmental effect of agriculture-related manufactured nano-objects on soil microbial communities. Environ Int 173:107819

Ako-Adounvo AM, Marabesi B, Lemos RC, Patricia A, Karla K (2017) Drug and gene delivery materials and devices. Elsevier, pp 375–392

Ali MA, Rehman I, Iqbal A, Din S, Rao AQ, Latif A, Samiullah TR, Azam S, Husnain T (2014) Nanotechnology, a new frontier in agriculture. Adv Life Sci 1(3):129–138

Aliyari Rad S, Nobaharan K, Pashapoor N, Pandey J, Dehghanian Z, Senapathi V, Minkina T, Ren W, Rajput VD, Asgari Lajayer B (2023) Nano-microbial remediation of polluted soil: a brief insight. Sustainability 15(1):876

An C, Sun C, Li N, Huang B, Jiang J, Shen Y, Wang C, Zhao X, Cui B, Wang C, Li X (2022) Nanomaterials and nanotechnology for the delivery of agrochemicals: strategies towards sustainable agriculture. J Nanobiotechnol 20(1):1–19

Anjali CH, Sharma Y, Mukherjee A, Chandrasekaran N (2012) Neem oil (*Azadirachta indica*) nanoemulsion—a potent larvicidal agent against *Culex quinquefasciatus*. Pest Manag Sci 68(2):158–163

Anjum SA, Tanveer M, Ashraf U, Hussain S, Shahzad B, Khan I, Wang L (2016) Effect of progressive drought stress on growth, leaf gas exchange, and antioxidant production in two maize cultivars. Environ Sci Pollut Res 23:17132–17141

Anselmo AC, Mitragotri S (2019) Nanoparticles in the clinic: an update. Bioeng Transl Med 4(3):e10143

Apodaca SA, Tan W, Dominguez OE, Hernandez-Viezcas JA, Peralta-Videa JR, Gardea-Torresdey JL (2017) Physiological and biochemical effects of nanoparticulate copper, bulk copper, copper chloride, and kinetin in kidney bean (*Phaseolus vulgaris*) plants. Sci Total Environ 599:2085–2094

Attwood MM, Fabbro D, Sokolov AV, Knapp S, Schiöth HB (2021) Trends in kinase drug discovery: targets, indications and inhibitor design. Nat Rev Drug Discov 20(11):839–861

Avellan A, Yun J, Morais BP, Clement ET, Rodrigues SM, Lowry GV (2021) Critical review: role of inorganic nanoparticle properties on their foliar uptake and in planta translocation. Environ Sci Technol 55(20):13417–13431

Babadi FE, Yunus R, Rashid SA, Salleh MAM, Ali S (2015) New coating formulation for the slow release of urea using a mixture of gypsum and dolomitic limestone. Particuology 23:62–67

Baker S, Kumar KM, Santosh P, Rakshith D, Satish S (2015) Extracellular synthesis of silver nanoparticles by novel *Pseudomonas veronii* AS41G inhabiting *Annona squamosa* L. and their bactericidal activity. Spectrochim Acta A Mol Biomol Spectrosc 136:1434–1440

Bala M, Bansal SK, Fatima F (2023) Nanotechnology: a boon for agriculture. Mater Today Proc 73:267–270

Bardgett RD, Van Der Putten WH (2014) Belowground biodiversity and ecosystem functioning. Nature 515(7528):505–511

Baveye C, Baveye J, Gowdy J (2016) Soil "ecosystem" services and natural capital: critical appraisal of research on uncertain ground. Front Environ Sci 4:41

Berendsen RL, Pieterse CM, Bakker:A. (2012) The rhizosphere microbiome and plant health. Trends Plant Sci 17(8):478–486

Biswal T (2023) Nano-phytoremediation and its applications. In: Phytoremediation: management of environmental contaminants, vol 7. Springer International Publishing, Cham, pp 335–364

Bobbin ML, Rossi JJ (2016) RNA interference (RNAi)-based therapeutics: delivering on the promise? Annu Rev Pharmacol Toxicol 56:103–122

Brevik EC, Cerdà A, Mataix-Solera J, Pereg L, Quinton JN, Six J, Van Oost K (2015) The interdisciplinary nature of SOIL. Soil 1(1):117–129

Bünemann EK, Bongiorno G, Bai Z, Creamer RE, De Deyn G, De Goede R, Fleskens L, Geissen V, Kuyper TW, Mäder P, Pulleman M (2018) Soil quality—a critical review. Soil Biol Biochem 120:105–125

Chareonviriyaphap T, Bangs MJ, Suwonkerd W, Kongmee M, Corbel V, Ngoen-Klan R (2013) Review of insecticide resistance and behavioral avoidance of vectors of human diseases in Thailand. Parasit Vectors 6:1–28

Chaudhary S, Umar A, Bhasin KK, Baskoutas S (2018) Chemical sensing applications of ZnO nanomaterials. Materials 11(2):287

Chugh G, Siddique KH, Solaiman ZM (2021) Nanobiotechnology for agriculture: smart technology for combating nutrient deficiencies with nanotoxicity challenges. Sustainability 13(4):1781

Corradini E, De Moura MR, Mattoso LHC (2010) A preliminary study of the incorporation of NPK fertilizer into chitosan nanoparticles. Express Polym Lett 4(8):509

Cuevas J, Daliakopoulos IN, del Moral F, Hueso JJ, Tsanis IK (2019) A review of soil-improving cropping systems for soil salinization. Agronomy 9(6):295

de Vries W, Kros J, Voogd JC, Ros GH (2023) Integrated assessment of agricultural practices on large scale losses of ammonia, greenhouse gases, nutrients and heavy metals to air and water. Sci Total Environ 857:159220

Díaz-Blancas V, Medina DI, Padilla-Ortega E, Bortolini-Zavala R, Olvera-Romero M, Luna-Bárcenas G (2016) Nanoemulsion formulations of fungicide tebuconazole for agricultural applications. Molecules 21(10):1271

Dominati E, Mackay A, Green S, Patterson M (2014) A soil change-based methodology for the quantification and valuation of ecosystem services from agro-ecosystems: a case study of pastoral agriculture in New Zealand. Ecol Econ 100:119–129

Doran, J.W., Coleman, D.C., Bezdicek, D.F. and Stewart, B.A., 1994. Defining soil quality for a sustainable environment

Douroumis, D., Onyesom, I., Maniruzzaman, M. and Edwards, M., 2011. Mesoporous silica nanoparticles as a drug delivery system

Elramady H, Brevik EC, Elsakhawy T, Omara AED, Amer MM, Abowaly M, El-Henawy A, Prokisch J (2022) Soil and humans: a comparative and a pictorial mini-review. Egypt J Soil Sci 62(2):101–122

Fatima F, Hashim A, Anees S (2021) Efficacy of nanoparticles as nanofertilizer production: a review. Environ Sci Pollut Res 28:1292–1303

Fu B, Chen L, Huang H, Qu: and Wei, Z. (2021) Impacts of crop residues on soil health: A review. Environ Pollut Bioavail 33(1):164–173

Gandhi G, Girgila S, Aggarwal RK, Buttar BS (2010) Propensity for DNA damage in psoriasis patients genotyped for two candidate genes. J Carcinogene Mutagene 1(112):2

Ghidan AY, Al-Antary TM, Salem NM, Awwad AM (2017) Facile green synthetic route to the zinc oxide (ZnONPs) nanoparticles: effect on green peach aphid and antibacterial activity. J Agric Sci 9(2):131–138

Gomes Martins CH, Carvalho TC, Mendes Souza MG, Ravagnani C, Peitl O, Zanotto ED, Panzeri H, Casemiro LA (2011) Assessment of antimicrobial effect of biosilicate (R) against anaerobic, microaerophilic and facultative anaerobic microorganisms. J Mater Scie Mater Med 22(6):1439

Gouin S (2004) Microencapsulation: industrial appraisal of existing technologies. Food Sci Technol 15:330–347

Gupta P, Khanday WA, Majid SA, Kushwa V, Tomar SS, Tomar R (2013) Study of sorption of metal oxoanions from waste water on surfactant modified analog of laumontite. J Environ Chem Eng 1(3):510–515

Heredia-Guerrero JA, Benítez JJ, Domínguez E, Bayer IS, Cingolani R, Athanassiou A, Heredia A (2014) Infrared and Raman spectroscopic features of plant cuticles: a review. Front Plant Sci 5:305

Hu R, Huang X, Huang J, Li Y, Zhang C, Yin Y, Chen Z, Jin Y, Cai J, Cui F (2015) Long-and short-term health effects of pesticide exposure: a cohort study from China. PLoS One 10(6):e0128766

Hubbard JD, Lui A, Landry MP (2020) Multiscale and multidisciplinary approach to understanding nanoparticle transport in plants. Curr Opin Chem Eng 30:135–143

Ingle AP, Duran N, Rai M (2014) Bioactivity, mechanism of action, and cytotoxicity of copper-based nanoparticles: a review. Appl Microbiol Biotechnol 98:1001–1009

Jakhar AM, Aziz I, Kaleri AR, Hasnain M, Haider G, Ma J, Abideen Z (2022) Nano-fertilizers: a sustainable technology for improving crop nutrition and food security. NanoImpact 27:100411

Jian J, Du X, Stewart RD (2020) A database for global soil health assessment. Sci Data 7(1):16

Johnston AE, Poulton R, Coleman K (2009) Soil organic matter: its importance in sustainable agriculture and carbon dioxide fluxes. Adv Agron 101:1–57

Kah M, Kookana RS, Gogos A, Bucheli TD (2018) A critical evaluation of nanopesticides and nanofertilizers against their conventional analogues. Nat Nanotechnol 13(8):677–684

Kaiser H (2014) Stomatal uptake of mineral particles from a sprayed suspension containing an organosilicone surfactant. J Plant Nutr Soil Sci 177(6):869–874

Khan MR, Ahamad F, Rizvi TF (2019) Effect of nanoparticles on plant pathogens. Academic Press, pp 215–240

Khodakovskaya M, Dervishi E, Mahmood M, Xu Y, Li Z, Watanabe F, Biris AS (2009) Carbon nanotubes are able to penetrate plant seed coat and dramatically affect seed germination and plant growth. ACS Nano 3(10):3221–3227

Kolenčík M, Ernst D, Urík M, Ďurišová Ľ, Bujdoš M, Šebesta M, Dobročka E, Kšiňan S, Illa R, Qian Y, Feng H (2020) Foliar application of low concentrations of titanium dioxide and zinc oxide nanoparticles to the common sunflower under field conditions. Nano 10(8):1619

Kumar A, Sharma S, Mishra S (2010) Influence of arbuscular mycorrhizal (AM) fungi and salinity on seedling growth, solute accumulation, and mycorrhizal dependency of *Jatropha curcas* L. J Plant Growth Regul 29:297–306

Kumar GD, Raja K, Natarajan N, Govindaraju K, Subramanian KS (2020) Invigouration treatment of metal and metal oxide nanoparticles for improving the seed quality of aged chilli seeds (*Capsicum annum* L.). Mater Chem Phys 242:122492

Large DE, Abdelmessih RG, Fink EA, Auguste DT (2021) Liposome composition in drug delivery design, synthesis, characterization, and clinical application. Adv Drug Deliv Rev 176:113851

Li Z, Zhang Y, Li Y (2013) Zeolite as slow release fertilizer on spinach yields and quality in a greenhouse test. J Plant Nutr 36(10):1496–1505

Lichtfouse E, Navarrete M, Debaeke P, Souchère V, Alberola C, Ménassieu J (2009) Agronomy for sustainable agriculture: a review. In: Sustainable agriculture. Springer, pp 1–7

Liu R, Lal R (2015) Potentials of engineered nanoparticles as fertilizers for increasing agronomic productions. Sci Total Environ 514:131–139

Lobo AO, Marciano FR, Regiani I, Matsushima JT, Ramos SC, Corat EJ (2011) Influence of temperature and time for direct hydroxyapatite electrodeposition on superhydrophilic vertically aligned carbon nanotube films. J Nanomed Nanotechnol 8(2). https://doi.org/10.4172/2157-7439.1000277

Lv J, Christie P, Zhang S (2019) Uptake, translocation, and transformation of metal-based nanoparticles in plants: recent advances and methodological challenges. Environ Sci Nano 6(1):41–59

Madanat HM, Al Antary TM, Zarqa M (2016) Toxicity of six ethanol plant extracts against the green peach aphid *Myzus persicae Sulzer* (Homoptera: Aphididae). Fresenius Environ Bull 25(3):706–718

Magniont C, Escadeillas G, Coutand M, Oms-Multon C (2012) Use of plant aggregates in building ecomaterials. Eur J Environ Civ Eng 16(sup1):s17–s33

Medfu Tarekegn M, Zewdu Salilih F, Ishetu AI (2020) Microbes used as a tool for bioremediation of heavy metal from the environment. Cogent Food Agric 6(1):1783174

Medina-Pérez G, Fernández-Luqueño F, Campos-Montiel RG, Sánchez-López KB, Afanador-Barajas LN, Prince L (2019) Nanotechnology in crop protection: status and future trends. In: Nano-biopesticides today and future perspectives. Academic Press, pp 17–45

Mirza AZ, Siddiqui FA (2014) Nanomedicine and drug delivery: a mini review. Int Nano Lett 4:1–7

Mittal D, Kaur G, Singh P, Yadav K, Ali SA (2020) Nanoparticle-based sustainable agriculture and food science: recent advances and future outlook. Front Nanotechnol 2:579954

Montanarella L, Pennock DJ, McKenzie N, Badraoui M, Chude V, Baptista I, Mamo T, Yemefack M, Singh Aulakh M, Yagi K, Young Hong S (2016) World's soils are under threat. Soil 2(1):79–82

Necula AM, Dunca S, Stoica I, Olaru N, Olaru L, Ioan S (2010) Morphological properties and antibacterial activity of nano-silver-containing cellulose acetate phthalate films. Int J Polym Anal Charact 15(6):341–350

Nene A, Galluzzi M, Hongrong L, Somani P, Ramakrishna S, Yu XF (2021) Synthetic preparations and atomic scale engineering of silver nanoparticles for biomedical applications. Nanoscale 13(33):13923–13942

Nuruzzaman MD, Rahman MM, Liu Y, Naidu R (2016) Nanoencapsulation, nano-guard for pesticides: a new window for safe application. J Agric Food Chem 64(7):1447–1483

Oliver MA, Gregory J (2015) Soil, food security and human health: a review. Eur J Soil Sci 66(2):257–276

Omara AED, Elsakhawy T, Alshaal T, El-Ramady H, Kovács Z, Fári M (2019) Nanoparticles: a novel approach for sustainable agro-productivity. Environ Biodivers Soil Secur 3(2019):29–62

Ortiz A, Sansinenea E (2021) Recent advancements for microorganisms and their natural compounds useful in agriculture. Appl Microbiol Biotechnol 105:891–897

Parisi C, Vigani M, Rodríguez-Cerezo E (2015) Agricultural nanotechnologies: what are the current possibilities? Nano Today 10(2):124–127

Pasquoto-Stigliani T, Guilger-Casagrande M, Campos EV, Germano-Costa T, Bilesky-José N, Migliorini BB, Feitosa LO, Sousa BT, de Oliveira HC, Fraceto LF, Lima R (2023) Titanium biogenic nanoparticles to help the growth of *Trichoderma harzianum* to be used in biological control. J Nanobiotechnol 21(1):1–21

Patil A, Chirmade UN, Trivedi V, Lamprou DA, Urquart A, Douroumis D (2011) Encapsulation of water insoluble drugs in mesoporous silica nanoparticles using supercritical carbon dioxide. J Nanomed Nanotechnol 2(3). https://doi.org/10.4172/2157-7439.1000111

Peng C, Tong H, Shen C, Sun L, Yuan P, He M, Shi J (2020) Bioavailability and translocation of metal oxide nanoparticles in the soil-rice plant system. Sci Total Environ 713:136662

Pimentel D (1995) Amounts of pesticides reaching target pests: environmental impacts and ethics. J Agric Environ Ethics 8:17–29

Pulit J, Banach M, Szczyg Owska R, Bryk M et al (2013) Nanosilver against fungi. Silver nanoparticles as an effective biocidal factor. Acta Biochim Pol 60:795–808

Rajput VD, Singh A, Minkina TM, Shende SS, Kumar P, Verma KK, Bauer T, Gorobtsova O, Deneva S, Sindireva A (2021) Potential applications of nanobiotechnology in plant nutrition and protection for sustainable agriculture. In: Nanotechnology in plant growth promotion and protection: recent advances and impacts. Wiley, pp 79–92

Raliya R, Nair R, Chavalmane S, Wang WN, Biswas P (2015) Mechanistic evaluation of translocation and physiological impact of titanium dioxide and zinc oxide nanoparticles on the tomato (Solanum lycopersicum L.) plant. Metallomics 7(12):1584–1594

Reddy KM, Feris K, Bell J, Wingett DG, Hanley C, Punnoose A (2007) Selective toxicity of zinc oxide nanoparticles to prokaryotic and eukaryotic systems. Appl Phys Lett 90(21):2139021

Rosarin FS, Mirunalini S (2011) Nobel metallic nanoparticles with novel biomedical properties. J Bioanal Biomed 3(4):85–91

Sahu JN, Karri RR, Zabed HM, Shams S, Qi X (2021) Current perspectives and future prospects of nano-biotechnology in wastewater treatment. Separ Purif Rev 50(2):139–158

Sangeetha J, Mundaragi A, Thangadurai D, Maxim SS, Pandhari RM, Alabhai JM (2019) Nanobiotechnology for agricultural productivity, food security and environmental sustainability. In: Nanotechnology for agriculture: crop production & protection. Springer, pp 1–23

Sansinenea E (2021) Application of biofertilizers: current worldwide status. Woodhead Publishing, pp 183–190

Sasson Y, Levy-Ruso G, Toledano O, Ishaaya I (2007) Nanosuspensions: emerging novel agrochemical formulations. In: Insecticides design using advanced technologies. Springer, pp 1–39

Sato Y, Murase K, Kato J, Kobune M, Sato T, Kawano Y, Takimoto R, Takada K, Miyanishi K, Matsunaga T, Takayama T (2008) Resolution of liver cirrhosis using vitamin A–coupled liposomes to deliver siRNA against a collagen-specific chaperone. Nat Biotechnol 26(4):431–442

Schreefel L, Schulte RPO, De Boer IJM, Schrijver AP, Van Zanten HHE (2020) Regenerative agriculture–the soil is the base. Glob Food Sec 26:100404

Servin A, Elmer W, Mukherjee A, De la Torre-Roche R, Hamdi H, White JC, Bindraban P, Dimkpa C (2015) A review of the use of engineered nanomaterials to suppress plant disease and enhance crop yield. J Nanopart Res 17:1–21

Seth A, Banyal A, Kumar P (2023) Commercialization and technology transfers of bioprocess. In: Basic biotechniques for bioprocess and bioentrepreneurship. Academic Press, pp 455–469

Shih MF, Wu CH, Cherng JY (2015) Bioeffects of transient and low-intensity ultrasound on nanoparticles for a safe and efficient DNA delivery. J Nanomed Nanotechnol 6(2):1

Singh P, Hussain T, Patel S, Akhtar N (2018) Impact of climate change on root–pathogen interactions. In: Root biology. Springer, pp 409–427

Singh A, Sengar RS, Singh AK, Singh S, Kumar A, Ahmad K (2019) Impact of genetic engineering on agriculture food crops. Biotech Today 9(1):17–24
Singh RP, Handa R, Manchanda G (2021) Nanoparticles in sustainable agriculture: An emerging opportunity. J Control Release 329:1234–1248
Solanki P, Bhargava A, Chhipa H, Jain N, Panwar J (2015) Nano-fertilizers and their smart delivery system. In: Nanotechnologies in food and agriculture. Springer, pp 81–101
Tahat MM, Alananbeh KM, Othman YA, Leskovar DI (2020) Soil health and sustainable agriculture. Sustainability 12(12):4859
Tothill IE (2011) Biosensors and nanomaterials and their application for mycotoxin determination. World Mycotoxin J 4(4):361–374
Tran S, DeGiovanni J, Piel B, Rai P (2017) Cancer nanomedicine: a review of recent success in drug delivery. Clin Transl Med 6:1–21
Tripathi A, Prakash S (2023) Nanobiotechnology: emerging trends, prospects, and challenges. Agric Nanobiotechnol 1:1–21
United Nations Department of Economic and Social Affairs Population Division (2022) World population prospects 2022. https://population.un.org/wpp/Graphs/DemographicProfiles/Line/900. Accessed on 21 Mar 2023
Usharani KV, Roopashree KM, Naik D (2019) Role of soil physical, chemical and biological properties for soil health improvement and sustainable agriculture. J Pharmacogn Phytochem 8(5):1256–1267
Vamvakaki V, Chaniotakis NA (2007) Pesticide detection with a liposome-based nano-biosensor. Biosens Bioelectron 22(12):2848–2853
Verger J, Boobis AR (2013) Reevaluate pesticides for food security and safety. Science 341(6147):717–718
Verma ML, Kumar P, Sharma D, Verma AD, Jana AK (2019) Advances in nanobiotechnology with special reference to plant systems. In: Plant nanobionics, Advances in the understanding of nanomaterials research and applications, vol 1. Springer, pp 371–387
Wang L, Hu C, Shao L (2017) The antimicrobial activity of nanoparticles: present situation and prospects for the future. Int J Nanomedicine 12:1227
Xiong T, Dumat C, Dappe V, Vezin H, Schreck E, Shahid M, Pierart A, Sobanska S (2017) Copper oxide nanoparticle foliar uptake, phytotoxicity, and consequences for sustainable urban agriculture. Environ Sci Technol 51(9):5242–5251
Yadav A, Yadav K, Ahmad R, Abd-Elsalam KA (2023) Emerging frontiers in nanotechnology for precision agriculture: advancements, hurdles and prospects. Agrochemicals 2(2):220–256
Zambrano-Mendoza JL, Sangoquiza-Caiza CA, Campaña-Cruz DF, Yánez-Guzmán CF (2021) Use of biofertilizers in agricultural production. In: Technology in agriculture, vol 193. InTech Open
Zhao L, Huang Y, Hu J, Zhou H, Adeleye AS, Keller AA (2016) 1H NMR and GC-MS based metabolomics reveal defense and detoxification mechanism of cucumber plant under nano-Cu stress. Environ Sci Technol 50(4):2000–2010
Zhu J, Li J, Shen Y, Liu S, Zeng N, Zhan X, White JC, Gardea-Torresdey J, Xing B (2020) Mechanism of zinc oxide nanoparticle entry into wheat seedling leaves. Environ Sci Nano 7(12):3901–i3913

Chapter 15
Advanced and Emerging Techniques in Soil Health Management

Sunita K. Meena, Ajeet Kumar, Khem R. Meena, Sanjeew K. Sinha, Lalita Rana, Navnit Kumar, Anil K. Singh, Hanuman P. Parewa, and Vijay S. Meena

15.1 Introduction

The world's many complex issues, including resource depletion and degradation, food and nutritional insecurity, climate change, and resource scarcity, are most likely caused by the rise of the human population. The global agricultural system must strike a delicate balance in order to satisfy a variety of human needs by 2050, when the world's population is projected to increase by 34% to 9.6 billion people. It also needs to create more opportunities for the hundreds of millions of rural poor people who depend on agriculture for a living while minimising negative environmental effects like ecosystem degradation and high greenhouse gas emissions. The world's population is currently estimated to be 7.24 billion, and it is increasing by around 80 million people annually or roughly the population of Germany. Although this growth is anticipated to significantly slow between 2050 and 2100, the world's population is still forecast to increase throughout the twenty-first century (OECD and FAO 2016).

FAO reported that by 2050, the demand for food is expected to double as a result of population increase and economic expansion. This rate of growth will require a

S. K. Meena (✉) · A. Kumar · S. K. Sinha · L. Rana · N. Kumar · A. K. Singh
Sugarcane Research Institute (SRI), Dr. Rajendra Prasad Central Agricultural University (RPCAU), Samastipur, Bihar, India
e-mail: sunita.meena@rpcau.ac.in

K. R. Meena
Department of Biotechnology, Central University of Rajasthan, Ajmer, Rajasthan, India

H. P. Parewa
College of Agriculture, Agriculture University Jodhpur, Jodhpur, Rajasthan, India

V. S. Meena
ICAR-Vivekananda Parvatiya Krishi Anusandhan Sansthan, Almora, India

R. K. Bhatia, A. Walia (eds.), *Advancements in Microbial Biotechnology for Soil Health*, Microorganisms for Sustainability 50,
https://doi.org/10.1007/978-981-99-9482-3_15

70% increase in food production. Croplands and water resources, which are strained due to rising human populations, are necessary for food production. Cropland expansion may be necessary due to population pressure on the planet's finite land resources. This frequently entails overusing arable land or destroying important forest resources. Despite slower rates of increase in food consumption, there will still be enough food to feed everyone on the planet in 2050.

Humans and their health have traditionally depended on soil, which offers resources for building homes and growing food. Since Thomas Malthus published his first well-known article in 1798, there has been concern about the relationship between an increasing global population and the soil's capacity to sustain an expanding food supply. Although the issues of human population growth and the maintenance of a sufficient food supply are undoubtedly old ones, finding a solution is becoming more and more difficult as the world's population is expected to reach 10 billion people before stabilising. Human interventions that are intended to maximise production functions and that, to some extent, always lead to a loss of other ecosystem functions have changed all agricultural soils from their original state. Significant changes may also be made to the soil food chain.

In order to produce more food, the world's population had to quadruple over the previous 100 years, which necessitated a fundamental shift in land and crop management. Mineral fertilisers are thought to have contributed 40% of the growth in food output over the previous 40 years (FAO 2017). Traditionally, agriculture has been thought of as delivering only one service, namely, the production of food. Primary and secondary productions are reliant on ecosystem services performed by the soil, including nutrient cycling, soil structure preservation, and biotic population control. The preservation of the whole spectrum of ecosystem services for future generations must be a fundamental goal of sustainable agriculture, necessitating the maintenance of the multifunctional capability of agricultural soils (Rattan et al. 2009; Bhattacharyya et al. 2015). The ability of soil to provide a variety of various ecosystem functions and services is referred to as soil health. According to Kibblewhite et al. (2008), soil health is an integrative property and the ability of soil to function as a living system. In order to assure productivity, profitability, and global food security, managing soil health is a daunting challenge. Soil health is strongly tied to maintaining food and nutritional security in the nation since improving soil health was one of the five recommendations made on hunger by the United Nations Millennium Development Task Force for raising agricultural productivity and combating hunger in India (United Nations 2015). The importance of soil health can be attributed to the fact that productive, high-performing soils lower production costs, boost profits, safeguard natural resources both on and off the farm, decrease nitrogen loading and sediment run-off, boost productivity, and maintain animal habitat.

The foundation of the food chain is healthy soil. In order for plants and animals to be productive, for water and environmental quality to be maintained or improved, and for plants and animals to be healthy, the soil must be able to function within the constraints of the ecosystem and the land use as a vital living system. To maintain agricultural production and environmental safety, soil health is crucial. A variety of environmental services, such as the provision of habitat, water infiltration, and profitable and sustainable agriculture, are provided by healthy soils (Costanza et al. 1997; Baveye et al. 2016). An urgent ecological problem in the recent past has been anthropogenic declines in soil health and of certain soil quality components. Increased public understanding of the value of regulating soil health for a sustainable agricultural production system is desperately needed. Combining several techniques in a creative way to improve the soil's physical, chemical, and biological compatibility for crop production is the management of soil health. The best nutrient and pesticide management techniques, which include crop rotations, conservation tillage, mulching with living and dead residue, reducing compaction, and addition of large amount of crop residues, manures, and composts, could be the key strategies for improving soil health (Das et al. 2020). There are very few case studies on the use of various nutrient management practices that analyse and summarise similar soil health information. The genesis of soil health deterioration, quantification and evaluation of soil health indicators, and the contribution of ecosystem services to soil health will all be covered in this book chapter. Future and emerging techniques will also be necessary to improve soil health.

15.2 Soil Health: A Big Concept

According to the NRCS (2018), "soil health is the soil's ability to function as a vital living ecosystem to sustain plants, animals, and humans." The ability of a particular soil to carry out a variety of tasks, such as nutrient cycling, maintaining productivity, preserving biodiversity, controlling water dynamics, and moderating climate, is referred to as soil health (Karlen et al. 1997). The realisation that anthropogenic influences are increasingly overlaid on natural soil quality is what sparked the current interest in soil health. Intensive agriculture, which is practised to meet the food needs of an ever-growing population, has a negative impact on soil health. A balanced microbiome supports the productivity of plants and the health of the soil. Complex communication takes place between soil microorganisms and plants. There is still a lot of translation needed in this discourse, which is mediated by root exudates. To enhance soil nutrient status and agricultural output, this plant and soil use biofertilizers, which are economical and environmentally benign. They can also increase productivity in a sustainable way. By tying water to organic matter, healthy soil stores more water and loses less to evaporation and run-off. As tillage decreases and plants and debris cover the soil, organic matter grows. In the top 6 in. of soil, 1% organic matter may hold about 27,000 gal of water per acre (Taylor 2014). According to USDA (US Department of Agriculture), if farmers are committed to undertake conservation practices to reach this aim, they may typically enhance soil organic matter in 3–10 years. Figure 15.1 illustrates the four fundamental principles for improving soil health and sustainability.

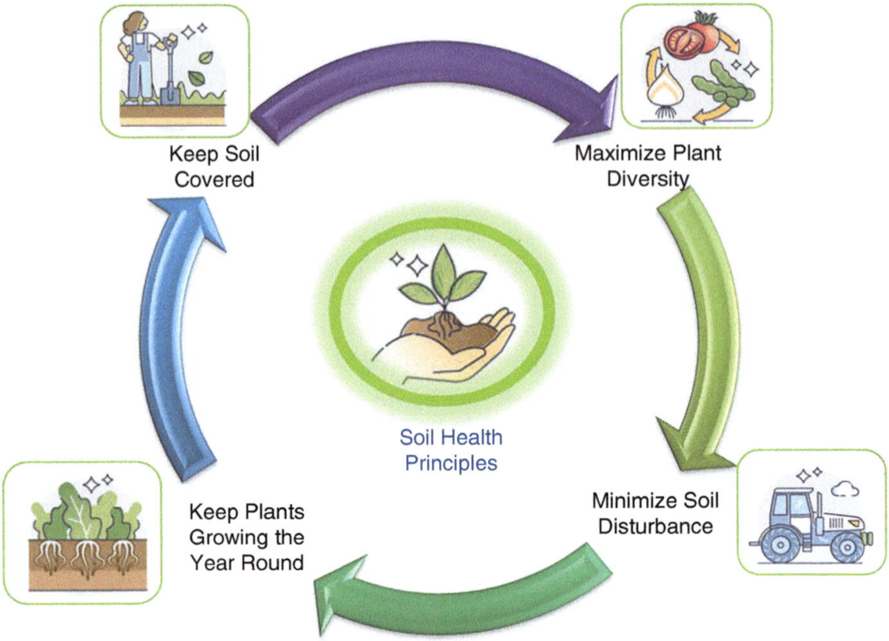

Fig. 15.1 The four key concepts of soil health to promote sustainability and soil health

15.3 Soil Health Principles

Due to the high cost and spatial and temporal variability of ecosystem services, it is inherently difficult to measure them directly. However, a wide range of soil health indicators, including chemical, physical, and biological soil properties, have been identified as proxy measurements (Doran and Parkin 1996; Stott 2019). Among these, soil organic carbon (SOC) is a keystone indicator that reflects a variety of ecosystem services and soil activities. Land use and management practices have a substantial impact on soil C dynamics, despite the fact that it is commonly acknowledged that climatic and edaphic conditions are determinants of soil C content (Nunes et al. 2020a). When soils are healthy, they have the ability to offer civilisation a variety of "ecosystem services" resources or natural processes that the environment provides that are beneficial to humans (Fig. 15.2). The soil functions that control ecosystem services have distinct features and can be measured using several soil test techniques (Table 15.1).

Our idea of soil health encompasses not only the soil biota and the plethora of biotic interactions that take place but also the soil as a habitat (Young and Ritz 2005). The ecosystem services that soil provides are driven by soil biological processes. Optimising agricultural yields while preserving ecosystem service delivery is the main problem of sustainable soil management. In order to control plant diseases, insects, and weeds, recycle vital plant nutrients, form advantageous

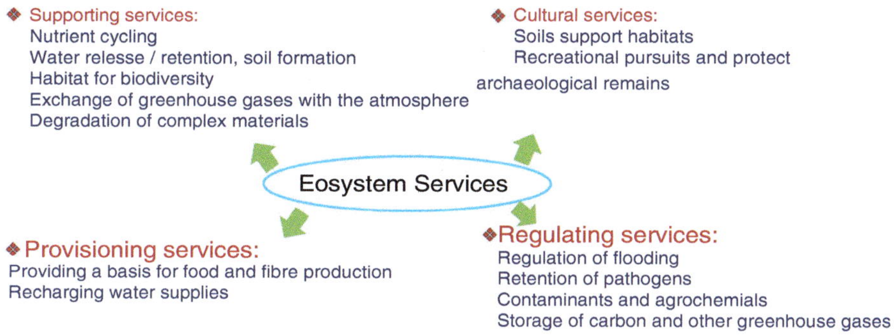

Fig. 15.2 Significance of ecosystem services in socio-economic sustainability

Table 15.1 Significance of different soil characteristics and soil health test relation to functions of soil

Soil functions	Soil health tests
Soil nutrient availability	• Soil nutrient availability test
	• Cation exchange capacity (CEC)
	• pH
Soil organic matter	• Total soil organic matter
	• Active soil organic matter
	• Particulate organic matter
	• Permanganate oxidizable carbon
Microbial diversity and habitat	• Soil respiration
	• Mineralizable carbon and nitrogen
	• Phospholipid fatty acid (PLFA)
	• β-Glucosidase (enzyme)
	• ACE soil protein/glomalin
Physical characteristics	• Bulk density
	• Water infiltration
	• Soil texture
	• Surface and subsurface hardness
	• Aggregate stability

symbiotic relationships with plant roots, and improve soil structure, which has a positive impact on the soil's ability to hold water and nutrients, healthy soils must be maintained (FAO 2008). It is reasonable to infer that a soil is healthy if the organic matter is added or maintained at an acceptable level for the growth of productive crops. Pest outbreaks are less likely to spread in healthy soil. In fertile soils, even damage from pests that are not present in the soil, like maize stem borers, is diminished (Chabi-Olaye et al. 2006). It is crucial to guarantee nutrient availability in soils and to apply a balanced amount of nutrients from organic sources and from mineral fertilisers, if necessary, in order to attain the increased productivity required meeting the current and future food demand. The work of monitoring soil health is highly difficult, and attempts are being made to apply it at the national, regional, and global levels (Steiner et al. 2000). A collection of techniques and instruments for conducting evaluations and monitoring tasks has been created by FAO and its partners (FAO 2010). Functional capacity rather than actual service outputs is a better indicator of soil health. As previously argued, an effective method for assessing the performance of soil systems appears to be the use of a set of diagnostic tests that have been selected to be representative of habitat condition, including physical and chemical factors like pH and salinity, energetic reservoirs like soil organic matter content, and important organisms and community structure like earthworms.

Because most adaption strategies for the climate change scenario rely more on current practices and sustainable agriculture than new technologies, the importance of soil health is rising (Jarvis et al. 2011). Agroforestry is one of several adaptation methods that is helpful for mitigation, while the exact balance of benefits varies greatly on local circumstances. For the majority of farming systems, adaptation will be centred on changes to the management of water and soil. Management of diseases and pests will be essential. An essential adaptation response is the selection or production of new crop varieties, which requires selecting for or breeding for particular features in response to localised changes in climatic conditions (such as resistance to heat, water stress, salt, or water logging, for example) (Thornton and Cramer 2012). When climate change is linked to numerous, interrelated environmental pressures (drought, heat, and low soil fertility), breeding, such as breeding beans for drought resistance, may be a limited choice for adaptation (Beebe et al. 2008). Breeding for future climates requires access to enough genetic diversity, which can be found in genebanks, the wild, and farmer's fields. In genebanks around the world, about 90% of all genetic features for rice, wheat, and maize are accessible, but only a significantly smaller percentage are available for several non-staple crops that provide essential micronutrients. Even though many crops, including the major cereals, show significant losses in yields when exposed to high temperatures within the range of present and near-term climate change, more research has been done on developing drought-tolerant crops than heat-tolerant crops (Lobell 2009). Water scarcity, which is the main factor restricting wheat growth, might be lessened with additional irrigation. Early planting would be possible, preventing (terminal) heat stress during the grain-filling season. To meet the fundamental crop water requirements, more irrigation water will be needed in the future—an average of 181 mm every season from 2080 to 2099 compared to only 134 mm historically (Thornton and Cramer 2012). Long-term climatic trends might require different adaptation measures than rising climate variability. In general, anticipatory legislation, long-term forecasts, and breeding for future climates are all necessary forms of preparation for long-term climate change. Increasing climatic variability necessitates the use of risk management instruments like crop insurance and better seasonal climate forecasts (Vermeulen et al. 2012). Crop residues, which are produced at a rate of roughly 336 million tonnes annually, can provide about 5.1 million tonnes of K in addition to organic matter.

15.4 Quantification and Assessment of Soil Health Indicator

An interpretative framework that offers a variety of indicator options is required to support these quantifications and evaluation of different land use patterns and management practices (Wander et al. 2019). During the evaluation of soil health and complex interactions among various soil properties like the inherent and dynamic, which include biological, physical, and chemical processes of soil evaluated, this interpretive framework is having soil health index with multi-indicator. In order to provide sustainable land management, various assumptions should be considered like site-specific intrinsic attributes, anthropogenic activities, and large-scale monitoring. Numerous indices have been created for soil health assessment by the several researchers; however few of them are widely acknowledged (Bunemann et al. 2018; Andrews et al. 2004; De Paul Obade and Lal 2016; Haney et al. 2018; Idowu et al. 2009; Rinot et al. 2019). Andrews et al. (2004) had created the Soil Management Assessment Framework (SMAF). By converting measured values into unitless (0–1) scores, the Soil Management Assessment Framework for soil health assessment incorporates many soil health indicators (biological, chemical, and physical) affecting soil's capacity to support vital functions, and developed algorithms had been sensitive to a number of regionally specific variables, including soil taxonomy, terrain, crop type, season, and climate. The conceptual SMAF for indication of very small dataset interpretation was established and has been applied widely. In addition to SMAF, another tool, the Comprehensive Assessment of Soil Health (CASH), is also used to evaluate various soil health metrics (Idowu et al. 2009). The Midwest and Mid-Atlantic soils have just been added to the CASH algorithms (Fine et al. 2017), although the spatial scope is still small, and the curves do not represent the national or international distribution of soils.

It is crucial to provide indexing techniques that can handle different soil properties and represent relevant soil health status for various types of soil in order to enhance scientifically based soil health evaluations (Bunemann et al. 2018). SMAF and CASH both used rather tiny and geographically constrained datasets, despite the fact that they were created to account for underlying site features. Additionally, a number of analyses identified potential issues with the scoring curves at the regional level, including both overestimations and underestimations of various characteristics and scores that most likely reflected dataset limitations and the need for more effective use of edaphic and climate data (Nunes et al. 2020b). SMAF and CASH were initially created and adopted on a global scale for scoring soil health indicators. The interpretation curves for the SMAF and CASH frameworks, however, were not validated at the national level because they were both built using a small dataset. Nunes et al. (2020b) suggested the Soil Health Assessment Protocol and Evaluation (SHAPE) tool as an extension of these ideas. The SHAPE, which takes into consideration edaphic and climatic parameters at the continental scale, was built using several numbers of soil organic carbon (SOC) data from across the United States. SHAPE was found significantly larger and representable dataset than

SMAF and CASH due to its flexible and quantitative nature that offers a spatial pertinent interpretation of different soil attributes.

15.5 Emerging Techniques to Improve of Soil Health

15.5.1 Spectroscopy Techniques

It is difficult and expensive and takes a lot of time and effort to identify and determine SOM and other soil physicochemical parameters using typical soil testing techniques in laboratories. Soil characterisation methods that are complementary or alternative are required for both in situ and laboratory settings. According to Lazaar et al. (2019), Vis-NIR visible-near-infrared and shortwave-infrared (SWIR) spectroscopy is a current alternative method for quantifying and analysing soil. It is quick, non-destructive, environmentally friendly, and repeatable and only needs a small amount of sample preparation. Key soil characteristics include soil organic carbon (SOC), calcium carbonates (CaCO3), texture, and total or organic nitrogen that are all commonly measured using this technique. Given that SOM significantly influences the shape and make-up of soil reflectance spectra (Gholizadeh et al. 2013; Lazaar et al. 2020), it is one of the key soil factors that can be predicted with reflectance spectroscopy. The basic standard measurements of soil as well as more recent methods like visible and near-infrared reflectance spectroscopy, a smart phone app, and metagenomics are all included in the evaluation of soil health indicators. Machine learning and regression algorithms, among others, are utilised to extract valuable and quantitative information from soil spectra. As they take into account the non-linear spectrum responses, numerous studies have shown that machine learning techniques are better to regression techniques in soil analysis (Nawar and Mouazen 2017; Viscarra Rossel and Behrens 2010). The following three primary elements, in addition to the modelling approach used, are related to the quality of prediction: (1) the method used to prepare the soil sample (e.g. whether it had been processed or fresh), (2) the measuring tool (e.g. the kind of detector, the spectral range, the spectral resolution, and the light source utilised), and (3) the applied laboratory methods (Fig. 15.3).

According to Coûteaux et al. (2003), Vis-NIR-SWIR proximal sensing relies on the creation of empirical models that predict soil constituent concentrations from complex spectral data, which includes different wavelength like visible (350–780 nm), near-infrared (780–1100 nm), and short-wave infrared (1100–2500 nm). In the past 40 years, visible and near-infrared reflectance spectroscopy has originated for advanced soil research (Nocita et al. 2014). Vis-NIR has numerous definite advantages over conventional methods of soil analysis (Viscarra Rossel et al. 2006), including the following:

- It is quick and non-destructive, to start. A Vis-NIR scan can be obtained in a matter of seconds without consuming any dirt.

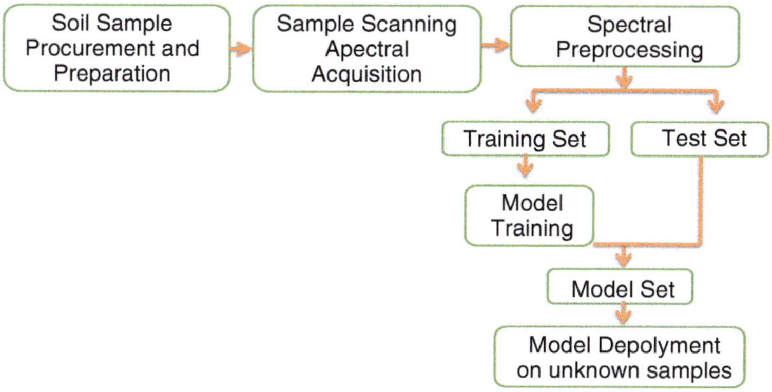

Fig. 15.3 The schematic workflow of soil analysis from the visible and near-infrared reflectance spectroscopy. (Source: Ge et al. 2019)

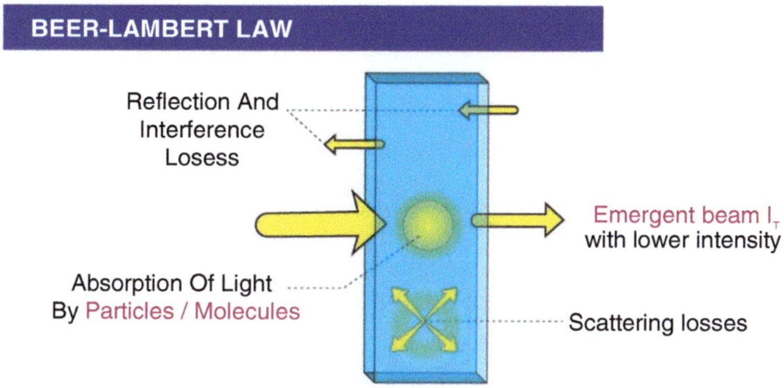

Fig. 15.4 Schematic diagram of Beer-Lambert law

- A single scan's spectrum makes it possible to simultaneously characterise several soil parameters on the same volume of soil material. Unlike standard assessments, it does not examine the soil properties independently on various sub-samples.
- Finally, it is inexpensive because there is small quantity preparation of sample required without involvement of any chemical extractant.

The initial cost of a Vis-NIR device is high, but when large quantity of soil samples are scanned and analysed, this cost quickly reduces. The foundation of Vis-NIR spectroscopy is the idea that particular soil components interact and absorb electromagnetic light, allowing for their quantification (Fig. 15.3).

According to the Beer-Lambert rule, spectral absorption is inversely proportional to the constituent's concentration (Fig. 15.4). Individual absorptions are challenging to distinguish, however, because spectral absorptions in the Vis-NIR spectral area are weak and soil absorption sources overlap across spectral bands. In soil analysis, the Beer-Lambert law is therefore rarely directly utilised. Rather, empirical, multivariate statistical models encompassing all wavebands are typically used to accomplish quantitative assessments.

In the Vis-NIR spectrum, different soil components (clay minerals, organic matter, iron oxides and carbonates, and water) are used which actively absorb light, used as the first class that Vis-NIR may effectively be used to measure, and active absorptions of spectral that are detectable in Vis-NIR region are not produced by these secondary characteristics. Furthermore, Vis-NIR has proven beneficial for measuring other soil chemical properties like CEC, salinity level, and pH, since different soil basic attributes are associated with other soil properties. In more recent times, assessing soil contaminants such as hydrocarbons and heavy metals with Vis-NIR has also yielded good results (Horta et al. 2015). The physical state of soil, particularly its aggregate size distribution, affects the penetration and scattering of Vis-NIR energy and consequently the samples' Vis-NIR spectra. In present era, researchers showed significant interest on its application on field; the majority of Vis-NIR spectroscopy examination is still carried out on soil samples (ai-dried and ground).

15.5.2 *Remote Sensing and Geostatistical Approaches*

The geographical variability of soil nutrients can be investigated using remote sensing and geostatistical approaches (Zhang et al. 2010). Geo-statistics is a natural phenomenon that describes spatial continuity which focuses mostly on spatially auto-correlated data using standard regression techniques; in addition to this, it is the study of events with varying geographical and temporal bases. Various fundamental elements of geo-statistics are explained in Fig. 15.5.

Mapping and monitoring of soil and land degradation, soil moisture, fertility evaluation, soil and water conservation strategies, suitability investigations of soil,

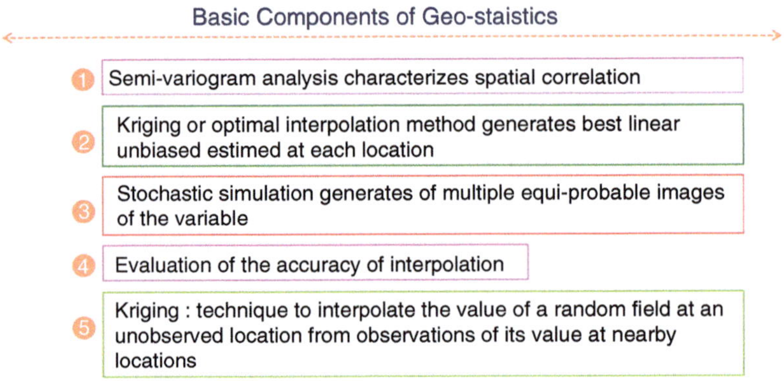

Fig. 15.5 The fundamental elements of geostatistics

and remote sensing (RS) techniques play a significant role. According to Ravisankar and Sreenivas (2011), the main use of RS is the development of soil nutrient availability maps as well as the soil fertility geographical assessment.

15.5.3 Nanotechnology: Boosting Soil Health and Plant Resilience

Nanotechnology, emerged as a promising technology, has the potential in twenty-first century to advance conventional methods of agriculture, enhancing management and conservation strategies and reducing agricultural input waste for sustainability (Shang et al. 2019; Dubey and Mailapalli 2016) (Fig. 15.6). Agrochemicals are often sprayed or broadcasted onto crops using conventional methods. As a result, relatively few agrochemicals—far less than the minimal effective concentration needed for healthy plant growth—reach the target areas of crops. Chemical leaching, photolysis, hydrolysis, and microbiological degradation are all factors contributing to the losses (Nair et al. 2010; Yang et al. 2016). The development of slow or controlled release fertilisers, insecticides, and herbicides based on nanotechnology has so attracted further attention in agriculture farming in order to assure environmentally friendly agricultural practises (Panpatte et al. 2016). Nanotechnology has steadily made its way from experimental trials conducted in laboratories to real-world uses. According to Shojaei et al. (2019), the goal of controlled delivery approaches is to release measured amounts of necessary and sufficient amounts of agrochemicals throughout time in order to achieve full biological competency while minimising loss and negative consequences. Due to their huge surface area, ease of attachment, and quick mass transfer, nanoparticles have the benefits of effective pesticide delivery (Ghormade et al. 2011).

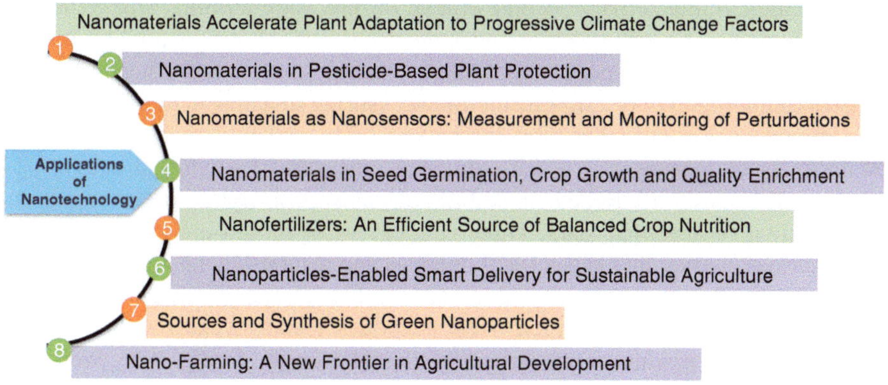

Fig. 15.6 Applications of nanotechnology in agriculture for achieving sustainability of agriculture and environment

The ultimate concentration of the fertilisers when they reach the target sites, whether they are sprayed on the leaves or administered directly to the soil, has a significant impact on how well they absorb nutrients (Solanki et al. 2015). According to Sabir et al. (2014), a very small amount—far below the minimum necessary concentration—reaches the intended site as a result of chemical leaching loss, drift, run-off, hydrolysis, evaporation, photolytic, or even microbiological destruction. Consequences of frequent overuse of fertilisers include a negative impact on the soil's natural nutrient equilibrium. In addition to this, toxic compounds have seriously damaged water ecosystems due to seeping into rivers and water reservoirs, which also contaminates drinking water. According to Godfray et al. (2010), the application of engineered nanomaterials in the context of sustainable agriculture has demonstrated an entirely new method of food production that has the ability to overcome the uncertainty in the crop sector with the limited resources at hand. Nanomaterials and nano-fertilizers have made promises to meet projected increases in both sustainable agriculture and global food demand as a result of the green nanotechnology revolution, which has fundamentally altered the global agriculture landscape. Nano-fertilizers may be the greatest option for reducing macro- and micronutrient deficiencies through improved nutrient usage efficiency and for resolving the long-term issue of eutrophication (Shukla et al. 2019) (Table 15.2). The potential for nano-fertilizers, which were specifically designed to control nutrient release according to crop needs while minimising differential losses, is enormous.

15.5.4 Biosensors or Nanobiosensors Used in Soil Health Analysis

Soil health is dynamic in nature, so its monitoring with many soil parameters (including pH, nutrients, temperature, moisture content, etc.) on-site before beginning any agricultural activity become crucial for precision farming. Due to heterogenic nature of soil, numerous nutrients are present in soil having varied forms, and each form sends out a unique signal. Here, nanobiosensors and other innovative modern technologies play a significant role in ensuring sustainable farming, because they provide a low-cost detection instrument with increased sensitivity (Kumar et al. 2021). A biosensor is a self-contained, integrated system that uses a bioreceptor with sensitivity to detectable biomolecules, or analytes, to generate accurate quantitative analytical data. Three generations of biosensors mostly reported under various studies are (1) electrical operates and responsive, (2) improvement in responsiveness between reaction and transducer through specific mediators, and (3) application of the reaction which itself generates responsiveness without direct involvement from product or mediator diffusion (Nelson and Griffin 1916; Weinstein et al. 2007). Sensitivity, selectivity, multiplexed detection capabilities, and portability of sensors are all greatly enhanced by the incorporation of nanoparticles and

Table 15.2 Efficacy and importance of nano-fertilizers as a source of balanced crop nutrition

Sl. no.	Nano-fertilizers	Effect on plant nutrition	Reference
1.	Nanomaterials—zeolites and chitosan	Controlling the need-based release and improving the crop uptake process significantly minimise nitrogen losses	Millan et al. (2008) and Panpatte et al. (2016)
2.	Ammonium-charged zeolites	Possibility to make phosphate minerals more soluble, resulting in better phosphorus availability and crop uptake	Dwivedi et al. (2016)
3.	Carbon-based nanomaterial/ graphene oxide	Prolonging the release pattern of potassium nitrate, enhancement of the function duration, and minimising leaching	Shalaby et al. (2016)
4.	Nanocalcite ($CaCO_3$ 40%), nano-SiO_2 (4%), MgO (1%), and Fe_2O_3 (1%)	Boosted P adsorption with the micronutrients Zn and Mn but also significantly improved the Ca, Mg, and Fe uptake	Sabir et al. (2014)
5.	Carbon nanoparticles	Improvement of the grain yields of rice (+10.29%), spring maize (+10.93%), soybean (+16.74%), winter wheat (+28.81%), and vegetables (+12.34–19.76%)	Liu et al. (2009)
6.	Chitosan-NPK fertilizer	Compared to control yield variables, significant increase in the harvest index, crop index, and mobilisation index of wheat yield variables	Abdel-Aziz et al. (2016)
7.	Nano-zinc oxide	Positively affects several plant species' development and physiological processes	Ali et al. (2019) and Asl et al. (2019)
8.	Zinc oxide nanoparticles	Boosts barley productivity (91%)	Kale and Gawade (2016)

nanostructures (Sharma et al. 2021). For the detection of biological products, physical or chemical transducers are integrated with identification components (Bhalla et al. 2016). A customised biosensor known as a nanobiosensor can be thought of as a little analytical tool/unit with a physico-chemical transducer connected to a physiologically or biologically generated sensitised element (Turner 2000; Rai et al. 2012). The first biosensor was created by Updike and Hicks (1967), and its invention sparked the creation of several altered biosensors. Fascinatingly, the idea of biosensors has been around since the early twentieth century, but until the development of science, they were primarily used in laboratories. Numerous fertilisers, herbicides, pesticides, insecticides, diseases, moisture, and soil pH can all be detected by nanobiosensors (Mondal et al. 2022). With a detection limit of 2.1 g/L, scientists have created adaptable lipid-based nanobiosensors to identify nitrites in soil samples; in that sensors, the hydrophobic interactions hold together bilayer lipid membrane for retaining the thermodynamic stability (Siontorou and Georgopoulos 2016).

15.5.5 *Artificial Intelligence for Soil Health Monitoring*

The research community is focusing increasingly on the breadth of artificial intelligence technologies, which encompasses software and big data solutions that may enhance and automate decision-making processes. Artificial intelligence (AI) is defined by Jones et al. (2022) as "AI assessments made by using different algorithms that either follow rules and machine learning followed the huge data review to find and follow patterns". Machine learning is opaque in comparison to conventional rule-following computing because it has numerous levels and machines create their own learning and patterns. An ecosystem for smart, effective, and sustainable farming is emerging with the development of cutting-edge technologies like artificial intelligence (AI), cloud machine learning (ML), satellite imagery, and sophisticated analytics. Farmers are now able to increase their average production per hectare and have better control over the price of food grains, ensuring that they continue to make a profit (Dharmaraj and Vijayanand 2018). In soil research, artificial intelligence can be applied to a wide range of tasks, including soil analysis, remote sensing, and mapping soil/land cover (Fig. 15.7). Different algorithms are used for agricultural analysis in artificial intelligence (AI) and machine learning (ML) systems that monitor soil quality and fertility.

Machine learning applications employ a variety of strategies to assist with data analysis processes, producing enough data to offer a statistical answer to problems requiring these techniques. Artificial intelligence performed the soil quality analysis, and generalising, analysing, and processing data from numerous monitoring equipment and issuing recommendations based on them constitute the wider agricultural industry. Without the aid of computer systems like artificial intelligence

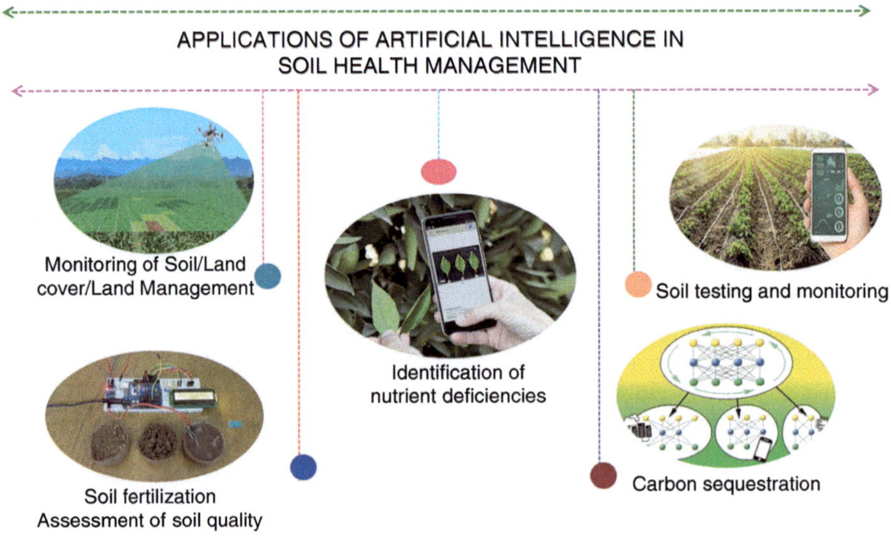

Fig. 15.7 Different applications of artificial intelligence for getting healthy soil

technologies, it is practically impossible to traverse this flood of information for big data analysis and its measures.

Exciting possibilities for sustainable food production are presented by the convergence of nanotechnology, artificial intelligence, and precision agriculture, which enables farmers to react in real time to changes in crop development (Zhang et al. 2021). Using artificial intelligence, especially electronic deep learning algorithms, farmers can spot potential nutrient deficiencies in soil quality. Farmers can now use a variety of agricultural technologies where all they need to do is to snap a photo with their smart phone and upload it to an AI development system. Farmers are given restoration methods, and other remedies after the issue has been evaluated in order to assist raise the quality of the soil and the yield of the crop. Previously inaccessible agricultural data sources, like satellite and unmanned aerial vehicle (UAV) readings, humidity sensor readings, and ground-based weather stations, are now available to agricultural producers for decision-making. Simultaneously, more individualised, precise analyses and forecasts of soil quality are made possible by the constant introduction of new monitoring and control technologies to the market. InCeres, a Brazilian agricultural start-up (2018), has created an app that uses soil application and nutrient uptake to forecast soil condition and fertility. The research is supported by information on crop types, weather patterns, soil chemical make-up, and satellite photos displaying plant growth rates. The programme created employs AI techniques to forecast soil fertility for each particular plot of farmland by analysing large amounts of data. Varatharajalu and Ramprabu (2018) have introduced an automated watering system that makes use of different sensor (temperature, moisture, molecular, and pressure regulator) related to soil to improve crop growth. This system was developed using various soil quality monitoring methodologies. The multiplexer receives the digital signals that have been converted from the sensor outputs over a wireless network like Zigbee or a hotspot. With the use of a standard RGB digital camera, Al-Naji et al. (2021) have developed a device that precisely analyses soil quality indicators including moisture. It analyses variations in soil colour using a standard video camera to find moisture content. An artificial neural network (ANN) that was trained to recognise various levels of soil moisture under various weather conditions was connected to the digital camera.

Forecasting and predicting variables like soil health are significantly hampered by the compilation of data (Ali 2022). Usually, reliable, reproducible, and representative data sets serve as the foundation for agricultural research. Because of this, even though we can develop a variety of Al models, weather and climate change can have an impact on temperature, soil moisture, photosynthetic rate, and ecological balance. In such circumstances, crop yield could drastically decline in a short period of time, necessitating frequent data collecting. Agriculture will inevitably experience an increase in digitisation across industries. Simple data inputs can be transformed into useful information by intelligent machines using artificial intelligence and machine learning. It has the potential to completely alter the world's food supply chain when used in the agricultural industry, even for monitoring factors like soil health.

15.6 Conclusion

A global audience has recently become more interested in soil health as it becomes a political, environmental quality, and human health issue. Traditionally, soil health has been evaluated in a form of output, but due to increasing awareness towards soil health, its quantification and monitoring become important. Soil health assessment is an interpretive framework of soil properties that are biological, physical, and chemical (which include inherent and dynamic soil functions). During soil analysis for different macro- and micronutrients, harsh chemical extractants have typically been used to assess the amount of nutrients that are accessible to plants. Additionally, complementary techniques such as soil texture, pH, and total soil organic matter do not assess biological soil characteristics.

So in the last four decades, various new methods have come into the frame which play an important role during soil analysis. Vis-NIR has several distinct advantages over traditional laboratory-based soil examination techniques, including speed, non-destructiveness, affordability, and multi-sensing capacity. In contrast to the abilities required for conventional soil analysis, the Vis-NIR technique implementation calls for knowledge and expertise in spectral data processing, multivariate statistical modelling, and advanced data analytics. For both laboratory and field Vis-NIR soil analyses, a strong basis has been established during the last 40 years of research. In the future, Vis-NIR is expected to become need of hour, increasingly significant for high-resolution and quantitative soil data, and become more crucial in numerous applications like environmental, ecological, hydrological, and agricultural. With lowered farming hazards, the potential of nanomaterials, biosensors, and artificial intelligence inspires a new green revolution. However, there are still a lot of unknown facts regarding the eco-toxicity, acceptable limit, and absorption capability of certain nanomaterials. More research is urgently required in these areas to better understand the behaviour and results of altered agricultural inputs as well as their interactions with bio-macromolecules present in living systems and the environment.

References

Abdel-Aziz HMM, Hasaneen MNA, Omer AM (2016) Nano chitosan-NPK fertilizer enhances the growth and productivity of wheat plants grown in sandy soil. Span J Agric Res 14:17

Ali O (2022) Artificial intelligence (AI) in soil quality monitoring. AZoRobotics. https://www.azorobotics.com/Article.aspx?ArticleID=482

Ali S, Rizwan M, Noureen S, Anwar S, Ali B, Naveed M, Abd-Allah EF, Alqarawi AA, Ahmad P (2019) Combined use of biochar and zinc oxide nanoparticle foliar spray improved the plant growth and decreased the cadmium accumulation in rice (Oryza sativa L.) plant. Environ Sci Pollut Res 26:11288–11299

Al-Naji A, Fakhri AB, Gharghan SK, Chahl J (2021) Soil color analysis based on an RGB camera and an artificial neural network towards smart irrigation: a pilot study. Heliyon 7(1):e06078. https://doi.org/10.1016/j.heliyon.2021.e06078

Andrews SS, Karlen DL, Cambardella CA (2004) The soil management assessment framework: a quantitative soil quality evaluation method. Soil Sci Soc Am J 68:1945–1962. https://doi.org/10.2136/sssaj2004.1945

Asl KR, Hosseini B, Sharafi A, Palazon J (2019) Influence of nano-zinc oxide on tropane alkaloid production, h6h gene transcription and antioxidant enzyme activity in Hyoscyamus reticulatus L. hairy roots. Eng Life Sci 19:73–89

Baveye PC, Baveye J, Gowdy J (2016) Soil ecosystem services and natural capital: critical appraisal of research on uncertain ground. Front Environ Sci 4:41. https://doi.org/10.3389/fenvs.2016.00041

Beebe SE, Rao IM, Cajiao C, Grajales M (2008) Selection for drought resistance in common bean also improves yield in phosphorus limited and favorable environments. Crop Sci 48(2):582–592

Bhalla N, Jolly P, Formisano N, Estrela P (2016) Introduction to biosensors. Essays Biochem 60:1–8

Bhattacharyya R, Ghosh BN, Mishra PK, Mandal B, Rao CS, Sarkar D et al (2015) Soil degradation in India: challenges and potential solutions. Sustainability 7:3528–3570. https://doi.org/10.3390/su7043528

Bunemann EK, Bongiorno G, Bai Z, Creamer RE, Deyn G, Goede R, Fleskens L, Geissen V, Kuyper TW, Mäder P, Pulle-man M, Sukkel W, van Groenigen JW, Brussaard L (2018) Soil quality: a critical review. Soil Biol Biochem 120:105–125. https://doi.org/10.1016/j.soilbio.2018.01.030

Chabi-Olaye A, Nolte C, Schulthess F, Borgemeister C (2006) Relationships of soil fertility and stem borers damage to yield in maize-based cropping system in Cameroon. Ann Soc Entomol (NS) 42(3–4):471–479

Costanza R, D'Arge R, de Groot R, Farberk S, Grasso M, Hannon B et al (1997) The value of world's natural ecosystem services and natural capital. Nature 387:253–260. https://doi.org/10.1038/387253a0

Coûteaux MM, Berg B, Rovira P (2003) Near infrared reflectance spectroscopy for determination of organic matter fractions including microbial biomass in coniferous forest soils. Soil Biol Biochem 35(12):1587–1600

Das A, Jayanta L, Ramkrushna I, Basavaraj S, Lal R, Rangappa K, Yadav G, Babu S, Ngachan SV (2020) Conservation tillage and residue management improves soil properties under upland rice–rapeseed system in subtropical eastern Himalayas. Land Degrad Dev 31. https://doi.org/10.1002/ldr.3568

de Paul Obade V, Lal R (2016) Towards a standard technique for soil quality assessment. Geoderma 265:96–102

Dharmaraj V, Vijayanand C (2018) Artificial intelligence (AI) in agriculture. Int J Curr Microbiol Appl Sci 7(12):2122–2128. https://doi.org/10.20546/ijcmas.2018.712.241

Doran JW, Parkin TB (1996) Quantitative indicators of soil quality: a minimum data set. In: Doran JW, Jones AJ (eds) Methods for assessing soil quality. Soil Science Society of America

Dubey A, Mailapalli DR (2016) Nanofertilisers, nanopesticides, nanosensors of pest and nanotoxicity in agriculture. In: Lichtfouse E (ed) Sustainable agriculture reviews, vol 19. Springer, Cham, pp 307–330

Dwivedi S, Saquib Q, Al-Khedhairy AA, Musarrat J (2016) Understanding the role of nanomaterials in agriculture. In: Singh DP, Singh HB, Prabha R (eds) Microbial inoculants in sustainable agricultural productivity. Springer, New Delhi, pp 271–288

FAO (2008) An international technical workshop investing in sustainable crop intensification: the case for improving soil health, 22–24 Jul 2008, Integrated crop management, vol 6. FAO, Rome

FAO (2010) Climate-smart agriculture: policies, practices and financing for food security, adaptation and mitigation. FAO, Rome

FAO (2017) The future of food and agriculture—trends and challenges. FAO, Rome

Fine AK, van Es HM, Schindelbeck RR (2017) Statistics, scoring functions, and regional analysis of a comprehensive soil health database. Soil Sci Soc Am J 81:589–601

Ge Y, Morgan CLS, Wijewardane NK (2019) Visible and near-infrared reflectance spectroscopy analysis of soils. Soil Sci Soc Am J 84:1495–1502

Gholizadeh A, Boruvka L, Saberioon M, Vasat R (2013) Mid-infrared spectroscopy applications for soil assessment with emphasis on soil organic matter content and quality. Appl Spectr OA 67(12):1349–1362. https://doi.org/10.1366/13-07288

Ghormade V, Deshpande MV, Paknikar KM (2011) Perspectives for nano-biotechnology enabled protection and nutrition of plants. Biotechnol Adv 29:792–803

Godfray HCJ, Beddington JR, Crute IR, Haddad L, Lawrence D, Muir JF, Pretty J, Robinson S, Thomas SM, Toulmin C (2010) Food security: the challenge of feeding 9 billion people. Science 327:812–818

Haney RL, Haney EB, Smith DR, Harmel RD, White MJ (2018) The soil health tool–theory and initial broad-scale application. Appl Soil Ecol 125:162–168. https://doi.org/10.1186/s13750-017-0108-9

Horta A, Malone B, Stockmann U, Minasny B, Bishop TFA, McBratney AB (2015) Potential of integrated field spectroscopy and spatial analysis for enhanced assessment of soil contamination: a prospective review. Geoderma 241–242:180–209. https://doi.org/10.1016/j.geoderma.2014.11.024

Idowu OJ, van Es HM, Abawi GS, Wolfe DW, Schindel-beck RR, Moebius-Clune BN, Gugino BK (2009) Useof an integrative soil health test for evaluation of soil management impacts. Renew Agric Food Syst 24:214–224. https://doi.org/10.1017/S1742170509990068

InCeres (2018) Artificial intelligence for soil fertility control. Available at: https://pesquisaparainovacao.fapesp.br/artificial_intelligence_for_soil_fertility_cotrol/747

Jarvis A, Lau C, Cook S, Wollenberg E, Hansen J, Bonilla O, Challinor A (2011) An integrated adaptation and mitigation framework for developing agricultural research: synergies and trade-offs. Exp Agric 47:185–203

Jones, K., Buchser, M. and Wallace, J. 2022. Challenges of AI. https://www.chathamhouse.org/2022/03/challenges-ai

Kale AP, Gawade SN (2016) Studies on nanoparticle induced nutrient use efficiency of fertilizer and crop productivity. Green Chem Technol Lett 2:88–92

Karlen DL, Mausbach MJ, Doran JW, Cline RG, Harris RF, Schuman GE (1997) Soil quality: a concept, definition, and framework for evaluation. Soil Sci Soc Am J 61:4–10

Kibblewhite MG, Ritz K, Swift MJ (2008) Soil health in agricultural systems. Philos Trans R Soc Lond Ser B Biol Sci 363(1492):685–701. https://doi.org/10.1098/rstb.2007.2178

Kumar A, Choudhary A, Kaur H, Mehta S, Husen A (2021) Metal-based nanoparticles, sensors and their multifaceted application in food packaging. J Nanobiotechnol 19:256

Lazaar A, Mahyou H, Gholizadeh A, Elhammouti K, Bilal M, Andich K et al (2019) Potential of VIS-NIR spectroscopy to characterize and discriminate top soils of different soil types in the Triffa plain (Morocco). Soil Sci Annu 70(1):54. https://doi.org/10.2478/ssa-2019-0007

Lazaar A, Mouazen AM, El Hammouti K, Fullen M, Pradhan B, Memon MS et al (2020) The application of proximal visible and near-infrared spectroscopy to estimate soil organic matter on the Triffa Plain of Morocco. Int Soil Water Conserv Res 8(2):195–204

Liu J, Zhang Y, Zhang Z (2009) The application research of nano-biotechnology to promote increasing of vegetable production. Hubei Agric Sci 1:41

Lobell D (2009) Climate extremes and crop adaptation. Summary statement from a meeting at the program on food security and environment held on June, 16–18, 2009. Stanford University, Stanford, CA

Millan G, Agosto F, Vazquez M, Botto L, Lombardi L, Juan L (2008) Use of clinoptilolite as a carrier for nitrogen fertilizers in soils of the Pampean regions of Argentina. Cienc Investig Agrar 35:293–302

Mondal R, Dam P, Chakraborty J, Paret ML, Katı A, Altuntas S, Sarkar R, Ghorai S, Gangopadhyay D, Mandal AK, Husen A (2022) Potential of nanobiosensor in sustainable agriculture: the state-of-art. Heliyon 8(12):e12207

Nair R, Varghese SH, Nair BG, Maekawa T, Yoshida Y, Kumar DS (2010) Nanoparticulate material delivery to plants. Plant Sci 179:154–163

Natural Resources Conservation Service. 2018. U.S. Department of Agriculture

Nawar S, Mouazen AM (2017) Predictive performance of mobile vis-near infrared spectroscopy for key soil properties at different geographical scales by using spiking and data mining techniques. Catena 151:118–129. https://doi.org/10.1016/j.catena.2016.12.014

Nelson JM, Griffin EG (1916) Adsorption of invertase. J Am Chem Soc 38(5):1109–1115. https://doi.org/10.1021/ja02262a018

Nocita M, Stevens A, Toth G, Panagos P, van Wesemael B, Montanarella L (2014) Prediction of soil organic carbon content by diffuse reflectance spectroscopy using a local partial least square regression approach. Soil Biol Biochem 68:337–347. https://doi.org/10.1016/j.soilbio.2013.10.022

Nunes MR, Karlen DL, Veum KL, Moorman TB, Cambardella CA (2020a) Biological soil health indicators respond to tillage intensity: a US meta-analysis. Geoderma 369:114335. https://doi.org/10.1016/j.geoderma.2020.114335

Nunes MR, Karlen DL, Veum KL, Moorman TB (2020b) A SMAF assessment of U.S. tillage and crop management strategies. Environ Sustain Indicators 369:100072. https://doi.org/10.1016/j.indic.2020.100072

OECD and FAO (Food and Agriculture Organization) (2016) OECD-FAO agricultural outlook 2016-25. OECD Publishing, Paris. https://doi.org/10.1787/agr_outlook-2016-en

Panpatte DG, Jhala YK, Shelat HN, Vyas RV (2016) Nanoparticles: the next generation technology for sustainable agriculture. In: Microbial inoculants in sustainable agricultural productivity. Springer, New Delhi, pp 289–300

Rai V, Acharya S, Dey N (2012) Implications of nanobiosensors in agriculture. J Biomater Nanobiotechnol 3:315–324. https://doi.org/10.4236/jbnb.2012.322039

Rattan RK, Kumar M, Narwal RP, Singh AP (2009) Soil health and nutritional security-micronutrients. In: Proceedings of platinum jubilee symposium of Indian Society of Soil Science, pp 249–265

Ravisankar T, Sreenivas K (2011) Soils and land degradation. In: Roy PS, Dwivedi RS, Vijayan D (eds) Remote sensing applications. NRSC Publication, Hyderabad

Rinot O, Levy GJ, Steinberger Y, Svoray T, Eshel G (2019) Soil health assessment: a critical review of current methodologies and a proposed new approach. Sci Total Environ 648:1484–1491

Sabir A, Yazar K, Sabir F, Kara Z, Yazici MA, Goksu N (2014) Vine growth, yield, berry quality attributes and leaf nutrient content of grapevines as influenced by seaweed extract (Ascophyllum nodosum) and nanosize fertilizer pulverizations. Sci Hortic 175:1–8

Shalaby TA, Bayoumi Y, Abdalla N, Taha H, Alshaal T, Shehata S, Amer M, Domokos-Szabolcsy É, El-Ramady H (2016) Nanoparticles, soils, plants and sustainable agriculture. In: Shivendu R, Nandita D, Eric L (eds) Nanoscience in food and agriculture, vol 1. Springer, Cham, pp 283–312

Shang Y, Hasan MK, Ahammed GJ, Li M, Yin H, Zhou J (2019) Applications of nanotechnology in plant growth and crop protection: a review. Molecules 24:2558. https://doi.org/10.3390/molecules24142558

Sharma P, Pandey V, Sharma MMM, Patra A, Singh B, Mehta S, Husen A (2021) A review on biosensors and nanosensors application in agroecosystems. Nanoscale Res Lett 16:1–24

Shojaei TR, Salleh MAM, Tabatabaei M, Mobli H, Aghbashlo M, Rashid SA, Tan T (2019) Applications of nanotechnology and carbon nanoparticles in agriculture. In: Suraya AR, Raja NIRO, Mohd ZH (eds) Synthesis, technology and applications of carbon nanomaterials. Elsevier, Amsterdam, pp 247–277

Shukla P, Chaurasia P, Younis K, Qadri OS, Faridi SA, Srivastava G (2019) Nanotechnology in sustainable agriculture: studies from seed priming to post-harvest management. Nanotechnol Environ Eng 4:11

Siontorou CG, Georgopoulos KN (2016) A biosensor platform for soil management: the case of nitrites. J Clean Prod 111:133–142

Solanki P, Bhargava A, Chhipa H, Jain N, Panwar J (2015) Nano-fertilizers and their smart delivery system. In: Rai M, Ribeiro C, Mattoso L, Duran N (eds) Nanotechnologies in food and agriculture. Springer, Cham, pp 81–101

Steiner K, Herweg K, Dumanski J (2000) Practical and cost-effective indicators and procedures for monitoring the impacts of rural development projects on land quality and sustainable land management. Agric Ecosyst Environ 81:147–154

Stott DE (2019) Recommended soil health indicators and associated laboratory procedures. Soil Health Technical Note No. 450-03. USDA–NRCS

Taylor, C. 2014. Natural Resources Conservation Service in Conservation

Thornton P, Cramer L (eds) (2012) Impacts of climate change on the agricultural and aquatic systems and natural resources within the CGIAR's mandate. CCAFS Working Paper 23. CGIAR Research Program on Climate Change, Agriculture and Food Security, Copenhagen

Turner AP (2000) Biosensors-sense and sensitivity. Science 5495(290):1315–1317

United Nations (2015) India and the MDGs towards a sustainable future for all. https://www.unescap.org/sites/default/files/India_and_the_MDGs_0.pdf

Updike SJ, Hicks GP (1967) The enzyme electrode. Nature 214:986–988. https://doi.org/10.1038/214986a0

Varatharajalu K, Ramprabu J (2018) Wireless irrigation system via phone call & SMS. Int J Eng Adv Technol 8:397–401. Retrieved from https://www.ijeat.org/wp-content/uploads/papers/v8i2s/B10821282S18

Vermeulen SJ, Aggarwal PK, Ainslie A, Angelone C, Campbell BM, Challinor AJ, Hansen JW, Ingram JSI, Jarvis A, Kristjanson P, Lau C, Nelson GC, Thornton PK, Wollenberg E (2012) Options for support to agriculture and food security under climate change. Environ Sci Policy 15(1):136–144

Viscarra Rossel RA, Behrens T (2010) Using data mining to model and interpret soil diffuse reflectance spectra. Geoderma 158(1–2):46–54. https://doi.org/10.1016/j.geoderma.2009.12.025

Viscarra Rossel RA, Walvoort DJJ, McBratney AB, Janik LJ, Skjemstad JO (2006) Visible, near infrared, mid-infrared or combined diffuse reflectance spectroscopy for simultaneous assessment of various soil properties. Geoderma 131(1–2):59–75

Wander MM, Cihacek LJ, Coyne M, Drijber RA, Grossman JM, Gutknecht JLM, Horwath WR, Jagadamma S, Olk DC, Ruark M, Snapp SS, Tiemann LK, Weil R, Turco RF (2019) Developments in agricultural soil quality and health: reflections by the research committee on soil organic matter management. Front Environ Sci 7:109

Weinstein RL, Bugler JR, Schwartz SL, Pey TA, Brazg RL, Mc-Garraugh GV (2007) Accuracy of the 5-day freestyle navigator continuous glucose monitoring system-comparison with frequent laboratory reference measurements. Diabetes Care 30(5):1125–1130

Yang H, Xu M, Koide RT, Liu Q, Dai Y, Liu L, Bian X (2016) Effects of ditch-buried straw return on water percolation, nitrogen leaching and crop yields in a rice-wheat rotation system. J Sci Food Agric 96:1141–1149

Young IM, Ritz K (2005) The habitat of soil microbes. In: Bardgett RD, Usher MB, Hopkins DW (eds) Biological diversity and function in soils. Cambridge University Press, Cambridge, pp 31–43

Zhang M, He PF, Chen WQ (2010) Spatio-temporal variability analysis of soil nutrients based on GIS and geostatistics. J Northeast Agric Univ 41(3):53–58

Zhang P, Guo Z, Ullah S, Melagraki G, Afantitis A, Lynch I (2021) Nanotechnology and artificial intelligence to enable sustainable and precision agriculture. Nat Plants 7:864–876

Correction to: Bioengineered Microbes for Restoration of Soil Health

Shweta Sharma, Amit Kumar Kesharwani, and Aditya Kulshreshtha

Correction to:
Chapter 3 in: R. K. Bhatia, A. Walia (eds.), Advancements in Microbial Biotechnology for Soil Health, Microorganisms for Sustainability 50, https://doi.org/10.1007/978-981-99-9482-3_3

The original version of the book was published with errors. The following corrections have been made after publication.

The affiliation of the authors for Chapter 3, A. K. Kesharwani and A. Kulshreshtha, has been changed to read as ICAR-Indian Agricultural Research Institute, New Delhi, India.

The updated version of this chapter can be found at https://doi.org/10.1007/978-981-99-9482-3_3

R. K. Bhatia, A. Walia (eds.), *Advancements in Microbial Biotechnology for Soil Health*, Microorganisms for Sustainability 50,
https://doi.org/10.1007/978-981-99-9482-3_16

MIX
Papier aus verantwortungsvollen Quellen
Paper from responsible sources
FSC® C105338

If you have any concerns about our products,
you can contact us on
ProductSafety@springernature.com

In case Publisher is established outside the EU,
the EU authorized representative is:
Springer Nature Customer Service Center GmbH
Europaplatz 3, 69115 Heidelberg, Germany

Printed by Libri Plureos GmbH
in Hamburg, Germany